CLIMATIC HAZARDS IN COASTAL BANGLADESH

CLIMATIC HAZARDS IN COASTAL BANGLADESH

Non-Structural and Structural Solutions

BIMAL KANTI PAUL
Professor, Department of Geography
Kansas State University
Manhattan, KS, USA

HARUN RASHID
Emeritus Professor, Department of Geography and Earth Science
University of Wisconsin - La Crosse
La Crosse, WI, USA

ELSEVIER

Amsterdam • Boston • Heidelberg • London
New York • Oxford • Paris • San Diego
San Francisco • Singapore • Sydney • Tokyo
Butterworth-Heinemann is an imprint of Elsevier

BH

Butterworth-Heinemann is an imprint of Elsevier
The Boulevard, Langford Lane, Kidlington, Oxford OX5 1GB, United Kingdom
50 Hampshire Street, 5th Floor, Cambridge, MA 02139, United States

Library of Congress Cataloging-in-Publication Data
A catalog record for this book is available from the Library of Congress

British Library Cataloguing-in-Publication Data
A catalogue record for this book is available from the British Library

ISBN: 978-0-12-805276-1

For information on all Butterworth-Heinemann publications
visit our website at https://www.elsevier.com/

Publisher: Joe Hayton
Acquisition Editor: Ken McCombs
Editorial Project Manager: Peter Jardim
Production Project Manager: Mohana Natarajan
Cover Designer: Greg Harris

Typeset by TNQ Books and Journals

DEDICATION

The risk of cyclone deaths in coastal Bangladesh due to storm surge inundation "intensifies as global warming causes the sea to rise... This means [says Dick Warrick, at East Anglia University's climate research unit] that future tidal waves will come from a higher sea level and will submerge more land" (M. Simon, *The Guardian Weekly*, 12 May 1991).

This book is dedicated to all past and future climate change victims of coastal Bangladesh.

CONTENTS

AUTHOR BIOGRAPHY

Bimal Kanti Paul, PhD (Kent State University, Ohio, USA) is a Professor of Geography at Kansas State University, Manhattan, KS, USA, where he has been teaching for last 26 years. He specializes in geography of environmental hazards and disasters, health/medical geography, population geography, and quantitative analysis in geography. Dr. Paul has a very impressive publication record—to date he has published 91 papers in nearly 40 referred journals, two books (one co-author with Harun Rashid), 25 book chapters, 17 book reviews, and 32 other papers. Most of these publications are the outcomes of externally funded research grants, including three from the National Science Foundation. In four different times, Dr. Paul received fellowships from the American Institute of Bangladesh Studies. He was immediate past editor of the *Geographical Review*, a publication of the American Geographical Society. Prior to that, he was book review editor of *The Professional Geographer*, a quarterly journal of the American Association of Geographers (AAG). He is currently a member of the editorial boards of several professional journals, including *The AAG Review of Books*. He also served as a member of the Editorial Board of several professional social science journals.

Harun Rashid, PhD (University of Saskatchewan, Canada) is an Emeritus Professor at the University of Wisconsin-La Crosse, USA. Earlier, he taught at the University of Wisconsin-La Crosse (2004–2009); Lakehead University, Canada (1975–2004); University of Benin, Nigeria (1981–1982); and University of Dhaka, Bangladesh (1965–1969). Dr. Rashid has published peer-reviewed journal articles and book chapters on floodplain management and urban floods in Bangladesh, Canada, USA, and Nigeria. His other peer-reviewed articles include such diverse environmental topics as shoreline management on the Great Lakes, morphologic effects of dams and flood control channels (channelization projects), remedial action plans for pollution alleviation in the Great Lakes, choice modeling in water resources management, and media discourse on floods and climate change. He is the co-author of two books on climate change in Bangladesh (Rashid and Paul 2014;

Paul and Rashid 2017). Dr. Rashid is a member of the editorial board of four international journals: *Disasters*, *International Journal of Disaster Risk Reduction* (IJDRR), *Environmental Management*, and *The Arab World Geographer*. He had also a keen interest in research administration and served as the Acting Associate Vice President Research at Lakehead University for two and half years (2002–2004).

PREFACE

For more than last three decades, we have conducted and published many studies on environmental issues and natural hazards and disasters in Bangladesh, Canada, Nepal, Nigeria, and the United States. We conducted research on different types of disasters, including river and flash floods, tornadoes, cyclones/hurricanes, blizzards, earthquakes, tsunamis, droughts, and forest fires. We dealt with a wide range of issues associated with these disasters, such as hazard preparedness, public and household response, individual perception, awareness and interpretation of hazards, compliance with hazard warnings, evacuation behaviors, emergency relief, recovery and rebuilding efforts, disaster-induced mortality and morbidity, disaster damage and impacts, and individual and household coping mechanisms.

Disasters can strike any place. Therefore, our initial interest was in the place affected by such extreme events. Our focus started to shift to a specific place, i.e., Bangladesh, since the early 2000s. By that time interests in global warming and climate change, which were initially confined primarily among climate specialists and scientists, had received significant attention of policy makers, politicians, mass media, and many concerned groups, such as nongovernmental organizations. As our ancestral homeland is in Bangladesh, naturally we became very much concerned about potential impacts of global warming–induced sea level rise, which is likely to inundate large parts of coastal Bangladesh, potentially displacing millions of people from their homes. Since the publication of the first report of the Intergovernmental Panel on Climate Change, estimates of potential loss of land due to sea level rise in coastal Bangladesh have been revised several times. According to the most conservative estimates, about one-fifth of the land area of the country would be lost due to a sea level rise of 3 feet (1 m) by the middle of this century with a corresponding displacement of at least 15 million people. Some other estimates projected much larger-scale population displacement from Bangladesh.

Our concerns about climate change impacts on the environment and livelihoods of the people of Bangladesh were reflected in a co-authored book entitled *Climate Change in Bangladesh: Confronting Impending Disasters*, which was published in 2014 by Lexington Books. We devoted a significant part of that book to sea level rise and its implications for Bangladesh. Although the book was on the entire country, it had a significant section on coastal Bangladesh. The current book may be considered as a sequel to this publication, thanks mainly to Kenneth P. McCombs, Senior Acquisitions Editor of Science and Technology Books, Elsevier, who encouraged us to consider a more in-depth analysis of coastal hazards in Bangladesh. We are heavily indebted to him for prodding us to this project. He followed up the project at every stage of its progress with a professional zeal and inspiration. We missed several deadlines, but he never lost his confidence in us. His constructive suggestions and efficient handling of every aspect of the manuscript have helped us in completing this important task. We are also grateful to the team at Elsevier, particularly Jardim Peter, and Mohana Natarajan for their professionalism and patience.

Although this book is about climatic hazards in coastal Bangladesh, in each chapter of the book we have provided a global context before turning into the area of our interest. In writing this book, we used materials from numerous sources. However, we have drawn heavily from our own publications on natural/climatic disasters in Bangladesh, which span a period of at least three decades. After introducing climate change challenges for Bangladesh and the coastal region, we have covered in detail four climatic hazards that pose the greatest threats to human lives in coastal Bangladesh. Land use changes in the coastal region are then presented along with programs so far implemented to properly and efficiently manage the coastal zone of the country. The next two chapters are on nonstructural and structural adaptation to sea level rise. The concluding chapter provides a summary of the main findings and recommendations for confronting some of the main challenges of coastal Bangladesh.

For the successful completion of this project we have benefited from numerous contributions of many of our colleagues, friends, and students. Our colleagues both at Kansas State University and University of Wisconsin—La Crosse always provided support and encouragements in our academic endeavors, including writing of this book. We frequently consulted with them about our writings, and their suggestions and constructive criticisms helped us tremendously. We are also grateful to several international colleagues and friends who gave us permission to reproduce some of the photographs from their personal collections. Dr. Emdad Haque (Professor, NRI, University of Manitoba, Canada) provided some of the rare historical pictures of damages from the 1991 Cyclone Gorky. The credit for the picture on shrimp polders (*ghers*) goes to Dr. Salim Momtaz (Associate Professor, University of New Castle, Australia). Two pictures in chapter 2 were provided by Dr. M. Khaledur Rahman (Post-Doctoral

Fellow, Center for Sustainability, Saint Louis University, St. Louis, MO, USA), whereas pictures of cyclone shelters were provided by MD Nawfel Huda and Khairul Matin (consultants). We acknowledge Dr. Masud Karim (engineering consultant) for pictures on the Sundarbans mangrove forest. We are grateful to Dr. Mohammad Zaman (an international consultant for the World Bank, ADB, Government of Bangladesh and others), who delivered some of the above pictures to us through his contact with Dr. Karim and two of his colleagues (i.e., Huda and Matin). We would also like to take this opportunity to pay homage to two of our inspiring teachers, mentors, and esteemed colleagues, the late Professor Aminul Islam and the late Professor Moniruzzaman Miah, who inspired us in geographical studies on natural hazards and disasters in Bangladesh.

Finally, we are deeply indebted to our family members, particularly to Mohsina Rashid and Anjali Paul, both of whom have been an enduring source of inspiration and have sacrificed countless hours of family life for the successful completion of this book project. Our children (Anjana Paul, Archana Paul, Rahul Deb Paul, Moona Rashid-Tyres, Zaid Rashid, and son-in-law Steve Tyres) and grandchildren (Shaun Tyres and Raiyah Tyres) have been waiting for the completion of this project. Their love, encouragement, and support over the years and particularly during the time we were busy writing this book have been a source of constant encouragement and inspiration.

Bimal Kanti Paul, Manhattan, Kansas, USA
Harun Rashid, Maple Ridge (Greater Vancouver), BC, Canada
July 2016

CHAPTER ONE

Introduction

CLIMATE CHANGE AS A GLOBAL PROBLEM

Who Are Polluting Our Atmosphere?

Climate change has emerged as a pervasive environmental problem for the entire world. Increasing concentration of greenhouse gases in the atmosphere is the root cause of this problem. Since most of the countries of the world have contributed more or less certain amounts of greenhouse gases into the atmosphere it can be argued that climate change is a collective global responsibility. Yet, data in Table 1.1 show that only a handful of industrial nations have emitted disproportionate amounts of greenhouse gases into the global atmosphere. This is the global context for opening this chapter and this book with a provocative question "who are polluting our atmosphere?"

Our question is based on an assumption (or more like an assertion) that the atmosphere is a global common property (called "a global commons") that is shared by all citizens of the world. This assumption is rooted in one of the basic physical characteristics of the atmosphere that it is an open system that exchanges energy and matter freely throughout the entire global atmosphere. The exchange of energy takes place through different forms of radiation. Solar radiation is the most basic form of energy for the atmosphere. Earth radiation (also called thermal infrared radiation) is another form of energy that is radiated by the earth surface into the atmosphere. Matters in the atmosphere include gas molecules (such as nitrogen, oxygen, carbon dioxide, ozone, and others), water vapor, dust particles, aerosols, and different types of particles from volcanic eruptions. Differences in air temperature from one place to another result in differences in air pressure which generates winds. At the global scale, a number of major wind systems called prevailing winds, such as the southeasterlies, the northeasterlies, southwesterlies, northwesterlies, and polar easterlies, transport both energy and matter across the latitudes. Thus, colder winds from the polar region are transported to tropical areas, whereas warmer tropical winds are carried to higher latitudes. Obviously, these exchanges of energy and matter do not follow any political boundary. Thus, warmer air temperatures attributable to

Climatic Hazards in Coastal Bangladesh
ISBN 978-0-12-805276-1
http://dx.doi.org/10.1016/B978-0-12-805276-1.00001-6

Table 1.1 Geography of Carbon Dioxide Emissions in 1990, 2000, and 2010

1990 Total Global Emissions: 22 Gt[a]	2000 Total Global Emissions: 25 Gt	2010 Total Global Emissions: 33 Gt
USA (22.62% of global total)[b]	**USA** (23.13% of global total)	China (26.36% of global total)
China (11.21)	China (13.88)	**USA** (16.37)
Russia (11.06)	**Russia** (6.56)	India (5.38)
Japan (5.27)	**Japan** (5.03)	**Russia** (5.18)
Germany (4.63)	India (4.18)	**Japan** (3.75)
Ukraine (3.49)	**Germany** (3.42)	**Germany** (2.49)
India (2.99)	**Canada** (2.17)	South Korea (1.79)
UK (2.67)	**UK** (2.15)	**Canada** (1.68)
Canada (2.03)	**Italy** (1.81)	**UK** (1.54)
Italy (1.93)	South Korea (1.77)	Mexico (1.39)
Top 10 countries: 68%	**Top 10 countries: 64%**	**Top 10 countries: 66%**
France (1.78)	**France** (1.61)	Indonesia (1.37)
Poland (1.41)	Mexico (1.49)	Brazil (1.32)
Mexico (1.41)	**Australia** (1.41)	**Australia** (1.31)
Australia (1.24)	**Ukraine** (1.39)	Saudi Arabia (1.29)
South Africa (1.22)	Brazil (1.36)	**Italy** (1.27)
Kazakhstan (1.16)	Iran (1.35)	**France** (1.19)
South Korea (1.14)	South Africa (1.22)	Iran (1.17)
Spain (1.03)	**Spain** (1.21)	**Poland** (1.01)
Brazil (0.99)	Indonesia (1.16)	South Africa (1.00)
Iran (0.93)	**Poland** (1.14)	**Ukraine** (0.92)
Top 20 countries: 80%	**Top 20 countries: 77%**	**Top 20 countries: 78%**
Contributions by remaining 167 countries: 20% (each country contributed between 0.01% and 0.9%)	**Contributions by remaining 167 countries: 23%** (each country contributed between 0.01% and 0.9%)	**Contributions by remaining 167 countries: 22%** (each country contributed between 0.01% and 0.9%)[c]

[a]See endnote 2 for an explanation of the short form Gt.
[b]Countries in bold indicate Annex I countries (see endnote 1).
[c]In 2010, Bangladesh contributed 0.35% of global total greenhouse gas emissions.
Prepared by authors based on original data from EDGAR tables Olivier, J.G.L., Janssens-Maenhout, G., Muntean, M., Peters, J.H.A.W., 2014. Trends in Global CO_2 Emissions – 2014 Report. PBL Publications, The Hague.

greenhouse gas emissions in China or the United States may reach distant places like Siberia or Greenland or Bangladesh, crossing many countries in its path. Because of such free movement of air (and air pollution) throughout the global atmosphere it is not an unreasonable question to ask who are polluting our atmosphere. The answer may already be known in general, but we have prepared Table 1.1 to demonstrate that only a limited few industrialized countries are responsible for the bulk of these emissions.

Data in Table 1.1 indicate that only 10 countries (out of 187 ranked) had been responsible for about two-thirds of the total carbon dioxide emissions for past several decades (1990–2010). Of them, six to eight countries were Annex I nations.[1] One of the most challenging trends for this period had been that, despite UNFCCC's efforts to curb greenhouse gas emissions, total carbon dioxide emissions (by far the most voluminous anthropogenic greenhouse gas) had increased from 22 gigatonnes (Gt) in 1990 to 25 Gt in 2000 and 33 Gt in 2010.[2] Another recent change (since 2010) has been the meteoric rise of China to the top as the infamous polluter of the global atmosphere (contributing 26% of global total CO_2 emissions), outdistancing the United States to the second place (16% of global total). When we expand the list beyond the top 10 countries, the data reinforce staggering geographical disparity in CO_2 emissions. Out of 187 countries ranked, only top 20 countries were responsible for about four-fifths of the total global emissions. The remaining one-fifth was contributed by the bottom

[1] The United Nations Framework Convention on Climate Change (UNFCC) divides countries into three main groups according to differing commitments for greenhouse gas emission reductions. Annex I Parties include the industrialized countries that were members of the OECD (Organization for Economic Co-operation and Development) in 1992, plus countries with economies in transition (the EIT Parties), including Russian Federation, the Baltic States, and several Central and Eastern European States. http://unfccc.int/parties_and_observers/items/2704.php ....

[2] Water vapor is by far the most voluminous natural greenhouse gas (much more voluminous than the combined values of natural carbon dioxide and anthropogenic carbon dioxide). There is nothing called anthropogenic water vapor because human activities are not capable of increasing atmospheric vapor indefinitely because of the feedback of the hydrologic cycle. Any excess vapor simply returns back to the earth surface by precipitation. In contrast, carbon dioxide is the most voluminous anthropogenic greenhouse gas that has been increasing its atmospheric concentration. Although anthropogenic methane has also been increasing its atmospheric concentration and it has a much higher global warming potential, its total concentration is much smaller than carbon dioxide. For this reason, in Table 1.1 we have listed emission data for only carbon dioxide, the most important anthropogenic greenhouse gas. All data in Table 1.1 are in gigatonnes. The term gigatonne is used as a part of the new metric SI units though it is not a proper SI unit. In the United States, the term *tonne* is expressed as a metric ton which is equivalent to 1000 kg (2204 pounds). One gigatonne is equal to one billion tons, i.e., 1×10^9 tons.

167 countries. Bangladesh was one of these countries, contributing only 0.35% of global greenhouse gas emissions in 2010 (Table 1.1).

Local Impacts of a Global Problem

It appears from the preceding analysis that climate change is a global problem because increasing amounts of greenhouse gases in the atmosphere (responsible for global warming and climate change) are being circulated freely by prevailing winds of the global atmosphere. Consequently, the climate of every country of the world has been affected adversely by global warming, but the nature of resulting climate change impacts has been highly uneven (Houghton, 2006). Poorer third world countries have been impacted more adversely than the industrial nations. In other words, third world countries like Bangladesh, which contributed very little greenhouse gases into the global atmosphere, have been disproportionately vulnerable to climate change impacts because of their biophysical and socioeconomic characteristics. Based on some of these biophysical and socioeconomic attributes, Maplecroft (2010, 2016) has developed a Climate Change Vulnerability Index to rank different nations of the world by their varied physical exposure to climate change and their government's "ability to adapt to climate change over the next 30 years."[3] Bangladesh ranks as the "first and most at risk" to climate change among 32 extreme risk countries. While such ranking is helpful for identifying climate change–vulnerable countries of the world, for planning purposes there is now a growing demand "among stakeholders across public and private institutions for spatially-explicit information regarding vulnerability to climate change at the local scale" (Preston et al., 2011). In recent years, the importance of identifying climate change vulnerability at the local scale has been gaining increasing attention in both scholarly work and planning

[3] Maplecroft, a British global risk analysis company, has applied the concept of vulnerability to develop an aggregate Climate Change Vulnerability Index (CCVI) for 170 countries, to assess climate change vulnerability of each of the nations (Maplecroft, 2010). In developing this index, Maplecroft evaluated 42 social, economic, and environmental factors to assess national vulnerabilities across three core areas: (1) exposure to climate-related natural disasters and sea level rise; (2) human sensitivity, in terms of population patterns, development, natural resources, agricultural dependency, and conflicts; and (3) future vulnerability by considering the adaptive capacity of a country's government and infrastructure to combat climate change. In its 2010 CCVI Bangladesh ranked at the top (#1) among 16 countries who were at "extreme risk" for climate change impacts, "facing the greatest risks to their populations, ecosystems and business environments" (Maplecroft, 2010). India ranked second. Other South Asian countries ranked in the highest category were Nepal, Afghanistan, and Pakistan (Rashid and Paul, 2014, p. 16). See endnote 1 in Chapter 3 for the 2015 update of this index.

initiatives. As an example of the latter, in a recent Website of the Columbia Basin Trust (CBT) (of the Canadian Columbia River Basin in southeastern British Columbia), its message urges its readers to "Start Thinking About Local Impacts" of climate change, which are "either direct or indirect effects of a changing climate on natural, human or built systems" (CBT, 2016). The Website further expands that climate change is likely to impact its local water supply, hydroelectric power, municipal and regional infrastructure (including roads and rail lines that are expected to be affected adversely by increased flooding), smoke from wildfire and related public health problems, and so on (http://www.cbt.org/uploads/pdf/LocalImpactsFactSheet.pdf).

While mountain ecosystems, typified by the Columbia Basin, are highly sensitive to climate change impacts, global warming–induced sea level rise poses a more direct threat to coastal ecosystems and coastal population in many low-lying coasts of the world. Climate change vulnerability of coastal zones has received significant attention in scholarly research. Hammer-Klose et al. (2003), for example, have assessed vulnerability of the Cape Cod National Seashore (in Massachusetts, USA) to sea level rise. Focusing mainly on coastal erosion, Alpar (2009) has assessed vulnerability of Turkish coasts to accelerated sea level rise, whereas Demirkesen et al. (2008) presented a method of "quantifying coastal inundation vulnerability of Turkey to Sea-Level Rise." Most of the smaller Pacific islands are extremely vulnerable to sea level rise. Gravelle and Mimura (2008) have analyzed how Viti Levu, one of the islands of Fiji, is vulnerable to sea level rise. Among a variety of publications on climate change and sea level rise in Bangladesh, Karim and Mimura (2008) assessed vulnerability of coastal Bangladesh to cyclone storm surge impacts with rising sea levels, using a numerical hydrodynamic simulation. Further, using a digital elevation model, they simulated potential depths of seawater flooding in low-lying coastal areas due to sea level rise. In contrast to such hydrodynamic modeling, we have conducted comprehensive analyses of different types of climatic hazards in coastal Bangladesh, using a variety of empirical data (both primary and secondary data) and an extensive literature review.

In the global context, Bangladesh ranks as the most vulnerable country (MVC) for climate change impacts because it has a very high population density, a climate-sensitive subsistence economy (mostly rice monoculture), and relatively low levels of technological amenities to deal with climate change impacts. Bangladesh is one of the smallest countries in the world in terms of its area (56,900 mi^2 or 147,500 km^2), which is about the size of Iowa, USA (Fig. 1.1).[4] However, in terms of its population (160 million in 2015) Bangladesh ranks as the eighth largest country in the world. Because of high density of population—2812 persons per square mile or 1085 persons per square kilometer—large numbers of people have no choice but to continue to live in different types of hazard-prone lands. Among these, coastal Bangladesh, constituting about 30% of the land area of Bangladesh, represents the most hazard-prone area of the country as it confronts the impending disasters of rising sea levels and increased magnitudes of tropical cyclones and storm surges due to climate change impacts.

Climate Change in Bangladesh: A Synopsis

Compared to the nature of climate change vulnerability of coastal Bangladesh, most of the climate change impacts in the rest of the country have been much more subtle. Some of the details of climate change in Bangladesh have been presented elsewhere (see Shamsuddin et al., 2015; Brammer, 2014; Rashid and Paul, 2014). As a context of our study, here we have summarized (below) climate change impacts on air temperature and rainfall in Bangladesh, the two most basic parameters of climate.

Temperature Trends

The *average (mean) annual* temperature in Bangladesh has increased from 75.24°F (24.02°C) in 1900–1930 to 77.58°F (25.32°C) in 1990–2012, i.e., an increase of 2.34°F (1.3°C) in 112 years (Table 1.2). These rates are

[4] Bangladesh is the eastern part of Greater Bengal of British India prior to their departure in 1947. Bengal had a tumultuous relationship with the British who landed on the coast of West Bengal in 1757 in guise of merchants but outmaneuvered the local *nawab* (governor of the Mughal rulers of India) in the historic Battle of Plassey and laid the foundation of an expanding colonial rule in India within a relatively short period of time. During the peak of the British rule (1757–1947) Bengal was the largest and most prosperous province of India—a far cry from the current impoverished Bangladesh. Calcutta, the initial capital of British India (before it was relocated to Delhi), was the largest metropolis of the British Empire outside London. In 1947 when the British left India Bengal was partitioned into two parts. West Bengal (Hindu majority area) became a state/province of India with Calcutta (later renamed as Kolkata) as its capital. East Bengal (Muslim majority area and later renamed as East Pakistan) became the western wing of Pakistan. The union of East Bengal/East Pakistan with West Pakistan, which was separated from each other by about 1000 miles, was a geographical and geopolitical oddity from the beginning of the creation of Pakistan. Following a 9-month long bloody civil war over East Pakistan's demand for complete regional autonomy, East Pakistan seceded from West Pakistan and became the independent nation of Bangladesh in December 1971. Dhaka (earlier known as Dacca), the capital of Bangladesh, has a population of at least 14 million and is one of the largest and most crowded metropolises of the Indian subcontinent.

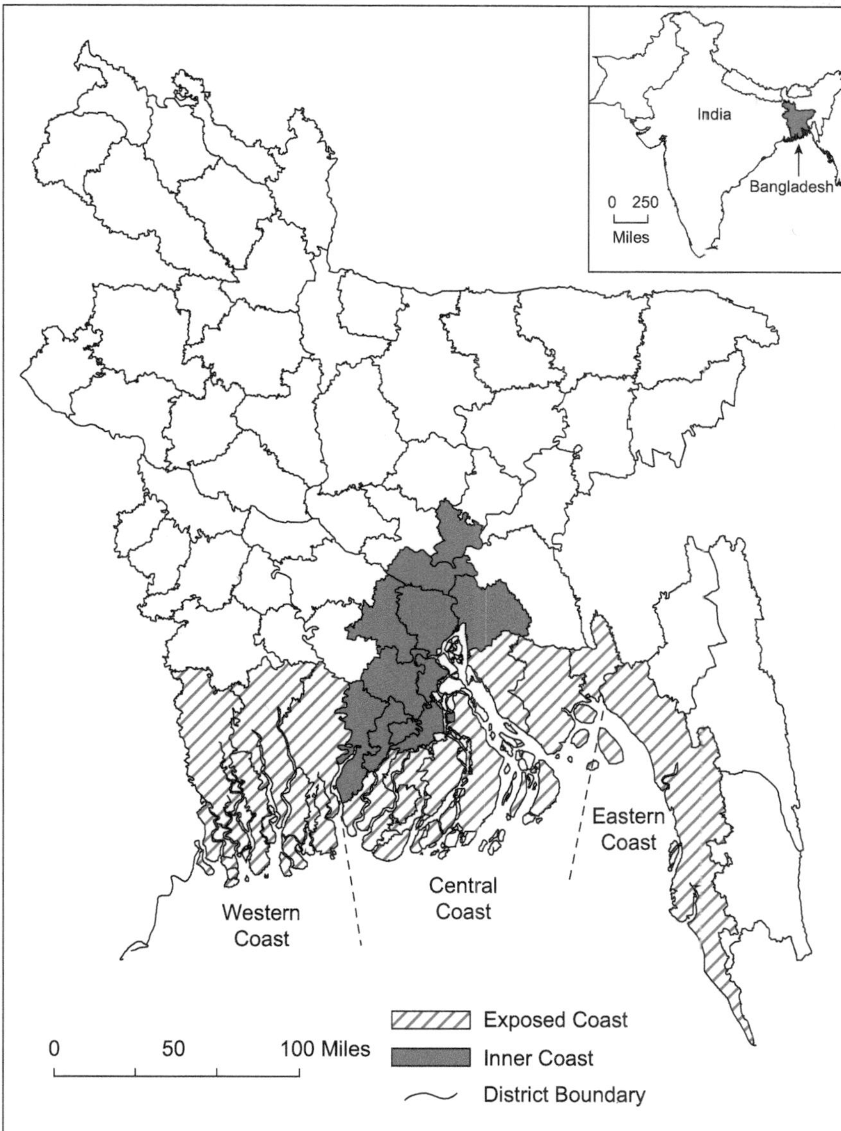

Figure 1.1 Map of Bangladesh highlighting the area of coastal Bangladesh.

Table 1.2 Average (Mean) Monthly Temperatures and Rainfall in Bangladesh, 1900–2012

Month	Mean Monthly Temperature in °F (°C)[a] 1900–1930	Mean Monthly Temperature in °F (°C) 1990–2012	Mean Monthly Rainfall in Inches (mm)[b] 1900–1930	Mean Monthly Rainfall in Inches (mm) 1990–2012
January	63.86 (17.7)	63.3 (18.5)	0.2 (5.3)	0.24 (6.1)
February	67.28 (19.6)	70.16 (21.2)	0.56 (14.3)	0.62 (15.7)
March	74.66 (23.7)	77.54 (25.3)	1.4 (35.6)	2.1 (53.4)
April	79.34 (26.3)	81.68 (27.6)	4.77 (121.4)	5.9 (149.9)
May	80.42 (26.9)	82.76 (28.2)	11.88 (294.2)	10.9 (276.9)
June	80.42 (26.9)	82.94 (28.3)	20.37 (517.5)	18.73 (475.8)
July	80.60 (27)	82.40 (28)	19.92 (506.2)	20.97 (532.8)
August	80.42 (26.9)	82.76 (28.2)	17.46 (443.7)	16.94 (430.3)
September	80.60 (27)	82.40 (28)	10.42 (264.8)	13.2 (335.4)
October	78.08 (25.6)	80.60 (27)	6.13 (155.9)	7.37 (187.1)
November	71.78 (22.1)	74.66 (23.7)	0.94 (24)	0.99 (25.1)
December	63.3 (18.5)	67.64 (19.8)	0.23 (5.9)	0.24 (6.1)
Year	**Annual average: 75.24 (24.02)**	**Annual average: 77.58 (25.32)**	**Annual total: 94.04 (2388.80)**	**Annual total: 98.21 (2494.6)**

[a]Original temperature data in °C in parenthesis; °F units are converted data.
[b]Original rainfall data in mm in parenthesis; inches are converted data.
Prepared by authors based on SWMC (Surface Water Modelling Centre) and BWDB (Bangladesh Water Development Board). 2000. Late Monsoon Flood in the Southwest Region of Bangladesh 2000. Dhaka: Government of Bangladesh, Ministry of Water Resources. *Climate Change Knowledge Portal* (CCKP). See endnote 5.

based on a dataset presented in World Bank's (2016) *Climate Change Knowledge Portal* (CCKP). Although these data are helpful for determining long-term temperature trends in Bangladesh, CCKP data are aggregates only without specifying the details of its source.[5] In contrast, the Climate Change Cell (CCC, 2008) of the Government of Bangladesh has provided a comprehensive analysis of changes in air temperature and rainfall for a period

[5] The World Bank's Climate Change Knowledge Portal (CCKP) is supported by the Global Facility for Disaster Reduction and Recovery. It is an online tool (a web platform) for access to comprehensive global, regional, and country data related to climate change and development. The CCKP dataset was produced by the Climate Research Unit (CRU) of the University of East Anglia, UK, but the portal does not specify the primary source of its dataset. http://sdwebx.worldbank.org/climateportal/.

of 60 years (1948–2007), based on data from 34 meteorological stations. According to this source, the *mean annual temperature* in Bangladesh during this period (1948–2007) increased at a rate of 2.34°F/100 years (1.3°C/100 years). Perhaps, more significantly, warming rates accelerated in the later part of the record (1980–2007), at about 3.85°F/100 years (2.14°C/100 years) (CCC, 2008). This rate was about 65% higher than the 60-year trend. Temperature trends for the second part of the record (i.e., 1980–2007) also showed significant seasonal differences in warming. The mean monthly temperatures in the monsoon season (June–October) registered an increase of +4.39°F/100 years (+2.44°C/100 years). The summer temperatures (March–May) increased at a rate of +4.32°F/100 years (+2.15°C/100 years), whereas winter temperatures increased at about +2.39°F/100 years (+1.33°C/100 years) (CCC, 2008).

The long-term (1948–2007) temperature trends also showed significant spatial variations in warming, irrespective of the relatively small size of Bangladesh. In a supposedly uniform hot and humid tropical climate, it is difficult to explain widespread station-to-station differences in warming trends. At best, the differences seem to be random as there are no systematic spatial patterns. The largest amount of warming occurred at Bogra and Faridpur (+4.96°F/100 year), both of which are interior stations. Dhaka and Jessore, which are also interior stations, registered significant warming (+3.78°F/100 year). Contrary to the general warming trend, three interior stations registered minor decreases (negative values) in warming, although these rates were not statistically significant. All of the remaining interior stations experienced either very low or moderate increases in mean annual temperatures (Table 1.3). Perhaps, the most significant finding from data in Table 1.3 is that all of the three coastal stations registered significant increases in mean annual temperatures, defying expected moderating effect of marine locations on air temperature. In particular, Cox's Bazar, situated literally on the coastline, registered the second highest increase in its mean annual temperature (+4.68°F/100 year). In short, thermal climate of coastal Bangladesh does not differ significantly from that of inland (interior) climate.

Rainfall Trends

Most of the climate change models predict increased rainfall in tropical monsoon climate as an outcome of global warming, assuming that higher temperatures are likely to increase evaporation, condensation, and cloud

Table 1.3 Temperature Trends at Selected Stations in Different Regions of Bangladesh, 1948–2007

Region	Station	Years of Record	Temperature Increases/ Decreases[a] °F/100 years (°C/100 years)
Northwest	Dinajpur	52	−0.72 (−0.4) (NS)
	Rangpur	45	0.18 (0.1) (NS)
	Bogra	56	4.96 (2.8)[a]
	Rajshahi	43	2.16 (1.2)[a]
West	Jessore	59	3.78 (2.1)[a]
Central	Dhaka	55	3.78 (2.1)[a]
	Faridpur	57	4.96 (2.8)
Northeast	Sylhet	51	1.98 (1.1) (NS)
	Srimangal	58	2.16 (1.2)[a]
East	Comilla	56	0.36 (0.2) (NS)
Southeast (inland)	Rangamati	51	−1.98 (−1.1)[a]
Southwest	Khulna	57	−0.18 (−0.1) (NS)
	Satkhira	58	1.44 (0.8)[a]
Coastal	Bhola (south-central)	41	3.86 (2.2)[a]
	Chittagong (southeast)	59	2.88 (1.6)[a]
	Cox's Bazar (southeast)	60	4.68 (2.6)[a]

[a]Changes are significant at 95% level. NS means not significant at 90% level.
Prepared by authors based on selected data from CCC (Climate Change Cell), 2008. Characterizing Long-term Changes of Bangladesh Climate in Contexts of Agriculture and Irrigation. Government of Bangladesh, Ministry of Environment and Forests, Department of Environment, Component 4b, and Ministry of Food and Disaster Management, Comprehensive Disaster Management Programme, Dhaka. http://www.climatechangecell-bd.org/.

formation leading to increased rainfall. The long-term data for Bangladesh (1900–2012) confirms this assumption as its mean annual rainfall has increased by about 4.4% (+4 inches) (Table 1.2). However, in-depth analyses by the Climate Change Cell (CCC, 2008), using monthly data for a period of 52 years (1960–2001), indicate that long-term rainfall trends are characterized by significant decadal, seasonal, and spatial divergences, implying that Bangladesh does not necessarily have a homogenous tropical monsoon climate (Table 1.4). Even within this compact country regional variations in rainfall are striking. We have analyzed these variations by two timelines: (1) by decades between 1960 and 2001 and (2) by averaging data for two decades, between 1960–1980 and 1980–2001.

Table 1.4 Rainfall Trends in Bangladesh, 1960–2001

Rainfall Increases or Decreases (−) Per Decade: 1960–2001	**Bogra (Northwest) Rainfall in Inches (mm)**[a]	**Dhaka (Central) Rainfall in Inches (mm)**	**Sylhet (Northeast) Rainfall in Inches (mm)**	**Chittagong (East Coast) Rainfall in Inches (mm)**
Summer (MAM)	1.24 (31.51)	1.1 (27.9)	3.25 (82.45)	2.52 (63.89)
Monsoon (JJASO)	1.74 (44.28)	−1.06 (−26.9)	−0.25 (−6.32)	−1.13 (−28.82)
Winter (NDJF)	0.14 (3.49)	0.52 (13.24)	−0.8 (−20.33)	0.78 (19.84)
Rainfall increases or decreases (−) per **two decades**	Bogra (northwest) rainfall in inches (mm)	Dhaka (central) rainfall in inches (mm)	Sylhet (northeast) rainfall in inches (mm)	Chittagong (east coast) rainfall in inches (mm)
Summer 1960–1980	9.9 (251.6)	18.84 (478.6)	36.35 (923.2)	15.68 (398.3)
Summer 1981–2001	12.66 (321.6)	20.84 (529.3)	45.22 (1148.6)	19.8 (503)
Change (%)	+28%	+11%	+24%	+26%
Monsoon 1960–1980	53.81 (1367)	60.07 (1526)	117.2 (2977)	92.13 (2340)
Monsoon 1981–2001	54.68 (1389)	57.64 (1464)	118.34 (3006)	91.46 (2323)
Change (%)	+1.6%	−4.23%	+0.97%	−0.73%
Winter 1960–1980	1.5 (38.1)	1.9 (48.2)	5.51 (139.9)	3.07 (77.9)
Winter 1981–2001	1.81 (46.1)	3.13 (76.9)	3.57 (90.7)	4.59 (116.7)
Change (%)	+21%	+65%	−35%	+50%

[a]Original rainfall data are in mm in parenthesis; inches are converted data.

Prepared by authors based on selected data from CCC (Climate Change Cell), 2008. Characterizing Long-term Changes of Bangladesh Climate in Contexts of Agriculture and Irrigation. Government of Bangladesh, Ministry of Environment and Forests, Department of Environment, Component 4b, and Ministry of Food and Disaster Management, Comprehensive Disaster Management Programme, Dhaka. http://www.climatechangecell-bd.org/.

Rainfall trends by decades differed significantly among four sample stations and during three different rainfall seasons.[6] These may be summarized as follows:

- *Pre-monsoon* (March–May) rainfall increased at all of the four sample stations, ranging from an increase of 1.1 inches/decade at Dhaka to 2.52 inches/decade at Sylhet (Table 1.4). The CCC has characterized this period as summer rainfall.
- *Monsoon* (June–October) rainfall: Contrary to the assumptions of most of the climate change models, three out of the four stations (Dhaka, Sylhet, and Chittagong) registered decreases in monsoon rainfall, ranging from −0.25 inches/decade to −1.3 inches/decade.
- *Winter* (November–February) rainfall increased in three out of four stations, ranging from +0.14 inches/decade at Bogra to +0.78 inches/decade at Chittagong. Paradoxically, Sylhet, the rainiest of the four stations, registered a decline of winter rainfall by −0.8 inches/decade.

When data were averaged by two decades for determining changes from 1960–1980 to 1980–2001, most of the decadal trends were retained with only minor modifications:

- *Pre-monsoon (summer) rainfall* increased at all of the four stations in the second part of the record (i.e., in 1980–2001), by 11%–28%.
- *Monsoon rainfall*: Decline in monsoon rainfall was evident from two stations: at Dhaka by −4.23% and at Chittagong by −0.73%.
- *Winter rainfall*: Most of the climate change models predict increasing droughts in the dry season. Contrary to such projections, winter (dry season) rainfall increased significantly in three stations (by 21%–65%), whereas Sylhet registered a 35% decline.

[6] Rainfall seasons in Bangladesh may be classified into three seasons: (1) premonsoon (March, April, and May), (2) monsoon (June–October), and (c) winter (November–February). The winter has the lowest rainfall and is normally considered as the dry season. We have used the term premonsoon rainfall in relation to the monsoon season. There is no popular term for it in Bangladesh, although Johnson (1982, p. 30) has used a vernacular term *chota barsat* (literally meaning "little rains"), which may be used in some areas of the country. He has also provided a tentative explanation of this rainfall in terms of the spring convection rainfall (associated with thunderstorm squalls), often called the norwesters (again this is not a popular term in the current literature). The main summer season in Bangladesh coincides with the monsoon season, although April is the warmest month at many stations. Somewhat paradoxically, the GoB Climate Change Cell (CCC) has used the term summer rainfall to refer to rainfall in the premonsoon season (March, April, and May).

COASTAL BANGLADESH

Definitions

Coastal Bangladesh is a frequently used term for the *coastal zone* of Bangladesh, which is difficult to define precisely. Emphasizing that the term coastal zone is a fuzzy concept, Gattuso and Smith (2011) have made an attempt to define it as a zone that includes "the coastal ocean as well as the portion of the land adjacent to the coast" (Gattuso and Smith, 2011). Based on these criteria, we define coastal Bangladesh as the coastal waters of the Bay of Bengal and a strip of land that extends inland from its coastline up to a distance that may be estimated based on different criteria.

Physical Characteristics of a Coastal Zone

While it is difficult to define coastal zone precisely some of its physical features, especially along the edge of water, can be defined objectively. The coastline (also called shoreline) is the line of separation between land and water. This line fluctuates with the diurnal regime of high and low tides. The sloping surface between the high tide and the low tide is called shore (or shore surface), which may be composed of bare rock or may be covered with beach materials, such as gravel or sand. Seaward of the coastline the continental shelf slopes gently toward the ocean terminating at the steep continental slope. Landward the beach may extend further up from the high tide level by wave action, especially during high storm waves. Coastal dunes behind most of the sandy beaches are one of the common morphological features of marine and aeolian processes (i.e., wave and wind actions). All coastal zones do not necessarily have beaches and coastal dunes because these are characteristics of sandy shores only. The main ecosystems of the coastal zone are "estuaries, macrophyte communities, mangroves, coral reefs, salt marshes and the rest of the continental shelf" (Gattuso and Smith, 2011).

Physical Characteristics of Coastal Bangladesh

The physical characteristics of coastal Bangladesh are highly diverse incorporating many of the morphological and ecological features reviewed above. There are also significant morphological differences between the western and the eastern sections of the coast. The western section, situated in the active delta of the Ganges, consists of an intricate network of river floodplains associated with the distributaries of the Ganges, the Brahmaputra, and the Meghna (GBM) Rivers. Closer to the coastline, large sections of these floodplains are inundated by daily regimes of astronomical tides. Therefore, these are often called tidal plains. To minimize confusion of terms for different segments of the floodplains, we may refer to all of the plains in the coastal zone as coastal plains. Overall, topographic slopes of all coastal plains are extremely low, ranging from about 1.5 inches/mile in the west to 1 inch/mile in the central section of the plains. In contrast, topographic slopes in the east coast are much steeper, since a narrow strip of coastal plain rises relatively steeply towards the north-south trending hills paralleling the coast, called Chittagong Hill Tracts.[7] The 25 ft contour, for example, runs parallel to the coastline within a distance of only 10–20 miles from the coast. Because of such topographic differences between the two sections of the coast, the coastal zone extends inland up to only 20–25 miles (30–40 km) in the east coast compared to a maximum of 120 miles (195 km in the west coast).

The tidal ranges are high along the coast. They vary within the range of 6.5–13 ft (2–4 m) in the western and central coastal zones. The tidal rage is more than 13 ft (4 m) in the northern part of the eastern zone (MOA, 2012). Intertidal areas are found almost all along the coastline. They comprise of alluvial soils and have low slope. These areas are inundated during diurnal high tides while falling dry at low tides. On many of these intertidal areas, new alluvial land is formed continuously, creating new cultivable land. Because alluvial soils are rich in nutrients (algae, for example) fertile intertidal areas play an important role in crop production in the coastal region (MOA, 2012).

Estimated Areas and Administrative Districts

Against this backdrop, the term coastal Bangladesh has been used in the literature to refer to about one-third of the land area of the country without providing the details of such estimates (see, for example, Shansuddoha and Chowdhury, 2007). Others have listed names of coastal and several adjacent districts as the coastal districts of Bangladesh without specifying the details of such a designation (see Alam and Ahmed, 2015).[8] For refining such

[7] Chittagong Hill Tracts consist of a set of north-south trending parallel ridges of sedimentary rocks of Miocene and Pleistocene geologic ages (Johnson, 1982). The formations are mainly sandstones, sandy clays, shales, and siltstones. The ridge crests range from about 1000 ft (300 m) in the district of Chittagong to 3300 ft (1000 m) along the Bangladesh-Myanmar border (Johnson, 1982).

[8] A district is the second largest administrative unit in Bangladesh, with an average population of 2.5 million.

estimates, we have employed two specific criteria in this study for identifying administrative areas of coastal Bangladesh:

- *Direct location on the coastline*: Eleven administrative districts are situated directly on the Bay of Bengal coastline, that is, each of these districts shares a segment of the coastline (Table 1.5a).
- *Estuarine and coastal influences*: Eight districts adjacent to some of the above districts do not share the Bay of Bengal coastline but are under significant estuarine and coastal influences, such as the daily regimes of astronomical tides and periodic impacts of cyclones and storm surges (Table 1.5b).

Estimated in this manner, about 30% of the total area of Bangladesh covering entire or parts of 19 districts (out of 64 districts), may be classified as coastal Bangladesh (Table 1.5). Based on the 2013 census update, coastal Bangladesh had a population of 39.41 million accounting for about 27% of the total population of the country.

Official Boundaries of Coastal Bangladesh

Several attempts had been made in the past to delineate boundaries of coastal Bangladesh but it was only in 2005 that the Government of Bangladesh (GoB) formally addressed this issue in its Coastal Zone Policy (CZP). The GoB formulated the CZP in 2005 to develop an integrated management plan for reducing coastal vulnerabilities, improving livelihoods of coastal people, ensuring effective use of coastal resources, and creating an enabling institutional environment. The CZP had been approved by the cabinet and subsequently published in the official Gazette (MWR, 2005). As per the Gazette notification, coastal Bangladesh includes 19 of the 64 districts of the country and the exclusive economic zone (EEZ) (Fig. 1.1). Three indicators have been considered for determining the landward boundaries of the coastal zone. These include (very similar to our criteria presented above) influences of (1) tidal waters, (2) salinity intrusion, and (3) cyclones/storm surges (MWR, 2005).

Among these 19 districts, 11 are coastal districts (Satkhira, Khulna, Bagerhat, Barguna, Patuakhali, Bhola, Lakshmipur, Noakhali, Feni, Chittagong, and Cox's Bazar), that is, each one of these districts contains a segment of the Bay of Bengal coastline (Fig. 1.1) (identical with our data in Table 1.5). The remaining eight districts (Pirojpur, Jhalokati, Barisal, Gopalganj, Jessore, Narail, Shariatpur, and Chandpur) do not directly share coastline with the Bay of Bengal (MWR, 2005). The first group of districts has been designated as "exposed coast," while the remaining eight districts have been characterized as "interior coast" (an oxymoron). Almost all of the districts of the

Table 1.5 Coastal Bangladesh: Districts, Areas, and Population

Coastal Zone	Districts	Area in mi^2 (km^2)[a]	Population[b] (millions)	Density/mi^2 (/km^2)[c]
(A) Districts on the Bay of Bengal Coastline				
Southwest	Satkhira	1474 (3817)	1.98	1347 (520)
	Khulna	1697 (4394)	2.31	1366 (528)
	Bagerhat	1529 (3959)	1.47	966 (373)
Central	Barguna	707 (1831)	0.89	1263 (488)
	Patuakhali	1244 (3221)	1.53	1235 (477)
	Bhola	1314 (3403)	1.77	1352 (522)
	Lakshmipur	556 (1440)	1.72	3109 (1200)
	Noakhali	1423 (3686)	3.10	2184 (843)
	Feni	382 (990)	1.43	3759 (1451)
Eastern	Chittagong	2040 (5283)	7.61	3734 (1442)
	Cox's Bazar	962 (2492)	2.28	2380 (919)
Subtotal: % of Bangladesh	Eleven districts on coastline	13,328 (34,518) 23%	26.16 18%	2063 (797)
(B) Inland Districts With Significant Coastal and Estuarine Influences				
Central	Pirojpur	493 (1278)	1.1	2256 (871)
	Jhalokati	273 (707)	0.68	2502 (966)
	Barisal	1075 (2785)	2.32	2162 (835)
	Madaripur	435 (1126)	1.16	2683 (1036)
	Shariatpur	453 (1174)	1.15	2550 (984)
	Munshiganj	388 (1004)	1.44	3728 (1439)
	Narayanganj	264 (684)	2.94	11,157 (4308)
	Chandpur	635 (1645)	2.41	3803 (1468)
Subtotal: % of Bangladesh	Eight inland districts with coastal influences	4017 (10,403) 7%	13.25 9%	3803 (1468)
Bangladesh total: % of Bangladesh	Coastal Bangladesh (19 districts)	17, 344 (44,921) 30%	39.41 27%	2959 (1143)

[a]Area in square miles (mi^2) and square kilometers (km^2).
[b]Population from the Bangladesh Bureau of Statistics (BBS) March 2013 updates.
[c]Density as persons per square mile (/mi^2) and per square kilometer (/km^2).
Prepared by authors based on data in Rashid, H., Paul, B., 2014. Climate Change in Bangladesh: Confronting Impending Disasters. Lexington Books, Lanham, MD, Boulder, CO, New York, Toronto, Plymouth, UK.

exposed coast contain one or more offshore islands. More than 70 islands have been identified in the coastal waters; most of these islands are small in size. The largest four islands are: Bhola (1702 mi^2 or 4408 km^2) Hatia (582 mi^2 or 1508 km^2), Swandip (294 mi^2 or 762 km^2), and Moheshkhali (103 mi^2 or 268 km^2). Some islands are either uninhabited or inhabited only during a particular season. Over time, new islands are formed by the process of land accretion, while old ones disappear due to erosion caused by strong winds, tidal activities, and storm surges. A former small island called Bholar Dweep, situated between Teknaf and St. Martin Island, disappeared by 1861. Similarly, South Talpatti Island, a disputed island along the India–Bangladesh international boundary, submerged under water by March 2010 (Rahman et al., 1993). This reflects the dynamic morphologic characteristics of the coastal waters.

Initially, newly formed coastal islands are inhabited by temporary settlers who are hired seasonally by absentee landlords for raising cattle, oxen, and buffaloes, or for lime production from shells and salt production. They build temporary shelters for the particular period of their stay in the islands. In course of time, some of these islands turned into permanent settlements (Elahi et al., 1995). The population density of the coastal zone is lower than the rest of Bangladesh. Within coastal Bangladesh there is no permanent settlement in the Sundarbans mangrove forests, which occupy a large part of the southwestern coastal zone (Paul, 2009).

Management of Coastal Zones

For management purposes, the coastal zone extends further inland beyond the direct influence of the coastal processes (MOA, 2012). In order to formulate an operational definition of coastal zone for management purposes, most countries of the world use some or a combination of four main elements: physical features, administrative boundaries, arbitrary distances from coastline, and selected environmental units (Ruddle and Manshard, 1981). In demarcating landward boundary, coastal managers and public authorities usually follow administrative boundaries. With climate change and sea level rise emerging as a major environmental concern of many countries, coastal zones across the globe have been receiving increasing attention of researchers, policy makers, general public, and many others. This is because the residents of coastal zones are directly vulnerable to global climate change, epitomizing "local impacts of a global problem," as we have alluded to it earlier. They are at the forefront of global warming and climate change. To confront impending disasters of sea level rise and other climate change impacts, many counties have already adopted different types of mitigation and adaptation measures. As indicated above, the GoB's response has been the formulation and adoption of the 2005 Coastal Zone Policy (CZP). Its overall goal is to create conditions for reductions of poverty, development of sustainable livelihoods, and integration of the coastal zone into national processes. The policy emphasizes preservation and enhancement of critical ecosystems and ecological processes (MWR, 2005). In 2012, the GoB formulated another Master Plan for Agricultural Development "to provide a road map for an integrated rehabilitation and development effort in Bangladesh's coastal zone aiming at sustainable food security, poverty reduction and livelihood development for poor" (MOA, 2012, p. 11). According to official poverty indicators, about 52% of the population of coastal Bangladesh lives below the poverty line, compared to the national average of 49% (Ahmad, 2005).[9] The 10-year plan (2012–2021) hopes to enhance productivity and balanced growth, increase employment, and provide access to food and nutrition for residents of coastal districts.

Coastal Resources

Economic Sectors

Climate change is likely to have direct and indirect impacts on coastal resources. Most vulnerable sectors are coastal agriculture, fisheries (including shrimp cultivation), forestry, salt industries, tourism, settlement, urban development, and infrastructures. Degradation of coastal land has already been reported due to the effects of increased salinity, flooding, and erosion. This has adversely affected lives and livelihoods of coastal residents as well as slowed down the pace of social and economic development in the region (MWR, 2005). Not only the coastal zone, the national economy and food security, biodiversity, ecosystem, and other sectors will also be adversely affected by climate change impacts. In addition to the diversity of natural resources, coastal region contains Export Processing Zones (EPZs), harbors, airports, land ports, and tourism complexes, and opportunities for other industries. This region has high potential for exploration of both onshore and offshore natural gas.

[9] Absolute poverty is defined in Bangladesh in terms of minimum dietary energy consumption per day per person. It is disaggregated into two levels: upper poverty line and lower poverty line. The former indicates a level of minimum dietary energy consumption of 2122 kcal/capita/day, while the latter indicates "extreme or hardcore poverty" and corresponds to dietary consumption of 1805 kcal/capita/day (BBS, 2010).

Coastal Ecosystems and Common-Property Resources

Coastal Bangladesh contains several ecosystems that have important conservation values. Most notably, the Sundarbans, the world's largest single tract of mangrove forest, lies along the southwestern coast of Bangladesh.[10] This World Heritage Site is known for its wide range of flora, fauna, and aquatic life. It is also an important common-property resource for local residents, especially for those whose subsistence depends on the forest from collection of fish, wood, and honey (Sarwar, 2005). Coral reefs along the St. Martin Island, the only coral island of the country—located near the southern end of the east coast, about six miles (10 km) offshore—provide another important ecosystem and an important tourist attraction (see section on southwestern coastal zone). Coastal fisheries are another important common-property resource, which has been providing food, employment, and income for coastal residents of Bangladesh for centuries.

Soils

In general, soils of the coastal region are less fertile and poor in organic matter content, ranging from less than 1%–1.5% (SRDI, 2001). The low organic content indicates poor physical condition of the coastal soils. The total nitrogen contents are generally low, mostly around 0.1%. This low nitrogen content may be attributed to low organic matter contents of most of the soils. Coastal soils are also deficient in phosphorous. Widespread zinc and copper deficiencies have been observed in the coastal regions (Karim et al., 1990). Overall, most of the top soils are acidic and subsoils are neutral to mildly alkaline. The Sundarbans soil is slightly alkanie to neutral with a pH of 5.0–6.8 (Rahman and Asaduzzaman, 2010). The pH values of parts of coastal soils range from 6.0–8.4 (SRDI, 2001). Clay loam (i.e., loam with a major proportion of clay) is the dominant soil texture, followed by clay (mostly clay with minor proportions of sand and

[10] Sundarbans mangrove forest, located in southwestern coastal zone of Bangladesh (21°30′–22°30′N and 89°12′–90°18′E), has an area of about 3860 mi^2 (10,000 km^2 or 1 million hectares). Approximately, two-thirds of the forests are located within Bangladesh; the remaining one-third is in the adjacent Indian State of West Bengal. The United Nations Education, Scientific and Cultural Organization (UNESCO) listed it as a World Heritage Site in 1997. In 1999, the Government of Bangladesh followed the suit, i.e., designated it as a World Heritage Site (Roy et al., 2012). Around 17% of the Sundarbans will go under water if the sea level rises by 3 ft (1 m). Because of its extensive cover and thick and sturdy composition the forest serves as a buffer for tropical cyclones that often make landfall along the Sundarbans coast. It also creates barrier against storm surge penetration.

silt). Soils in the southwestern areas are mostly clay to clay loam and in southeastern areas they are mostly clay loam to loamy in texture.

Land Types for Agricultural Land Use

Although coastal plains have relatively low elevations (as we have emphasized earlier) there are significant micro-topographic variations for the purpose of land use. Based on seasonal flooding, the GoB has divided land types in coastal regions into four categories: highland (16%), medium highland (76%), medium lowland (7%), and lowland (1%) (MOA, 2012, p. 38). These data indicate that, contrary to a broad generalization of low-lying topography, land types of coastal plains are dominated by highlands and medium highlands, which favor agricultural land use. Their combined percentages (92%) far exceed comparative data for the entire Bangladesh (75%). During annual normal floods both medium lowlands and lowlands are inundated up to depths of 3–6 ft (0.9–1.8 m) and 6–10 ft (1.8–3 m), respectively (MOA, 2012). Highlands are used for growing *kharif* crops (e.g., broadcast *aman* and *aus* rice) which are sown at the beginning of the summer/monsoon rains and harvested at the beginning of winter.[11] Medium highlands are suitable for crops that can tolerate shallow flooding, such as broadcast *aus* and jute and transplanted *aman*. Medium lowlands are used for growing transplanted *aman* and mixed broadcast *aus* and transplanted *aman*. Winter or *rabi* crops, such as *boro* rice, and vegetables and pulses, are widely grown on soils which drain in October or November. Deepwater broadcast *aman* is typically grown in lowland areas. Winter crops can only be grown in this land type if floodwater recedes before December (MOA, 2012). In short, livelihoods in the coastal region are largely dependent on agriculture and as in the rest of Bangladesh agriculture is dominated by rice cultivation. Nearly 77% of crop area is devoted to rice cultivation in the coastal region.

Wetlands

Coastal region contains many types of wetlands, such as rivers, canals, ponds, *beels* (marshes), and *baors* (oxbow lakes). Ponds are manmade water bodies

[11] As elsewhere in Bangladesh, three distinct but partly overlapping cropping seasons can be recognized in the coastal zones. The *bhadoi* or *kharif* season begins in late March or early April and lasts until the end of August or early September. *Aus* rice and jute are the two important crops grown in this season. The *aghani* or *haimantic* season, which roughly corresponds with the late monsoon period, extends from August to November or December. The main crop of this season is rain-fed *aman* rice. The *rabi* crop season starts from November and continues up to February (Paul 1984).

that are used for fish culture, sources of drinking water, and other household uses. Coastal Bangladesh has large numbers of ponds covering a surface area of about 230,000 acres (93,000 hectares), accounting for about 40% of the total pond area of the country (MOA, 2012). Freshwater marshes *(beels)* are natural depressions in floodplains and are not embanked. Oxbow lakes *(baors)* are remnants of abandoned meandering channels. Active river channels, marshes, and oxbow lakes provide rich habitats for aquatic species and represent common properties for fishing and other purposes. Together they account for about 8% of the total area of the coastal region. Additionally, saltwater marshes and tidal channels provide opportunities for capture fishing and accounts for about 3% of the coastal land. Since late 1980s saltwater shrimp cultivation has become an important sector of the economy of coastal Bangladesh. Coastal waters of Bangladesh are also rich in marine fisheries. Salt production is another important coastal resource. It is concentrated mainly in the east coast.

Forests

One of the important natural resources of coastal Bangladesh is its forest resources. Forests occupy about 18% of the land area of coastal Bangladesh, which is significantly higher than the national average (10–11%). More importantly, nearly one-half of the total forests of Bangladesh are located in coastal Bangladesh (Banglapedia, 2016). Mangroves and tropical evergreen and semievergreen forests are the major forest types. As indicated above, the southwestern coast houses the world's largest mangrove forest called the Sundarbans. Homestead forests are also a common feature of rural homes.

Demographic and Socioeconomic Profile

Based on the updates of the 2011 census, coastal Bangladesh had a population density of 2083 persons per square mile (797 per square kilometer), compared to the national density of 2850 per square mile (1170 per square kilometer) (Table 1.5a). However, the average household size in coastal Bangladesh (4.7 per household) was slightly higher than the national average (4.4 per household) (MOA, 2012). Also, there were more women (50.4%) than men (49.6), reflecting outmigration of male labor force. As in the rest of the country, households are overwhelmingly headed by male in coastal Bangladesh; only 5.0% households are headed by women (BBS, 2010). A large proportion of the population of this region (26%) lives in several major urban centers, compared to the national average of 23.5% (MOA,

2012). Chittagong, situated on the east coast, is the second largest metropolis of Bangladesh and it is the largest sea port of the country. It is the gateway of Bangladesh to the Indian Ocean. Khulna, situated on the southwest coast, is the third largest city of Bangladesh. Other smaller cities of southwestern and central coasts include Satkhira, Bagerhat, Barguna, Patuakhali, and Bhola. These districts are also least urbanized, with urbanization rates of about 10−12% of the district population.

The overwhelming majority of coastal residents live in poverty. The poverty rate is at least 40% above the national average. More than half of coastal households are functionally landless. According to the 2008 Census of Agriculture, average ownership of land per household was 0.79 acre in Bangladesh, but in the coastal districts, the corresponding figure was 0.72 acre (BBS, 2010). Land is also highly concentrated in few hands. Data show that top 7% households own 36% of the land and top 1% households own 10%, while the bottom 46% households own only 13% of the total land (MOA, 2012). According to the same source (2008 agricultural census), there were about 3.1 million farm households in the coastal region as defined in the Master Plan for Agricultural Development (MOA, 2012). Among them 2.7 million or 46% were small farmers who owned less than 2.5 acres of agricultural land. As expected, their proportion was slightly higher (1%) than that of national figure. There were over 1.6 million households in the coastal region whose predominant source of livelihood was agricultural labor.

Besides farmers, landless agricultural laborers constituted another occupational group in rural areas. Other occupational groups included salt and shrimp farmers, fishers, and shrimp fry and honey collectors. Salt farming was overwhelmingly concentrated in the eastern coastal zone where 15% of the total rural households were salt farmers. They produced enough raw salt to meet the country's demand (MOA, 2012). Skilled labor force includes indigenous technical expertise in boat-building, carpentry, and net making (MOA, 2012).

In coastal Bangladesh fishing is the primary source of income for 14% households, compared to the national average of 8% (BBS, 2010). Fishermen operate in the estuaries, on coastal waters, and sometimes in the deep sea. Barguna in the central coastal zone had the highest proportion of fisher households (38%), followed by Khulna (29%) and Jhalokati (26%) (MOA, 2012). Monsoon (June to October) is the main fishing season. While shrimp farms are more concentrated in the greater Khulna district, most shrimp fry collectors are concentrated in Patuakhali and Barguna districts. Nearly one million people were engaged in this activity and a large number of them were children. The cycle of fry collection is from mid-February to mid-August. Besides agriculture and fishing, a significant section of the subsistence economy includes different types of activities for extracting resources from the Sundarbans mangrove forest.

Tourism is a growing attraction of the coastal region contributing significantly to the national economy. Cox's Bazar on the east coast, famed for its well-developed long beaches, attracts both international and internal tourists. St Martin's Island, as indicated above, is also an important tourist destination (see section on the east coast). Nature tourism in the Sundarbans forest has been gaining increasing popularity.

COASTAL ZONES OF BANGLADESH

Based on ecological and physiographic characteristics, coastal Bangladesh may be divided into three coastal zones (Fig. 1.1).

Western Zone

This zone covers the western and older parts of the Ganges delta. In the past, it was once a part of the active Ganges delta. Presently it is undergoing a process of transition from active delta to a semimoribund delta sustained only by the regional rivers. Of these regional rivers only the Gorai receives a significant water flow from the Ganges (SWMC and BWDB, 2000). The elevation of this zone ranges between 3 ft (0.9 m) and 7 ft (2.1 m) above the mean sea level (Iftekhar and Islam, 2004). The flat coastal topography is subjected to different types of coastal processes, which include wave actions along the coastline, storm surges associated with tropical cyclones, and saltwater intrusion in tidal channels. The average tidal range at Hiron Point along the southernmost coastline is approximately 10 ft (3 m) (see Chapter 3). Tidal effects in river channels are felt inland throughout the entire zone. The effects of rising sea levels are already evident in an increase in the monsoon season sea levels by about 2 ft (60 cm) as far inland as 200 miles (300 km). Sediments are carried inland by tidal currents. Sediments are also transported downstream by major rivers flowing into the Bay of Bengal. Tidal propagation causes intrusion of saline water. The level of salinity in the coastal rivers of this zone much depends upon freshwater discharges from upstream rivers (SWMC and BWDB, 2000).

The Sundarbans, world famous mangrove forest, is undoubtedly the most distinguishing physical feature and ecosystem of this zone (and perhaps of entire Bangladesh). Mangrove swamps, tidal flats, natural levees, and tidal creeks are some of the morphological features of the Sundarbans area. The mangrove forest provides significant protection from heavy wind and storm surges by retarding their flow. The forest stands also protect the land from coastal erosion and help in sediment accretion, thus building the forest floor (Warrick and Ahmad, 1996). Unlike the next two coastal zones, there are no large islands off the coast of the Sundarbans. Islands located south of the Sundarbans are smaller in size, fewer in number, and not permanently settled (Paul, 2009). People associated with the fishing industry often live on these islands, but only during the fishing season.

The Sundarbans occupy southern parts of the districts of Satkhira, Khulna, and Bagerhat extending about 50 miles (80 km) inland from the coastline of the Bay of Bengal. It is bounded on the east by the Baleswar River (at the border between Bagerhat and Barguna districts) and on the west by the international border with India (Paul, 2009). About 40% of the Sundarbans extend westward across the international border into the Indian state of West Bengal. The Sundarbans are famed for its biodiversity. The plant ecology of this forest is typified by its salt-tolerant mangrove species, which flourishes in tidal mudflats. At least 334 species of trees, shrubs, and other plants have been recorded (Rahman and Asaduzzaman, 2010; Kabir and Hossain, 2008). The forest has taken its name from the most abundant tree species, locally called *Sundri* (*Heretiera fomes*), which covers about three-quarters of the forest area. In terms of abundance *Gewa* (*Excoecariato agallocha*) is the second most important species covering about 16% of total forest area. In addition to trees, there are at least 35 legumes, 29 grasses, and 19 sedges (Rahman and Asaduzzaman, 2010). The mangroves also support over 400 species of fish, 270 species of birds, 35 species of reptiles, and 42 species of mammals (including the famous Royal Bengal tiger, which is near extinction) (Rahman and Asaduzzaman, 2010; Kabir and Hossain, 2008).

The Sundarbans are the dominant sources of hardwood timber for entire Bangladesh and it provides livelihood support for about 3.5 million local forest-dependent people living within or the vicinity of the forest. The main livelihood activities of these people include the following (Kabir and Hossain, 2008):

- *Bawalis* (wood collectors)
- *Mawalis* (honey collectors)
- *Golpata* (fonds) collectors
- Medicinal plant collectors
- Collectors of crabs, oysters, fish fries, jelly fish

According to a baseline survey conducted by the Sundarbans Biodiversity Conservation Project (SBCP), among forest-dependent people, 35% were fry collectors, 33% fishers, 22% *Bawalis*, 4% boatmen, 3% *Golpata* collectors, 2% shell/crab collectors, and about 1% *Mawalis*, and medicinal plant collectors (MOA, 2012). Some of these professions are high-risk jobs as many lost their lives due to attacks by tigers. Honey collectors are often injured by bee stings despite their protective devices. Many poor households also depend on recently planted forests in newly developed islands *(chars)* in Patuakhali, Bhola, and Noakhali for fuel wood and other materials for house construction (MOA, 2012).

Central Zone

The central zone consists of the active Ganges delta and the funnel-shaped estuary of the Ganges-Brahmaputra-Meghna (GBM) Rivers in the east. It is bounded by the Baleswar River in the west and by the Feni River in the east. It is by far the largest of the three coastal zones as it includes six coastal districts (from West to East: Barguna, Patuakhali, Bhola, Lakshmipur, Noakhali, and Feni) and parts of at least eight other districts that are under significant coastal influences (Table 1.5). The main geomorphological characteristic of the central zone is that it receives large amounts of discharges and sediment loads from the GBM Rivers. It has been estimated that at least 70% of the sediment load consists of fine silts and about 10% of sands (Allison et al., 2003). Most of the sediments are deposited in the GBM estuary building the delta front and resulting in formation of many unstable channel bars (sand bars) and larger islands. A comparison of Landsat images taken respectively in 1984 and 2007 showed a net land gain of 174 square miles (451 square km) in the estuary within that period, representing an average growth rate of 7.6 square miles/year (19.6 km^2/year) (Brammer, 2014). It should be stressed here that the bulk of the land gain has been in the form of beginning of land accretion (deposition) barely at the sea level and most of it has not yet been reclaimed for human settlement. Erosion and sedimentation along the edges of the newly formed land, especially along new islands, have been constantly changing their outlines. Channel instability and bank erosion of rivers and islands result from highly dynamic estuarine environments that include both strong river currents (fluvial processes) and wave actions (coastal processes) in the Bay of Bengal and in the GBM estuary. Strong winds of tropical cyclones and storm surges (forward surging high waves) associated with tropical cyclones also play significant roles in landform changes.

Many small and large islands are located in the central zone including Bhola, the largest island of Bangladesh, which is also an administrative district consisting of seven subdistricts. According to an estimate by Brammer (2014), erosion in the estuarine environments by both fluvial and coastal processes has been responsible for major land losses in both Sandwip Island (which is a part of the east coast as well as of the GBM estuary) and Hatiya Island, situated east of Bhola along the central line of the GBM discharges. Landward of the estuary, rates of coastal changes are small, excepting emergence of highly unstable new islands at the mouth of the GBM Rivers.

The GBM estuary is the most productive fishing grounds of Bangladesh. Besides agriculture, fishing is the second most important source of livelihood for most residents of this zone (Rashid and Paul, 2014). Overall, the topography of the central zone is extremely flat, as the 10-ft (3 m) contour above the sea level extends as far inland as 85 miles (about 135 km) in the west to 120 miles (almost 200 km) at the center.

The density of population in the central zone varies greatly between its western subsection (i.e., west of the GBM estuary) and its eastern subsection (east of the estuary). As data in Table 1.5 indicate, the density ranges from 1235/mi^2 in Patuakhali to 1352/mi^2 in Bhola (the western sub-section) to 2184/mi^2 in Noakhali to 3759/mi^2 in Feni (eastern subsections) (Table 1.5). Among the three coastal zones, the impacts of sea level rise will be most severe in the vast estuarine zone in the central part. Not only does this zone face the greatest impact of sea level rise and population displacement, it has already seen other impacts of a rising sea level: an alarming intrusion and spread of salinity, adversely affecting agriculture and coastal freshwater fishery (Paul, 2011; Yu et al., 2010; Shamsuddoha and Chowdhury, 2007; Houghton, 2006). A recent study by the Bangladesh Agricultural Development Corporation (BADC) has shown that salinity has spread up to 80 miles (120 km) inland from the coast during the past three decades (Rahaman, 2012). Spread of salinity is likely to decrease availability and productivity of agricultural land, increase food insecurity, and result in serious scarcity of safe drinking water and loss of biodiversity.

Eastern Zone

The eastern zone is the east coast of Bangladesh comprising of the coastal plain running in a north-south direction between the Feni River in the north (eastern boundary of the central zone) and the southeastern tip of Bangladesh (Dakhinpara village) along its international boundary with

Myanmar (Burma).[12] This is a relatively narrow and straight coastal zone between the Bay of Bengal coastline and a series of north-south trending parallel hills, called the Chittagong Hill Tracts. The east coast differs significantly from the remaining two zones in both its physiographic and ecological characteristics. It is likely to be least affected by sea level rise because it is protected by the Chittagong Hill Tracts. There are three large islands, namely Maheshkhali, Kutubdia, and Sandwip, but they seem to be unrelated to two major rivers of the area, the Karnafuli and the Matamuhiri. The Naf River, flowing into the Bay of Bengal in a north-south direction along the southern peninsula of Bangladesh (Teknaf peninsula), forms the international boundary between Bangladesh and Myanmar.

Unlike the other two zones, this coastline is shaped more by marine (coastal) processes than by fluvial processes of estuarine environments. This coast is protected along the sea by mudflats and submerged sand bars. These bars have formed two sandy beaches: one stretches for about 100 miles (160 km) from Cox's Bazar to Teknaf; the other lies immediately south of the Chittagong city in the north. The former beach is the longest beach of Bangladesh. Both beaches attract international and domestic tourists.

Like other two zones, fishing is an important economic activity of this zone but the nature of fishing is almost exclusively saltwater marine fishing. In other two zones, both freshwater coastal fishing near river mouths and salt water fishing in estuarine-marine environment are common. Unlike other two zones, the east coast provides far better economic opportunities for its residents because of two major metropolitan centers in this zone, namely Chittagong and Cox's Bazar. While Chittagong, the second largest city of

[12] Dakhinpara village (in Cox's Bazar district, Teknaf subdistrict) is situated on the southern tip of the Teknaf peninsula (20.76° North and 92.32° East). It is the southernmost village of mainland Bangladesh. When the Google map is zoomed at a scale of about 1 cm = 100 m (1 inch = 833 ft or 1.2 inches = 1000 ft) some of the details of its geographical positions are clearly visible. Dakhinpara village is located immediately south of the Teknaf-Shah Parir Dweep Road and about +500 m (1/3 of a mile) south of the Shah Parir Dweep Cyclone Shelter. The southernmost landmass of mainland Bangladesh lies further south of this village (about 2000 ft or 600 m south) in the form of a 1.2-mile long or 2-km long recurved (hooked) *sandspit* (an active sand bar) that curves into the wide estuary of the Naf River (international boundary between Bangladesh and Myanmar). Sandy beaches along the Cox's Bazar-Teknaf coast and the recurved sandspit at the tip of the peninsula are indications (evidence) of northwest-southeast orientation of the prevailing longshore transport (current) of sand along the east coast of the Bay of Bengal. The southernmost landmass of Bangladesh lies further south in St Martin's Island at the latitude/longitude coordinate of about 20.57° North and 92.33° East (this particular spot on the island is marked on the Google map as Cherar Dweep).

Bangladesh, is a port city and is also a major industrial center. Cox's Bazar has a vibrant tourist economy providing employment opportunities for a large working population. Because of these two metropolitan areas the density of population in this zone is significantly higher than in the remaining two zones (see Table 1.5). St. Martin Island, a small coral island, situated southwest of the southern tip of Bangladesh, also attracts both international and local tourists. The poverty level in this zone is significantly lower than in the other zones.

ORGANIZATION OF THIS BOOK

Objectives of This Book

The preceding review of the nature of physical and socioeconomic environments of coastal Bangladesh indicates that it is a high-density area that is largely dependent upon land and sea for an indigenous system of livelihoods. The area is now at risks of several climate change—related hazards. First, magnitudes of tropical cyclones and their related storm surges have already increased in recent years due to global warming and consequent warming of ocean waters. Second, storm surge impacts have been exacerbated by rising sea levels as surge levels have been penetrating further inland from the current coastline. Finally, overriding these impacts projected rise in sea levels is likely to inundate most of the areas of coastal Bangladesh, making this area unsuitable for continued living. To sum up what are at risks of inundation due to sea level rise in coastal Bangladesh, the following is an incomplete list:

- Approximately one-third of the land area of Bangladesh
- One-quarter to one-third of the population of Bangladesh
- At least 3 million rural homes and their agricultural lands
- Largest mangrove forest of the world, the Sundarbans, a World Heritage

Sea level rise is such a geophysical phenomenon that it evokes simplistic images of large-scale inundation of coastal areas and massive population displacements from coastal Bangladesh. The international news media have often painted a sensational scenario of millions of climate refugees lining up to flee coastal Bangladesh. One of the problems (i.e., inadequacy) of such scenarios is that time and space are blurred or condensed in such images. In most likely scenarios, sea level rise is expected to be relatively slow at least up to 2050, beyond which sea level rise is projected to accelerate. For the interim period, i.e., between now and 2050, people of coastal Bangladesh have some opportunities to live with rising sea levels by adopting a number of nonstructural and structural adaptation measures.

Our intent is neither to discredit international news media reports dealing with climate change impacts on coastal Bangladesh, nor to question the ranking of Bangladesh as the MVC for climate change impacts. Instead, we would like to demonstrate that the details of potential climate change impacts in coastal Bangladesh need greater attention. Some of the details of climate change vulnerabilities of coastal Bangladesh as we have presented in this book may confirm popular perceptions about the MVC status of Bangladesh, but very little is known about potential adaptation measures for dealing with such impacts. As the title of the book suggests, we have two main objectives of this study:

- First, to provide a comprehensive analysis of major climatic hazards in coastal Bangladesh based on our own interpretations of primary (original) and secondary data and an extensive literature review.
- Second, to explore different types of nonstructural and structural adaptation measures for dealing with impending climatic hazards that may allow the people of coastal Bangladesh to continue to live in this area for an interim period prior to complete inundation of the area by rising sea levels.

Climatic Hazards

In this study, we have defined climatic hazards broadly to include all hazards and their adverse impacts that may be related to climate change. Unlike tropical cyclones, sea level rise is not an atmospheric process. It is certainly a geophysical hazard (it may be considered as a marine or coastal process) but we have characterized it as a climatic hazard because sea level rise is being fuelled by global warming—induced warming of ocean waters and melting of land ice. For the same reason, we have characterized all other adverse impacts of rising sea levels as climatic hazards, such as increased salinity intrusion and coastal erosion and sedimentation.

Adaptation Measures

We have classified adaptation measures into two categories: (1) nonstructural adaptation and (2) structural adaptation. Among nonstructural adaptation measures (solutions) we have reviewed both indigenous measures and modern scientific measures. The emphasis of structural measures is on modern engineering interventions. The distinction between nonstructural and structural adaptation measures is not clear for two measures we have reviewed. Thus, we have placed coastal forestry and the triple-F model (for forest, fruit, and fish) under structural measures because these may or may not involve large-scale engineering interventions.

Chapter Arrangements

Chapter 1: We started this chapter (Chapter 1) with a set of primary data on sources of greenhouse gas emissions into the global atmosphere as the context for climate change–induced (climatic) hazards in coastal Bangladesh, suggesting that it is a case study of "local impacts of a global problem" and, thus, implying that there is a global responsibility for these hazards. In the remainder of the chapter we have provided a comprehensive review of physical and socioeconomic settings of coastal Bangladesh, implying that all of the resources of coastal Bangladesh, including its natural resources, human settlements, and agricultural lands are at risks of inundation due to sea level rise.

Chapter 2: Among all climatic hazards, undoubtedly tropical cyclones and storm surges associated with them pose the greatest threats to human lives and their shelters (homes). Climate change has intensified these storms and increased the magnitudes of storm surges, resulting in increasing fatalities and property damage. In Chapter 2, we have provided a comprehensive treatment of this topic, with different types of supporting data. River floods are also a major climatic hazard for the Ganges, the Brahmaputra, and the Meghna floodplains inland from the coastal belt but it is not a major problem for coastal Bangladesh.[13]

Chapter 3: In this chapter our main focus is on delineation of areas likely to be flooded by a 3-ft (1 m) rise in sea level. We have used topographic contours and a map of storm surge areas for determining that approximately one-third of Bangladesh is likely to be impacted by a 3-ft rise in sea level. In this chapter, we have also analyzed original tide gauge data and a least

[13] River floods in Bangladesh rank second only to tropical cyclones as a climatic hazard. Like cyclones, floods have also been exacerbated by climate change impacts. Based on historical data, there are some indications that flood magnitudes have increased during the last half of the 20th century, particularly during its last two decades (e.g., catastrophic floods in 1987, 1988, and 1998). Model simulations indicate that global warming is likely to increase flood magnitudes significantly in the Ganges, the Brahmaputra, and the Meghna floodplains (Mirza et al., 2003). These floodplains are located inland from coastal Bangladesh. Although some inland parts of coastal Bangladesh may be/are impacted by river floods, most of the coastal belt closer to the coastline does not experience catastrophic floods because floodwaters are drained into the Bay of Bengal. For this reason we have not included a chapter on river floods.

square regression to determine a 20-year trend in sea level rise along the Bangladesh coast.

Chapter 4: Accelerated coastal erosion and sedimentation are secondary effects of climate change because these may be induced by sea level rise. Similar erosion and sedimentation problems in coastal rivers may also be considered as secondary impacts of climate change because these may be induced by a combination of global warming-induced intensification of hydrologic regimes in rivers and effects of sea level rise on estuarine channels. Rapid morphologic changes due to erosion, accretion, and land subsidence are more prevalent in the western and central coastal zones than in the east coast. Specific reasons for these changes are explained in Chapter 4.

Chapter 5: Like coastal erosion and sedimentation problems, salinity intrusion in coastal soils (especially in agricultural lands) and into freshwater bodies (particularly in coastal aquifers) are also secondary impacts of climate change. These have already been induced by rising sea levels and are likely to increase in the future due to continued rise in sea levels. Chapter 5 deals with salinity problems in coastal Bangladesh based on an extensive literature review. We have explained causes of increased salinity, geographical extent of salinity intrusion, and potential measures for dealing with these problems.

Chapter 6: For confronting impending climatic disasters, our focus is on nonstructural measures. Land use and coastal zone management are examples of nonstructural measures that may go a long way to minimize adverse effects of climate change. Land use in coastal Bangladesh is diverse, competitive, and conflicting. With population increase the land is being converted from directly productive purposes, such as crop cultivation, to other uses, such as housing, roads, and urban settlement. Present land uses in coastal zones, their trend, and key drivers of land use changes are presented in Chapter 6. The Government of Bangladesh has already initiated integrated coastal zone management to enhance productivity, balanced growth, increased employment, and increased access to food and natural resources. Salient features of coastal management programs and their both actual and potential outcomes are evaluated in this chapter.

Chapter 7: Chapters 7 and 8 are our major contributions to nonstructural and structural adaptation to sea level rise. In Chapter 7, we started with a theoretical introduction to the concept of adaptation with definitions of relevant terms, especially mitigation, adaptation, structural and nonstructural adaptation, and several other relevant terms. A major theoretical emphasis is on adaptation and adaptive capacity. To demonstrate how some of these theoretical concepts can be applied in coastal Bangladesh, we have presented examples of cyclone adaptation features in indigenous houses in coastal Bangladesh. A major section of this chapter deals with potential green revolution in coastal Bangladesh due to development of large varieties of salt-tolerant rice by the Bangladesh Rice Research Institute (BRRI). We have also reviewed adaptation of coastal fisheries to climate change impacts. We concluded this chapter with a section on disaster management policies for coastal Bangladesh.

Chapter 8: In this chapter we have reviewed four major engineering projects, namely a World Bank infrastructure project, cyclone-resistant homes, cyclone shelters, and coastal polders, using both primary and secondary data and an extensive literature review. We have also included coastal forestry and the triple F model (for forest, fruit, and fish) in this chapter because these may (or may not) involve large-scale operations using heavy machinery.

Chapter 9: We conclude this book with a summary of the main findings of this study and some recommendations for short-term and medium-term management initiatives for confronting climatic hazards in coastal Bangladesh.

REFERENCES

Ahmad, M., 2005. Living in the Coast—People and Institution. WARPO, Ministry of Water Resources, Dhaka.

Alam, S., Ahmed, R., 2015. Tropical cyclones in Bangladesh. In: Shamsuddin, S.D., Ahmed, R., Jahan, R. (Eds.), Climate Variability: Issues and Perspectives for Bangladesh. Shaitya Prakash, Dhaka, pp. 151−182.

Allison, M.A., Khan, S.R., Goodbred, J.S.L., Kuehl, S.A., 2003. Stratigraphic evolution of the late Holocene Ganges−Brahmaputra lower delta plain. Sedimentary Geology 155, 317−342.

Alpar, B., 2009. Vulnerability of Turkish coasts to accelerated sea level rise. Geomorphology 107, 58−63.

BBS (Bangladesh Bureau of Statistics), 2010. Census of Agriculture 2008. Ministry of Planning, Dhaka.

Banglapedia, 2016. Forest and Forestry. http://en.banglapedia.org/.

Brammer, H., 2014. Bangladesh's dynamic coastal regions and sea-level rise. Climate Risk Management 1, 51−62.

CBT (Columbia Basin Trust), 2016. Start Thinking About Local Impacts. Local Impacts of Climate Change. Columbia Basin Trust: A Legacy for the People (Website Factsheet). http://www.cbt.org/uploads/pdf/LocalImpactsFactSheet.pdf.

CCC (Climate Change Cell), 2008. Characterizing Long-term Changes of Bangladesh Climate in Contexts of Agriculture and Irrigation. Government of Bangladesh, Ministry of Environment and Forests, Department of Environment, Component 4b, and Ministry of Food and Disaster Management, Comprehensive Disaster Management Programme, Dhaka. http://www.climatechangecell-bd.org/.

Demirkesen, A.C., Evrendilek, F., Berberoglu, S., 2008. Quantifying coastal inundation vulnerability of Turkey to sea-level rise. Environmental Monitoring and Assessmment 138, 101−106.

Elahi, K.M., Sultana, S., Das, S.C., 1995. Management of Coastal Environment in Bangladesh: A Pre-project Study of Selected Issues. Jahangirnagar University, Savar, Bangladesh.

Gattuso, J.F., Smith, S.V., 2011. Coastal zone. In: Duffy, J.E. (Ed.), The Encyclopedia of Earth. Electronic Publication. www.eoearth.org/view/article/151298.

Gravelle, G., Mimura, N., 2008. Vulnerability assessment of sea-level rise in Viti Levu, Fiji islands. Sustainability Science 3, 171–180.

Hammar-Klose, E.S., Pendleton, E.A., Thieler, E.R., Williams, S.J., 2003. Coastal Vulnerability Assessment of Cape Cod National Seashore to Sea-level Rise. Open-File Report 02–233. US Geological Survey, Woods Hole. Cited in/taken from: Preston et al. 2011.

Houghton, J., 2006. Global Warming: The Complete Briefing, third ed. Cambridge University Press, Cambridge, UK.

Iftekhar, M.S., Islam, M.R., 2004. Managing mangroves in Bangladesh: a strategy analysis. Journal of Coastal Conservation 10, 139–146.

Johnson, B.L.C., 1982. Bangladesh, second ed. Heinemann Educational Books, London.

Kabir, D.M.H., Hossain, J., 2008. Resuscitating the Sundarbans: Customary Use of Biodiversity and Traditional Cultural Practices in Bangladesh. Unnayan Onneshan, Dhaka. http://www.forestpeoples.org/.

Karim, M.F., Mimura, N., 2008. Impacts of climate change and sea-level rise on cyclonic storm surge floods in Bangladesh. Global Environmental Change 18, 490–500.

Karim, Z., Hussain, S.G., Ahmed, M., 1990. Salinity Problems and Crop Intensification in the Coastal Regions of Bangladesh. Soils publication No. 33. Bangladesh Agricultural Research Council (BARC), Dhaka.

Maplecroft V., 2010. Big Economies of the Future—Bangladesh, India, Philippines, Vietnam and Pakistan—Most at Risk from Climate Change. Maplecroft Company, UK. http://maplecroft.com/.

Maplecroft, V., 2016. Climate Change and Environmental Risk Atlas 2015. Verisk Maplecroft. Latest Products and Report. http://maplecroft.com/portfolio/new-analysis/analysis/2014/.

Mirza, M.M.Q., Warrick, R.A., Ericksen, N.J., 2003. The implications of climate change on floods of the Ganges, Brahmaputra and Meghna rivers in Bangladesh. Climatic Change 57, 287–318.

MOA (Ministry of Agricultre), 2012. Master Plan for Agricultural Development in Southern Region of Bangladesh. Government of Bangladesh, Ministry of Agriculture, Dhaka.

MWR (Ministry of Water Resources), 2005. The National Coastal Zone Policy. Government of Bangladesh, Ministry of Water Resources, Dhaka.

Olivier, J.G.L., Janssens-Maenhout, G., Muntean, M., Peters, J.H.A.W., 2014. Trends in Global CO_2 Emissions –Report. PBL Publications. The Hague.

Paul, B.K., 2011. Environmental Hazards and Disasters: Contexts, Perspectives and Management. Wiley-Blackwell, Hoboken, NJ.

Paul, B.K., 2009. Why relatively fewer people died? The case of Bangladesh's cyclone Sidr. Natural Hazards 50, 289–304.

Paul, B.K., 1984. Perception of and agricultural adjustments to floods in Jamuna floodplain, Bangladesh. Human Ecology 12 (1), 3–19.

Preston, B.L., Yuen, E.J., Westaway, R.M., 2011. Putting vulnerability to climate change on the map: a review of approaches, benefits, and risks. Sustainability Science 6 (2), 177–202.

Rahaman, P., July 13, 2012. A Briny Future for Bangladesh. The Third Pole. http://www.thethirdpole.net/a-briny-future-for-bangladesh.

Rahman, M.R., Asaduzzaman, M., 2010. Ecology of Sundarbans, Bangladesh. Journal of Science Foundation (1 and 2), 35–47.

Rahman, S.M.N., Gafoor, A., Hossain, T.I.M.T., 1993. Coastal Zone Monitoring Using Remote Sensing Techniques. Bangladesh Space Research and Remote Sensing Organization, Dhaka.

Rashid, H., Paul, B., 2014. Climate Change in Bangladesh: Confronting Impending Disasters. Lexington Books, Lanham, MD, Boulder, CO, New York, Toronto, Plymouth, UK.

Roy, A.K.D., Alam, K., Gow, J., 2012. A review of the role of property rights and forest policies in the management of the Sundarbans mangrove forest in Bangladesh. Forest Policy and Economics 15, 46–53.

Rudle, K., Manhard, W., 1981. Renewable Natural Resources and Environment. Tycooly International, Dublin.

Sarwar, M.G.M., 2005. Impacts of Sea Level Rise on the Coastal Zone of Bangladesh (Master's thesis). Lund University, Sweden.

Shamsuddin, S.D., Ahmed, R., Jahan, R. (Eds.), 2015. Climate Variability: Issues and Perspectives for Bangladesh. Shaitya Prakash, Dhaka.

Shamsuddoha, M., Chowdhury, R.K., 2007. Climate Change Impact and Disaster Vulnerabilities in the Coastal Areas of Bangladesh. COAST Trust, Dhaka.

SRDI (Soil Resource Development Institute), 2001. Soil and Water Salinity in the Coastal Region of Bangladesh. Soil Resource Development Institute, Dhaka.

SWMC (Surface Water Modelling Centre) and BWDB (Bangladesh Water Development Board), 2000. Late Monsoon Flood in the Southwest Region of Bangladesh 2000. Government of Bangladesh, Ministry of Water Resources, Dhaka.

Warrick, R.A., Ahmad, Q.K., 1996. Implications of Climate and Sea Level Change for Bangladesh. Kluwer Academic Publishers, Dordrecht, the Netherlands.

World Bank, 2016. Climate Change Knowledge Portal: For Development Practitioners and Policy Makers. The World Bank Group. http://sdwebx.worldbank.org/climateportal/.

Yu, W., Alam, M., Hasan, A., Ruane, A.C., Rosenzweig, C., Major, D.C., Thurlow, J., 2010. Climate Change Risk and Food Security in Bangladesh. World Bank, Washington, DC. http://documents.worldbank.org/.

CHAPTER TWO

Tropical Cyclones and Storm Surges

MEANING OF CYCLONES AND THEIR ALTERNATIVE TERMS

A *cyclone* is a low pressure system of the atmosphere in which air pressure has dropped below the standard (normal) atmospheric pressure (1013.2 millibar or hectopascal, short-formed as mb or hPa) and winds rotate inward in a counter-clockwise direction in the northern hemisphere and clockwise direction in the southern hemisphere. In meteorological terms, any low-pressure system may be characterized as a cyclone. Therefore, a distinction should be made between two different types of cyclones:

- *Mid-latitude cyclones*, also called *frontal cyclones*, are large traveling low-pressure systems that develop along weather fronts (lines of separation between a cold air mass and a warm air mass) in the mid-latitudes (35°—55° N and S). Most of them develop as weak low-pressure systems with pressure values barely below the standard atmospheric pressure. Only in exceptional cases, an intense mid-latitude cyclone may have a surface pressure as low as 970 mb or even lower.
- *Tropical cyclones*, which develop over tropical oceans, are commonly characterized by intense low-pressure systems (most often below 950 mb), high wind speeds, and storm surges, i.e., rapidly advancing (surging forward) high waves (often erroneously called tidal waves or tidal surges).

Tropical cyclones are known by many vernacular names depending on their geographical location, but the following English terms are commonly used in three major cyclone zones:

- Atlantic coasts of North America and Europe: Hurricane
- Pacific coasts of East Asia and Pacific Islands: Typhoon
- Indian Ocean coasts of South Asia and Australia: Cyclone

Climatic Hazards in Coastal Bangladesh
ISBN 978-0-12-805276-1
http://dx.doi.org/10.1016/B978-0-12-805276-1.00002-8

Thus, hurricanes in North America, typhoons in Japan, and cyclones in India and Bangladesh are identical storm systems.[1]

ORIGIN OF TROPICAL CYCLONES

A tropical cyclone begins from a low-pressure system (depression) but why only limited numbers of depressions convert into tropical cyclones? While the processes that lead to such a transformation are exceedingly complex, a brief review of the preconditions for cyclone formation may provide some insights into the genesis of such a severe storm system.

Threshold Sea Surface Temperature

A tropical cyclone is a large and deep low-pressure system that is sustained by large amounts of evaporation from a warm sea surface. A threshold sea surface temperature (SST) of 26–27°C (79–80°F) has been observed as a precondition for cyclone formation (Houze, 2014; Emanuel, 2003; Barry and Chorley, 2003). SST refers to the temperature of a layer of warm water that extends up to a depth of at least 50 m (165 ft) (www.bom.gov.au/cyclone/faq). Every kilogram of vapor transfers about 2257 kilojoules (kJ) of energy from the sea surface into the atmosphere in the form of *latent heat of evaporation* (see Barry and Chorley, 2003, p. 69 for an equation on latent heat).[2] This provides the initial source of energy for large numbers of convection cells, in which warm and moist air rises upward, expands, and cools, leading to the development of cumulus and cumulonimbus (thunderstorm) clouds. The *latent heat of condensation* released into the atmosphere following condensation, cloud formation, and rainfall, sustains the life of a tropical cyclone. Thus, the combination of latent heat of evaporation and condensation provides the fuel for the "heat engine" that drives a tropical cyclone.

[1] The vernacular name for cyclones in Bangladesh is *ghurni jhar*. This term is used in Bengali newspapers but it is not used commonly in conversational language. The average people would relate much better to the English term cyclone than to its vernacular equivalent (as is the case for telephone, television, chair, etc.).

[2] "The latent heat of vaporization is the amount of heat required to convert a unit mass of a liquid into vapor without a change in temperature". For water, this value is 2257 kJ/kg. http://www.engineeringtoolbox.com/fluids-evaporation-latent-heat-d_147.html.

Preexisting Disturbances

It is now widely recognized that some forms of preexisting disturbances in the tropical atmosphere, especially in the upper troposphere, are conducive to cyclone formation (Emanuel, 2003; Riehl, 1954). The key to this process is high-level outflow (divergence) which, in turn, allows the development of low-pressure and convergence of strong winds near the surface (Barry and Chorley, 2003). In tropical latitudes where cyclones originate the prevailing winds are easterlies, i.e., northeast and southeast trade winds. The easterlies experience frequent low-level disturbances, especially in the lower half of the troposphere (up to about 5—6 km or 3—4 miles). These promote widespread upward motion of warm and moist air and a lowering of the surface pressure. Conditions for cyclone formation are ripe when a diverging system at high altitudes (upper troposphere) moves over a converging system at low levels (Battan, 1961). Once upward air motion begins in a convection cell, it would be accompanied by inflow in the lower levels (convergence) and outflow in the upper levels (divergence) of the atmosphere. As moist and warm air from surrounding areas converges into the core of the low-pressure system near the sea surface, both latent heat of evaporation and sensible heat from the sea surface (i.e., direct heat transfer by conduction process) warm the core of the disturbance and lowers the central pressure that increases wind speeds. The warming of the core and the increased surface winds thus reinforce each other in "a positive feedback mechanism" (Zehnder, not dated).

Vortical Hot Towers

A distinctive feature of a tropical cyclone is its warm vortex, compared to other tropical depressions and incipient storms that have "a cold core area of shower activity" (Barry and Chorley, 2003, p. 271). A tropical cyclone is essentially a "mesoscale convective system," which may start as a "set of one or more isolated deep vortical hot towers. The vorticity of the low-level environment is stretched by convergence in the lower portions of the updraft and advected upward. The updrafts thus become upward extending centers of high positive vorticity, called vortical hot towers" (Houze, 2014, p. 289). The warm core of a cyclone is the "mesoscale convective vortex," which develops through the actions of hundreds (or even thousands) of convection cells, which develop into tall cumulonimbus clouds ("hot towers") releasing latent heat of condensation (Davis, 2015; Houze 2010, 2014; Hendricks et al., 2004; Barry and Chorley, 2003). Numerical simulations have revealed "the presence of vortical hot towers, or cores of deep cumulonimbus convection possessing strong vertical vorticity, that arise from buoyancy-induced stretching of local absolute vertical vorticity in a vorticity-rich prehurricane environment" (Hendricks et al., 2004). These simulations also demonstrated that the strong vertical vorticity in the hot towers, which effectively trapped the latent heat from moist convection, incorporated a "tangential momentum spinup" from local warm core formation (Hendricks et al., 2004, p. 1211). Thus, strong vertical vorticity of incipient convection cells seems to be one of the prerequisites of cyclone formation. Further, development of strong vorticity "the environmental winds must have very little shear to allow the cyclone to develop with vertical coherence" (Houze, 2014, p. 288).

Coriolis Effect

Another prerequisite for the formation of tropical cyclones is the effect of earth's rotation on air movement, an effect that may be expressed mathematically as the *Coriolis parameter.* Because of the influence of the earth's rotation, "the converging air will be turned and start to move in a circular path" (Battan, 1961, p. 102). Tropical cyclones generally form between 5° and 20° latitudes and they rarely form within 5° of the equator because "the Coriolis force is too weak for low-level convergence to be able to generate enough relative vorticity for storm formation. The strong background positive vorticity helps to trap energy released in clouds so that it contributes to strengthening of the cyclone" (Houze, 2014, p. 288) The magnitude of the Coriolis effect is zero (0) at the equator and reaches the maximum value of 1 at the poles, which can be translated mathematically into "sine of latitude 90° = 1" (Battan, 1961, p. 47). At intermediate latitudes, the Coriolis effect can be calculated from the following equation (Battan, 1961, p. 46):

$$F_c = 2 \times V \times A \times \text{sine}\, l \tag{2.1}$$

where F_c = Coriolis effect, V = wind velocity, A = angular velocity of the earth, and l = latitude.

Based on this equation, magnitude of the Coriolis effect at the poles "is equal to twice the product of wind velocity, V, and the angular velocity of the earth, A" (Battan, 1961, p. 46). The Coriolis values in the Bay of Bengal, situated approximately between 10°N and 20°N, range between 0.17 and 0.34, which are sufficient to provide significant deflections to large storm systems. Detailed storm tracks clearly show such deflections. The Bay of Bengal cyclone Sidr (2007), for example, headed northwestward from the Andaman Islands. By the time it hit the coast of Bangladesh it veered slightly to the northeast direction (Buis, 2007).

CHARACTERISTICS OF TROPICAL CYCLONES

Wind Speed and Air Pressure

Perhaps, the most defining characteristic of a tropical cyclone is its wind speed. For this reason, tropical cyclones are classified into different categories of strengths based on their wind speeds. According to the Saffir-Simpson Scale (Table 2.1), a widely used scale of storm intensity, a tropical cyclone is defined as an intense tropical storm that has reached a critical wind speed of at least 64 knots [74 miles per hour (mph) or 120 kilometer per hour (km/h)]. The Government of Bangladesh Meteorology Department uses a slightly more elaborate classification of tropical depressions, tropical

Table 2.1 Saffir-Simpson Scale of Tropical Cyclones

Storm Category	Wind Speed (knots)[a]	Wind Speed (mph)	Wind Speed (km/h)	Central Low Pressure (mb)
Tropical depression (TD)[a]	33 knots or lower	38 mph or lower	61 km/h or lower	
Tropical storm (TS)[a]	34–63	39–73	63–117	
Category 1	64–82	74–95	119–153	980 mb or higher
Category 2	83–95	96–110	154–177	965–979
Category 3	96–112	111–129	178–208	945–964
Category 4	113–136	130–156	209–251	920–944
Category 5	137 knots or higher	157 mph or higher	252 km/h or higher	Lower than 920 mb

1 knot (nautical mile) = 1.15 miles and 1.852 km.
Note: The Government of Bangladesh Meteorology Department uses an eight-point scale for all tropical disturbances, ranging from minor tropical disturbances to different categories of tropical cyclones. We have used the Saffir-Simpson scale since it is widely used for most of the global cyclone zones.
[a]Tropical Depression and Tropical Storm are not a part of the Saffir-Simpson Scale.
Prepared by the authors based on wind data from NOAA: http://www.nhc.noaa.gov/ and pressure data from the US Atlantic Oceanographic and Meteorological Laboratory in Miami, Florida: https://www.unc.edu/~rowletts/units/scales/saffir.html/.

storms, and tropical cyclones. We have used the Saffir-Simpson Scale for the sake of consistency of terms used in other international cyclone zones.

In meteorological terms, the main driving force for winds (of any speed) is the pressure-gradient force (PGF), which is responsible for "accelerating a parcel of air from a high atmospheric pressure region to a low pressure region, resulting in wind" (Kastendeuch, 2007, p. 2058). On weather charts (regularly displayed on television screens during weather reports) differences in pressure from one area to another are indicated by isobars, i.e., lines joining points of equal pressure. The PGF acts at right angles to isobars in the direction from high to low pressure. In satellite imagery most often tropical cyclones/hurricanes appear as almost perfectly circular bands of clouds. On weather charts of such a storm system the isobars are represented by closely packed circular lines, implying strong PGF and resulting strong wind speeds. Closer the isobars, stronger are the PGF values and corresponding wind speeds, which may be expressed mathematically by the following equation (Battan, 1961; Barry and Chorley, 2003):

$$\mathrm{PGF} = -1/\mathrm{D} \times (\mathrm{P_2} - \mathrm{P_1})/\mathrm{B} - \mathrm{A} \tag{2.2}$$

where D = density of air, P2 − P1 = pressure difference (from one isobar to another); and B − A = distance between the isobars.

The minus sign enters the equation because the force is directed from high to low pressure (Battan, 1961, p. 41). It follows from the above equation that the PGF is inversely related to the distance between the isobars and directly related to the pressure gradient, which may be measured as differences in pressure between successive isobars over a unit distance. Thus, the magnitude of a tropical cyclone is related to the pressure gradient between its outer margin and its center. As an approximation, an isobar with a value of 1010 mb or 1000 mb may be considered as the outer margin of a cyclone since the normal (standard) atmospheric pressure is 1013.2 mb. The steepness of the pressure gradient at a given location depends on the distances between successive isobars and the value of the central low pressure. The world record for the lowest central pressure of a hurricane/tropical cyclone is 882 mb, which was registered by hurricane Wilma, a 2005 Caribbean Sea super-cyclone. In Table 2.2, we have provided a limited amount of pressure-track data for Wilma to demonstrate how it transformed from a tropical storm to a category 1 hurricane within 6 h and then directly into a category 4 and category 5 super-cyclone within the next 12 h. The corresponding pressure changes and wind speeds from Category 1 to category 4, in particular, demonstrate how its central pressure and wind speed

Table 2.2 Transformation of the Caribbean Hurricane *Wilma* (October 2005) From a Tropical Storm Into a Category 5 Hurricane: Samples of Pressure Track Data

Date/Time	Latitude/Longitude Coordinate	Central Pressure (mb)	Wind Speed in Knot (Storm Category)
18 Oct/0600 h	15°42′N/79°54′W (a)	982	60 (TS)
18 Oct/1200	16°12′N/80°18′W (b)	979	65 (Category 1)
	Distance: (a)–(b): 37.79 knot (kt)	Pressure difference: 3 mb/ 37.79 kt = 0.079 mb/kt	
18 Oct/1200	16°12′N/80°18′W (a)	979	65 (Category 1)
18 Oct/1800	16°36′N/81°6′W (b)	975	75 (Category 1)
	Distance: (a)–(b): 51.84 kt	Pressure difference: 4 mb/ 51.84 kt = 0.077 mb/kt	
18 Oct/1800	16°36′N/81°6′W (a)	975	75 (Category 1)
19 Oct/0000	16°36′N/81°48′W (b)	946	130 (Category 4)
	Distance: (a)–(b): 40.19 kt	Pressure difference: 29 mb/ 40.19 kt = 0.72 mb/kt	
19 Oct/0000	16°36′N/81°48′W (a)	946	130 (Category 4)
19 Oct/0600	17°00′N/82°12′W (b)	892	150 (Category 5)
	Distance: (a)–(b): 33.18 kt	Pressure difference: 54 mb/ 33.18 kt = 1.63 mb/kt	
19 Oct/0600	17°00′N/82°12′W (a)	892	150 (Category 5)
19 Oct/1200	17°18′N/82°48′W (b)	882	160 (Category 5)
	Distance: (a)–(b): 17.97 kt	Pressure difference: 10 mb/ 17.97 kt = 0.56 mb/kt	

Prepared by the authors based on original pressure track data from NOAA *Tropical Cyclone Reports*: http://www.nhc.noaa.gov/data/tcr.

changed rapidly within relative short distances. The central pressure decreased from 975 mb to 946 mb, a drop of 29 mb, in only 6 h within a short distance of 40 knots (46 miles). The wind speeds increased from 75 knots (86 mph) to 130 knots (150 mph). Within the next 6 h, Wilma registered the historic record of 882 mb as its central low pressure—a drop of 131 mb from the standard atmospheric pressure. The corresponding wind speed of 160 knots (184 mph = 296 km/h) was among one of the highest wind speeds of all super-cyclones.

Thus, it appears from the data presented in Table 2.2 that, in addition to changing wind speeds, the central low-pressure values of Wilma demonstrated the dynamic nature of this rapidly evolving storm system. Such detailed data on pressure changes along a cyclone/hurricane track are not readily available for the Bay of Bengal cyclones. Instead, data on wind speeds monitored from permanent buoys in the Indian Ocean provide most of the information on cyclone tracks. Based on such data, the Bay of Bengal cyclone Sidr, a historic cyclone that hit the coast of Bangladesh in November 2007, transformed from a tropical storm directly into a category 3 cyclone within a period of 24 h (November 12–13) since its inception south of the Andaman Islands (India), about 500 miles (800 km) south of the Bangladesh coast (Table 2.3). It then strengthened successively into a

Table 2.3 Transformation of the Bay of Bengal Cyclone *Sidr* (November 10–16, 2007) From a Tropical Depression Into a Super-Cyclone, Categories 3–5: Samples of Wind Speeds Along the Storm Track[a]

Date/November	Latitude/ Longitude Coordinate[b]	Wind Speed in Knots (mph/km/h)	Storm Category
10	9.5°N/93.5°E	15 (17 mph/28 km/h)	Tropical depression
11	10°N/92.5°E	30 (35 mph/56 km/h)	Tropical depression
12	10.4°N/90°E	60 (69 mph/111 km/h)	Tropical storm
13	12°N/90.5°E	110 (127 mph/204 km/h)	Category 3
14	14.2°N/89.5°E	115 (132 mph/213 km/h)	Category 4
15	17.9°N/89.1°E	140 (161 mph/259 km/h)	Category 5
15 (landfall)	22°N/89.9°E	64 (75 mph/120 km/h)	Category 1
16 (inland)	24.2°N/91°E	60 (69 mph/111 km/h)	Tropical storm
16 (inland)		45 (52 mph/83 km/h)	Tropical storm

[a]Pressure data were not monitored on a continuous basis along the cyclone track.
[b]Latitude/longitude coordinates are approximate only as these were interpreted visually by the authors from a web-based chart by O'Hara and Falvey (2007).
Prepared by the authors based on original cyclone track data for *Sidr*, posted on websites of the US Naval Maritime Forecast Center/ Joint Typhoon Warning Center, Pearl Harbor, Hawaii (O'Hara and Falvey, 2007). www.usno.navy.mil/NOOC/nmfc-ph/RSS/jtwc/atcr.pdf.

category 4 and a category 5 super-cyclone within the next 48 h (Table 2.3) (Gutro and Olsen, 2007). By the time it hit the coast of Bangladesh on November 15 it weakened into a category 1 cyclone. Soon afterward, it dissipated inland as a tropical storm. The lack of detailed pressure data was compensated somewhat by simulation studies that suggested that a strong PGF in the southeastern quadrant of the storm along the rough terrain of the Bay of Bengal coast had evolved Sidr into one of the largest and one of the strongest tropical cyclones of the Bay of Bengal. Its diameter was approximately 280 miles (450 km) with a cloud height of 8–9 miles (13–15 km) (Akter and Tsuboki, 2012). The maximum sustained wind speeds of Sidr reached 140 knots (161 mph, 259 km/h)—a category 5 cyclone (O'Hara and Favley, 2007). As indicated above, only limited amount of pressure data were available for Sidr, but Hasegawa (2008, cited by Akter and Tsuboki) reported that its lowest central pressure before landfall was 944 mb.

Storm Surges

Storm surges are one of the most devastating hydrodynamic features of tropical oceans that strike low-lying coastal areas, often with great losses of human lives. Historically, these have proven to be the main agent of human casualties of cyclones, whereas strong wind velocities have been responsible for most of the property damage. Storm surges are rapidly advancing (surging forward) high waves that are propelled by an advancing cyclone toward the coastline. Strong winds blowing over sea surface for a prolonged period of time push sea water toward the shore resulting in higher water levels ahead of an advancing storm (also called wind set-up). Advancing high waves are further superimposed on high water levels. The latter are often mistaken as high tides and storm surges are often termed erroneously as tidal waves. However, if high astronomical tides coincide with storm surges, storm surge impacts would then incorporate three hydrodynamic features: higher-than-normal sea levels, high tides, and rapidly advancing high waves. If these three characteristics synchronize their combined effects could be devastating. This is exactly what happened during the 1970 November Bhola cyclone in East Pakistan (now Bangladesh), which has been labeled as the killer storm of the 20th century (and perhaps the killer storm of all historical records). An estimated 300,000 people (some references cite even higher casualties) perished under the crushing impacts of a 10 m (34 ft) high storm surge that struck the gently sloping coastal plain of Bhola Island (situated in the mouth of the Ganges estuary). Another catastrophic cyclone in April 1991 struck the southeastern coast of Bangladesh with a 6 m (20 ft) high storm surge that killed nearly 140,000 people. Table 2.4 presents data on wind speeds and storm surge heights associated with a selected number of high-magnitude cyclones that struck the Bangladesh coast during a period of 50 years (1960–2010), which may be characterized as a period of acceleration in high-magnitude cyclones.

Coastal Bangladesh is particularly vulnerable to high-magnitude storm surges because of the shallow bathymetry (depth) of its continental shelf. Surges are further magnified by the funneling effect of the highly indented

Table 2.4 Maximum Wind Speeds and Storm Surge Heights Associated With a Selected Number of High-Magnitude Cyclones That Struck the Bangladesh Coast: 1960–2010

Year (Date)	Coastal Area	Maximum Wind Speed in Knots (mph, km/h)	Storm Surge Height in Meter (ft)
1960 (30 Oct)	SE coast (Chittagong, Cox's Bazar)	112 knots (129 mph, 208 km/h)	6.1 (20)
1961 (9 May)	Central coast (Bhola, Noakhali)	86 knots (99 mph, 160 km/h)	3.0 (9.8)
1965 (11 May)	Barisal, Noakhali	87 knots (101 mph, 162 km/h)	4.0 (13.1)
1966 (23 October)	Noakhali, Chittagong	78 knots (90 mph, 145 km/h)	6.7 (22)
1970 (12 November): *Bhola Cyclone*	Bhola, Chittagong	120 knots (138 mph, 222 km/h)	10.6 (34.8)
1988 (29 November)	Khulna	86 knots (99 mph, 160 km/h)	4.4 (14.4)
1991 (29 April): *Cyclone Gorky*	Chittagong	121 knots (140 mph, 225 km/h)	6.1 (20)
1997 (19 May)	Chittagong, Feni	121 knots (140 mph, 225 km/h)	4.6 (15.1)
2007 (15 November): *Cyclone Sidr*	Barguna, Patuakhali	119–135 knots (137–155 mph, 220–250 km/h)	6.0 (19.7)
2009 (25 May): *Cyclone Aila*	Chittagong	65 knots (75 mph, 120 km/h)	2–3 (6.6–9.8)

Prepared by the authors based on selected data from Karim, M.K., Mimura, N., 2008. Impacts of climate change and sea level rise on cyclonic storm surge floods in Bangladesh. Global Environmental Change 18 (3), 490–500. Data on Cyclone *Aila* from: www.storm-surge.info/sev-data.

Table 2.5 Effect of Basin Bathymetry on Storm Surge Heights: Results of Simulation Experiments

Depth of Water in Meter (ft)	Surge Height in Meter (ft)
5 (16.40)	9.87 (32.40)
10 (32.81)	6.59 (21.6)
15 (49.21)	5.12 (16.79)
20 (65.62)	4.08 (13.40)

Prepared by the authors based on simulation data presented in Siddiqui, Z.A., 2009. Storm surge forecasting for the Arabian sea. Marine Geodesy 32 (2), 199–217, Table 1.

Ganges estuary. Simulation experiments by Siddiqui (2009) have demonstrated that storm surge height is inversely proportional to depth of water but directly related to atmospheric pressure drops. Thus, as surge water approaches the Bangladesh coastline over a relatively shallow continental shelf storm surges tend to magnify in amplitudes (Table 2.5). As explained in the previous section, the PGF is the driving force for strong cyclonic winds and, by proxy, for storm surges as well. Applying the mathematical principle of PGF, shorter the distances between successive isobars, larger are the PGF values reflecting drops in pressure. Consequently, storm surges respond to the rates of pressure drop: greater the drop, higher the storm surge heights (Table 2.6). Using real bathymetry and differences in pressure drops, Siddiqui's (2009) simulations showed that higher surges were basically associated with larger differences in pressure drop in a relatively shallow bathymetric environment.

Applying similar numerical simulations to a specific case study, for example, Cyclone Gorky (also called Cyclone 2B) (April 29, 1991), Murty

Table 2.6 Effect of Atmospheric Pressure Drop on Storm Surge Heights: Results of Simulation Experiments

Pressure Drop (hPa or mb)[a]	Surge Height in Meter (ft)
20	0.457 (1.5)
25	0.558 (1.83)
30	0.699 (2.29)
35	0.833 (2.73)
40	0.958 (0.14)

[a]hPa = hectopascal; mb = millibar.

Prepared by the authors based on simulation data presented in Siddiqui, Z.A., 2009. Storm surge forecasting for the Arabian sea. Marine Geodesy 32 (2), 199–217, Table 4.

and Flather (1994) demonstrated how pressure drop was directly related to storm surge heights along the Bangladesh coast. The contribution to the water level from atmospheric pressure variations was given by the inverse barometric effect (Murty and Flather, 1994, p. 153):

$$\eta_p = \Delta p/\rho g \tag{2.3}$$

where $\eta_p =$ storm surge elevation, $\Delta p =$ pressure deficit in the cyclone, $\rho =$ water density, and $g =$ gravitational acceleration.

"This relationship means that a decrease in pressure of 1 mb raises the water surface by 0.99 cm" (Murty and Flather, 1994, p. 153). Moreover, water surface slope due to wind stress (a friction factor) also plays an important role in surge elevation. The details of the effect of water surface slope, $d\eta_w/dx$, are given below (Murty and Flather, 1994, p. 153):

$$d\eta_w/dx = f_w(1 + \lambda)(\rho_a/\rho)\left(U^2/gh\right) \tag{2.4}$$

where f_w is the friction factor (~ 0.0026), $\lambda =$ ratio of the bottom friction to surface friction, $\rho_a =$ density of the air, $\rho =$ water density, $U =$ average wind speed, and $h =$ water depth.

The total storm surge elevation, η, is $\eta_p + \eta_w$. Using Eq. (2.4) (above) and assuming a change of still water depth from 10 m to 0 m over a distance of 53 miles (80 km), a pressure drop of 65 mb, and an average wind speed of 30 m/s, the maximum storm surge elevations at selected coastal stations along the Bangladesh coast during Cyclone Gorky (April 29, 1991) were as follows:

- Teknaf (surge amplitude): 4.1 m
- Cox's Bazar: 3.6 m
- Chittagong: 8.2 m
- Sandwip: 8.2 m

Most of these simulation data were similar to actual field measurements. Further, field evidence agreed that the highest surge fatality occurred in Chittagong and Sandwip, i.e., areas with higher storm surge amplitudes.

The geography of storm surge impacts in coastal Bangladesh is also dependent on the cyclone track orientation and the specific location of landfall of a given cyclone. Overall, the central coast, composed of a relatively flat delta plain, is more susceptible to storm surge damage when cyclone tracks are located west of the Meghna (GBM) estuary (Azam et al., 2004). The surge elevations are reduced as the cyclone tracks shift eastward from the estuary, but they still pose significant threats to the east coast, especially to two major metropolitan centers, namely, Chittagong and Cox's Bazar (Debsarma, 2009). Variations in track orientation are of little consequence for the north-south trending east coast since most of the cyclones approach the coast from the south quadrant. The track angle has greater effects on surge heights in southwestern and central coasts. "The height increases when the cyclone approach angle deviates more from the south to the west" (Azam et al., 2004, p. 285).

Types of Clouds

A tropical cyclone is a rapidly evolving mesoscale convection system that is accompanied by different types of clouds. The specific cloud types are contingent upon their relative vertical and horizontal positions. Clouds with vertical development are called *cumulus* and those associated with active thunderstorms are *cumulonimbus* (thunderstorm) clouds. A distinction should be made here between the rotating cumulonimbus clouds (the so-called "vortical hot towers") associated with the cyclone vortex and ordinary thunderstorm clouds without rotating vortices, which are normally situated away from the core. By far the most complex cloud structures are found along the vertical profile of the vortex. Besides the rotating cumulonimbus clouds that extend up to and above the top of the troposphere, lower level clouds include *nimbostratus* (low-level rain-bearing stratus) and *stratocumulus* (low-level cumulus) in the central zone and upper-level *cirrus* and *cirrostratus* spiraling anticyclonically outward (i.e., in the divergence zone) (Houze, 2014).

Eye of the Storm

One of the striking features of a tropical cyclone is the *eye of the storm*, which refers to a small circular opening in the middle of its spiraling cloud pattern. Typically, cyclone eyes cover relatively small areas, ranging from as small as 6 miles (10 km) in diameter in rare cases, to 25 miles (40 km) on average, and to as large as 60 miles (100 km). In satellite imagery the eye is visible as a small cloud-free circular area at the center of the spiral bands of clouds. Similarly, on weather maps the eye appears as the smallest circular isobar at the center of the tightly spaced isobars representing the storm. The eye is one of the most dynamic features of a cyclone. Whereas the center of the eye is a relatively cloud-free area with warm air temperatures and light winds, some of the strongest cyclone winds are experienced along the eye wall, as it is implied by tight spacing of isobars around the eye. For this reason, when the eye of the storm passes over land, the people residing inside the temporary eye area may be misled for a while by its cloud-free sky and light winds, but soon after the passage of the eye they would be subjected for the second time to the full fury of the storm. Two complementary drafts maintain the warm core of the storm vortex. The updraft is the dominant vertical motion that carries warm and moist air upward, which is a cooling process in a convection system. This is countered by a complementary downdraft that warms the descending air by adiabatic heating process (due to subsidence and compression). For this reason air temperature inside the eye is always warmer than outside, sometimes by as much as 5.5°C (10°F) (Zehnder, not dated).

Although the eye is cloud-free, the rotating vortex achieves its force balance via a vertically overturning "strong secondary circulation which produces an *eyewall cloud*" (Houze, 2014, p. 300). A zoomed-in satellite view of the eye of Hurricane Katrina (2005) provided a real-time glimpse of its eye-wall clouds. It showed that the cloud top surrounding the eye sloped downward and inward toward the ocean surface. "Seen from an aircraft flying inside the eye, the cloud surface bounding the eye region and sloping at an angle of 45° gives an observer on the plane the impression of being inside a giant circular sports stadium with the grandstand banking upward and outward" (Houze, 2014, p. 300).

Rainbands

Another product of secondary circulation along the cyclone vortex is the development of *rainband* clouds surrounding the eye, which produce a significant proportion of total rainfall of a tropical cyclone. Rainbands have a "spiral configuration distinct from the circular eyewall shape" (Houze, 2014). Multiple rainbands may develop around the eye. Inner rainbands lie close to the vortex center and are more convective than relatively motionless outer ranbands, which develop at distances of 50–90 miles (80–150 km) away from the center (Akter and Tsubuki, 2012). Cyclone Sidr, the Bay of Bengal cyclone that struck the coast of Bangladesh in November 2007, provides an ideal case study of well-developed rainband clouds. Using Meteosat 7 satellite imagery and Tropical Rainfall Measuring Mission radar data, Akter and Tsubuki (2012) have studied Sidr's rainband characteristics at some length. Cyclone Sidr contained two long, prominent rainbands: one on the eastern side of the eye elongated toward the south (the eastern band) and the other to the west of the eye extended to the north (the northern band). The northern band appeared early during the cyclone, whereas the eastern band developed with cyclone intensification (Akter and Tsuboki, 2012, Fig. 1). Consequently, the eastern band contained more intense convective cells and heavier precipitation than the northern band.

CLIMATE CHANGE IMPACTS ON BAY OF BENGAL CYCLONES

Assumptions About Climate Change Impacts on Cyclones

The Bay of Bengal is one of the leading breeding grounds for tropical cyclones, particularly because of its warm SSTs. More significantly, coastal

Bangladesh ranks as the most vulnerable geographical area for human fatalities due to storm surge impacts. Consequently, in recent years concerns about simultaneous impacts of global warming–induced sea level rise and storm surges associated with tropical cyclones have gained increasing attention from scientists, environmentalists, politicians, and the news media. These concerns are based on an assumption that absorption of increasing amounts of atmospheric heat by ocean water is likely to increase cyclonic activities. While most of the scientific analyses support this broad assumption, we have addressed below some of the specific questions related to this assumption, using data on (1) frequencies and magnitudes of tropical cyclones in the Bay of Bengal, (2) frequencies of high-magnitude cyclones striking the coast of Bangladesh, and (3) temperature anomalies in the Indian Ocean and the Pacific Ocean linking them to the Bay of Bengal cyclones.

Have Cyclones Become More Frequent and More Violent in the Bay of Bengal?

Theoretical Assumptions

Since tropical cyclones are fueled by warm SSTs, all model simulations predict that cyclone/hurricane intensity should increase with increasing global mean temperatures (Emanuel, 2005). By using an index of the "total dissipation of power," integrated over an area (radius) of a cyclone and over the lifetime of the cyclone, Emanuel has demonstrated that this index has increased since 1970s.[3] He found that this index, which integrates mathematically several pertinent parameters, such as maximum sustained wind speed over 10 m above the sea surface, surface air density, surface drag coefficient, etc., is "highly correlated with tropical sea surface temperature, reflecting well-documented climate signals, including multi-decadal oscillations in the North Atlantic and North Pacific, and global warming"

[3] Emanuel (2005) defines an index of the potential destructiveness of tropical cyclones/hurricanes based on the total dissipation of power, integrated over the lifetime of a cyclone. This index has increased since the mid-1970s. Hurricane power dissipation, integrated over the surface area affected by a storm and over its lifetime is given by (Emanuel, 2005, p. 685).

$$\mathrm{PDI} \equiv \int \mathcal{T}_0 v_{max}^3 \, dt$$

where PD represents the units of energy and reflects the total power dissipated by a storm over its lifetime; V_{max} is the maximum sustained wind speed at the conventional measurement altitude of 10 m; $\mathcal{T}$ represents the integral, which is over the radius to an outer storm limit and over the lifetime of the storm; and v_{max}^3 implies that the total power dissipation rises roughly as the cube of the wind speed.

Table 2.7 Numbers of Tropical Cyclones in the Bay of Bengal, 1891–2007

Decade	Numbers of Cyclones[a]	Numbers Per Decade
1891–1900	50	5
1801–1910	43	4.3
1911–1920	44	4.4
1921–1930	60	6
1931–1940	52	5.2
1941–1950	45	4.5
1951–1960	31	3.1
1961–1970	51	5.1
1971–1980	48	4.8
1981–1990	35	3.5
1991–2000	28	2.8
2001–2007	15	2.14
1891–2007 total	**502**	**4.29**

[a]Include tropical storms with wind speeds: 39–73 miles per hour/63–119 kilometer per hour.
Prepared by the authors based on data presented in Fig. 3 in: Mahala, B.K., Nayak, B.K., Mohanty, P.K., 2015. Impacts of ENSO and IOD on tropical cyclone activity in the Bay of Bengal. Natural Hazards 75, 1105–1125.

(Emanuel, 2005, p. 686). Cyclone frequencies, in contrast, do not show any detectable trends. Thus, the question posed above (in the heading) involves two separate issues. Using data in Tables 2.7 and 2.8, we would like to demonstrate how the answers differ significantly from one another, depending upon the nature of the questions. The language of communication (discourse) is very critical for describing changes in cyclone frequencies and their magnitudes in the Bay of Bengal, especially because we are dealing here with data for two different geographical situations. First, data in Table 2.7 refer to all cyclonic activities in the Bay of Bengal. Many of those storms did not strike Bangladesh. Some had landfalls over India and Myanmar but many simply dissipated over the ocean. Second, in contrast, when people talk about impacts of cyclones in coastal Bangladesh probably they imply impacts of storms that had struck the coast of Bangladesh. We will analyze this specific concept under a separate section below by using data in Table 2.7.

Frequencies of Cyclonic Storms

Dealing with the first question, we have summarized 117 years of data (1891–2007) on the numbers of cyclonic storms that had developed over the Bay of Bengal during this period (Table 2.7). These data include all

Table 2.8 Landfall of Tropical Cyclones Over Coastal Bangladesh, 1877–2003

Decade	TD[a]	TS[b]	Category 1[c]	Cat 2[c]	Cat 3[c]	Cat 4[c]	Cat 5[c]
1877–1888	7	3					
1889–1899	2	1	3[d]				
1900–1909	2	5	1[d]				
1910–1919	3	2	1[d]				
1920–1929	5	3	1[d]				
1930–1939	6	3	1[d]				
1940–1949	4	2	1[d]				
1950–1959	5	2					
1960–1969	5	10	2	1		2[e]	
1970–1979		8	3			1[e]	
1980–1989		5	1		1		
1990–2003		6	3		1	2 (1[e])	1[e]
1877–2003 total	**39**	**50**	**17**	**1**	**2**	**5**	**1**

[a]TD = Tropical depressions.
[b]TS = Tropical storms.
[c]Category represents for tropical cyclones.
[d]Exact wind data not available.
[e]Super-cyclones with wind speed exceeding 120 knots (128 mph/222 km/h).
Prepared by the authors based on data presented in Tables 4–15 in: Islam, T., Peterson, R.E., 2009. Climatology of landfalling tropical cyclones in Bangladesh 1877–2003. Natural Hazards 48 (1), 115–135.

categories of tropical cyclones (categories 1–5), in addition to tropical storms that start from a minimum wind speed of 39 mph (miles per hour) (63 km/h) (Table 2.1). According to the Saffir-Simpson scale, a tropical cyclone has a threshold wind speed of a minimum of 74 mph (120 km/h) (Table 2.1). The maximum speeds include category 5 cyclones (>157 mph or 252 km/h). We would like to stress here that our data in Table 2.7 are secondary in nature, as we have drawn them from a recent publication by Mahala et al. (2015), but the original data used by these authors are from a primary source, namely the cyclone e-Atlas of India Meteorological Department.

The very first finding from data in Table 2.7 seems to be contrary to our expectation, as there was a declining trend in cyclonic activities in the Bay of Bengal for the entire period of record (1891–2007). To verify this finding, let us look at the details of this trend by comparing the number of storms by successive decades. In all, there were 502 storms in 117 years with an average frequency of 4.29 storms per year. Out of the 11 decades (1891–2000), the numbers of storms in six decades were above the long-term average, but they all occurred in the first half of the record (i.e., between 1891 and 1950). The maximum frequency (6 per year) was in 1920–1930, whereas the minimum (2.8 per year) was in 1991–2000. With the exception of

1961–1980, when there were slight increases in frequencies of storms (with above-average values), overall, the declining trend started after 1950 and the latest data (2001–2007) seemed to continue this trend. Thus, we can conclude from these data that, contrary to popular expectation, frequencies of storms/cyclones in the Bay of Bengal have declined over a period of at least one century. It should be stressed here that the bulk of the data presented in Table 2.7 consists of less intense tropical storms and lower categories of tropical cyclones.

Frequencies of High Magnitude Cyclones

Table 2.8 addresses the question if tropical cyclones striking coastal Bangladesh have become more intense and more violent in recent years. The answer is a definite yes based on the existing literature (Wang et al., 2013). To test this assumption, we have analyzed 127 years of storm track information for the landfall of tropical cyclones along the Bangladesh coast. Although our data in Table 2.8 are again secondary in nature as we have obtained them from a publication by Islam and Peterson (2009), the authors of this article have used primary data from the *Global Tropical Cyclone Climatic Atlas*.[4]

In all, there were 115 storms in 127 years (1877–2003). This is a post-Industrial Revolution period during which carbon dioxide had been accumulating in the atmosphere for more than 100 years. For the first 82 years of record (1877–1959) there is no consistent pattern of increases in landfall of cyclones along the Bangladesh coast. However, starting from 1960, there was a striking increase in the landfall of tropical storms and tropical cyclones in coastal Bangladesh. There were 29 tropical storms and 18 tropical cyclones between 1960 and 2003. Out of the 26 tropical cyclones in 127 years, 18 of them, that is, more than two-thirds (69%) occurred between 1960 and 2003. Not only was there a significant spike in cyclone activities during this relatively short period, perhaps more significantly, all of the five super-cyclones (with wind speeds exceeding 138 mph, 222 km/h) occurred during this period, providing evidence of recent intensification

[4] The data for the *Global Tropical Cyclone Climatic Atlas* are based on a compilation of several sources of primary data. These include the following: (1) two historical digital tape deck data files archived at the US National Climatic Data Center (NCDC), (2) a Joint Typhoon Warning Center (JTWC) historical data file from Guam, and (3) data forwarded to NCDC by several specialized meteorological centers participating in a Tropical Cyclone Program, sponsored by the UN's World Meteorological Organization (WMO) (Islam and Peterson, 2009).

of tropical cyclones over northern Bay of Bengal. These storms have also become extremely destructive causing massive property damage and loss of lives. It has been estimated by the CRED EM-DAT International Disaster Database that between 1961 and 2014 storms and cyclones have been responsible for the loss of nearly 557,000 lives in coastal Bangladesh (see further on cyclone deaths in a later section).[5] Further, increasing numbers of coastal residents have been affected by these cyclones in many different ways. Most significantly, millions had been displaced temporarily during some of the catastrophic cyclones. Although there are no reliable data on population displacements due to cyclones, it has been widely reported that large numbers of coastal residents have also been displaced permanently (that is, uprooted from their homes).

Nature of Sea Surface Temperature Anomalies

One of the problems with SSTs is that they are not necessarily uniform throughout an entire ocean body. Instead, they tend to develop anomalies, called SST anomalies, in the form of large pools of abnormally warm or abnormally cold water in certain parts of the ocean. The air temperature, air pressure, and cyclonic activities of the atmosphere immediately above the sea surface are affected by the temperature characteristics of these pools. Since atmosphere is an open system warmer or cooler air associated with these SST anomalies may not necessarily remain confined within the boundaries of a given ocean. The general circulation of the atmosphere (major wind systems) may transport these warmer or cooler air masses from one ocean to another or from one area to another. Thus, cyclones over a given ocean may be affected not only by its own SST anomalies but also by such anomalies in other distant oceans. In short, cyclones are impacted by a chain of interrelated factors: (1) general warming of the atmosphere (i.e., global warming), (2) impact of global warming on SSTs, (3) mutual exchange of energy (feedbacks) between SSTs and the atmosphere, and (4) general circulation of the atmosphere redistributing impacts of SST anomalies from different oceans. Using long-term historical data on cyclone frequencies and magnitudes in the Bay of Bengal and timing of SST anomalies in the Indian Ocean and the Pacific Ocean, we have

[5] This database is maintained by the Center for Research on the Epidemiology of Disasters (CRED), Université Catholique de Louvain-Brussels, Belgium. It is also short-formed as EM-DAT: The OFDA/CRED International Disaster Database (www.emdat.be).

made an attempt (below) to isolate and analyze the main issues of these complex problems.

Teleconnections

Some of the latest research on Bay of Bengal cyclones suggests that recent intensification of these storms may be related to global warming–induced SST anomalies, not only in the Indian Ocean but also in distant Pacific Ocean. The SST anomalies in the Pacific Ocean are known as El Niño and La Niña, whereas their Indian Ocean counterpart is called the Indian Ocean Dipole (IOD). How could these distant phenomena be interrelated over distances of thousands of miles/kilometers? There are at least two potential interconnections. First, the prevailing winds (i.e., the dominant winds of the general atmospheric circulation) transport warmer/cooler air from one area to another, since the global atmosphere is an open system. Second, warmer/cooler ocean water might be transported from the Pacific to the Indian Oceans (and vice versa) through openings between Australia and Indonesia. Further, these SST anomalies (El Niño, La Niña, and IOD) are coupled ocean-atmosphere phenomena. The term "coupled ocean-atmosphere phenomena" means that changes in SSTs result in simultaneous changes in air temperature and air pressure over the sea surface. In short, SST, air temperature, and air pressure are interrelated. Thus, changes in these characteristics in either the Pacific Ocean or in the Indian Ocean may affect each other's climatic characteristics. Such distant connections of the global atmosphere are called teleconnections.

What are El Niño and La Niña?

These are SST anomalies in the Pacific Ocean, which develop in cycles of warmer and cooler SSTs in alternating geographical positions within the equatorial/tropical belt of the Pacific Ocean. To explain these anomalies, it is convenient to consider three distinct phases of SSTs in the Pacific Ocean.

Phase 1: The *normal* or *neutral* phase does not experience either El Niño or La Niña. During this phase, the SST in western Pacific is warmer than in eastern Pacific. The prevailing winds over the tropical Pacific are easterlies, originating over East Pacific and blowing toward West Pacific. As the easterlies blow constantly over the surface water toward the west, this has a "skimming effect," removing a relatively shallow surface layer of warm water toward West Pacific. The lost warm surface water is then replaced by cooler water from below by a process called *upwelling* (similar to bubbling up). This is a very significant process for fish stocks and other marine life in the upper parts of the ocean where life-sustaining nutrients are delivered from deeper water below by upwelling. Because of strong upwelling of cool water along the coasts of Ecuador, Peru, and northern Chile, this area enjoys a productive fishing industry during normal periods. Cool surface water also means very little evaporation. Consequently, the entire coastal area of Ecuador, Peru, and Chile experiences a dry desert climate.

Phase 2. El Niño is the name for the warmer phase of SSTs in East Pacific. By technical definition by the World Meteorological Organization, to qualify as an El Niño condition the SSTs must be warmer than normal by at least +0.9°F or +0.5°C for three consecutive months and lasting for a minimum of five months. Most of the El Niño (and alternating La Niña) episodes within the last 50 years occurred in cycles of 3–5 years. On average, each episode lasted between one and two years. Exceptionally long ones persisted for periods of up to 34 consecutive months with temperature anomalies reaching as much as ±2–3°C. During major El Niño episodes the easterlies (east to west airflows) weaken significantly and may even reverse their directions (flowing west to east). This allows accumulation of a large pool of warm water in East Pacific, extending along the equator all the way from Central Pacific (north of Tahiti) to the coasts of Ecuador and Peru. Accumulation of unusually warm water in East Pacific results in thickening of the surface layer of warm water (up to 1000 ft or 300 m, compared to the normal of 130 ft or 40 m). This, in turn, reduces upwelling of cooler water (resulting in lack of nutrient delivery to the surface layer from deeper water below) with devastating effects for the fish stock along the coasts of Ecuador, Peru, and Chile. Extensive fish kills and reductions of fish stocks occur during El Niño.

The climatic effects of El Niño have also been devastating for coasts of Ecuador and Peru. Excessive rainfall and stormy conditions often had resulted in flash floods and landslides. The El Niño effects could also be far-reaching beyond the equatorial belt reaching as far north as the subtropical and midlatitude belts of North America. Whereas winters in northern Mexico and California tend to be wetter during El Niño episodes, both northwestern and northeastern parts of North America experience drier and warmer winters. As expected, cyclonic activities increase over East Pacific. Most of the high-magnitude hurricanes in East Pacific also occur during the El Niño events.

Phase 3. La Niña is the name for the reverse condition of El Niño. The easterlies become stronger than normal. Strong easterlies drive warmer surface water toward West Pacific accumulating a large pool of unusually warm water northeast of Australia and Papua New Guinea and east of Indonesia. Cooler water develops in East Pacific (cooler at least by −0.9°F or −0.5°C for three consecutive months and lasting for at least five months). During La Niña, rainfalls and cyclonic activities increase in northern and northeastern Australia and in many parts of Southeast Asia. Even high-magnitude floods in Bangladesh have been attributed partly to the La Niña effect.

What is ENSO?

In teleconnection research, effects of SSTs on changing air temperatures and air pressures over the Pacific (responsible for its cyclonic activities) are attributed to ENSO (El Niño-Southern Oscillation), a classic example of a single large-scale coupled ocean-atmosphere phenomenon. The Southern Oscillation (SO) is measured as an index called Southern Oscillation Index (SOI). It measures the differences or fluctuations in air pressure between Darwin (northern Australia) and Tahiti (central Pacific). During negative SOI (called low phase) air pressure is high over Darwin and low over Tahiti and East Pacific, resulting in a westerly flow (west to east), that is, a reversal of the easterlies. The resulting accumulation of warmer SSTs in Central and East Pacific leads to El Niño. During positive SOI (high phase) air pressure is high over the Central Pacific (Tahiti) and low over Darwin, resulting in strengthening of the easterlies, promoting La Niña.[6]

[6] Air pressure varies widely from one place to another and from one latitude belt to another belt. At a global scale the general circulation of the atmosphere is based on a model of a semipermanent high pressure system over the subtropical belt (called subtropical high pressure belt) over 25°−35° N (north of the equator) and 25°−35° S (south of the equator). Prevailing winds (dominant winds) blow in a north-south direction (to begin with) along the meridians (longitudes) from the subtropical high-pressure belts to the equatorial low-pressure belt (between 10° N and 10° S). Earth's rotation (*Coriolis effect*) deflects these winds as northeasterlies (north of the equator) and southeasterlies (south of the equator). Since these winds blow along the meridian (in a north-south direction to begin with) these are also called meridian flows.

In contrast to meridian flows, the *Walker Circulation* (discovered by Sir Gilbert Walker, the Chief Meteorologist of British India in the 1920s) refers to zonal flows that develop as east-west winds over the Pacific. Conceptually, Walker tied these east-west flows to the Southern Oscillation, which is measured as the differences in air pressure between Darwin (northern Australia) and Tahiti (central Pacific). The Walker Circulation reinforces the easterlies (a limb of this zonal flow) during La Niña, whereas during El Niño the Walker Circulation is weakened. For an explanation of how El Nino and Southern Oscillation are coupled as ENSO, see above the section on ENSO.

Another feature of the Walker Circulation is that warm and moist air over the West Pacific rises up (to the top of the troposphere at about 10 miles/16 km above the sea level), exhausts all moistures by condensation and precipitation, and becomes cooler, then crosses the Pacific and sinks over East Pacific, strengthening the high pressure system. This reinforces the easterlies. Thus, a complete east-west vertical loop (cell) is created, which is called the Walker Circulation. More recent research has found that multiple Walker Circulation cells extend from the Pacific Ocean to the Indian Ocean to the Atlantic. Thus, the Walker Circulation plays an important role in transporting winds from surfaces of SST anomalies between the Pacific and the Indian Oceans.

What is IOD?

The Indian Ocean Dipole (IOD) is the Indian Ocean counterpart of the Pacific El Niño and La Niña. The term dipole means two "poles" or two areas of differences. The IOD measures differences in SSTs between the Arabian Sea (western pole) and the eastern Indian Ocean south of Indonesia (eastern pole). Both of these poles are situated within the equatorial belt of the Indian Ocean (i.e., between 10°N and 10°S) but they have a northwest-southeast diagonal orientation because of the physical configuration of the North Indian Ocean. The Arabian Sea is located north of the equator and the Indian Ocean south of Indonesia is located south of the equator.

Like ENSO, IOD is a coupled ocean-atmosphere phenomenon. The shifting pools of warm/cool water contribute to variations in rainfall and storm activities of many countries surrounding the Indian Ocean. It is postulated that the IOD is linked to the Pacific ENSO through the easterlies (considered as a limb/part of the *Walker Circulation*)[7] and through transport of warm waters from the Pacific into the Indian Ocean. During a positive IOD warmer SSTs develop over western Indian Ocean (Arabian Sea, in particular). During a negative IOD, the opposite happens, that is, the western Indian Ocean becomes cooler with higher air pressures resulting in westerly winds blowing toward the Indian subcontinent (i.e., reversing the prevailing easterlies). Positive IODs are often associated with El Niño and negative IODs with La Niña.

Effects of ENSO and IOD on Bay of Bengal Cyclones

Data in Table 2.9 provide evidence of simultaneous occurrence (co-occurrence) of tropical cyclones with ENSO and IOD events in the Bay of Bengal for a period of 117 years (1891–2007). These data are remarkable because they confirm several scientific and popular assumptions. First, conforming to the general observation stated above, positive IODs were associated more with El Niño events (42 events that also co-occurred with tropical cyclones) than with La Niña (8 events), whereas 63 La Niña events (and tropical cyclones) co-occurred with negative IOD, compared to only nine El Niño events (first row of the table). Second, the data confirm the assumption of ENSO effects on cyclonic activities in the Bay of Bengal. This is a significant finding because two-thirds of the tropical cyclones (333 out of 502) co-occurred with El Niño and La Niña during three

[7] See footnote 6.

Table 2.9 Simultaneous Occurrence of Tropical Cyclones With ENSO and IOD Events in the Bay of Bengal, 1891–2007

Nature of IOD	Number of Tropical Cyclones Co-occurred With El Niño	Number of Tropical Cyclones co-occurred With Neutral ENSO	Number of Tropical Cyclones Co-occurred With La Niña	Total Number of Tropical Cyclones
Negative IOD	9	9	63	81
No IOD	127	138	84	349
Positive IOD	42	22	8	72
Total	178	169	155	502

Prepared by the authors based on data in Table 3 in: Mahala, B.K., Nayak, B.K., Mohanty, P.K., 2015. Impacts of ENSO and IOD on Tropical Cyclone Activity in the Bay of Bengal. Natural Hazards 75, 1105–1125.

different phases of the IOD. This confirms the significance of teleconnection between SST anomalies in the Pacific and the Indian Oceans. Third, the dominant effect of ENSO events on tropical cyclones in the Bay of Bengal can be seen from the data indicating that, compared to 66% of total cyclones during ENSO events, only about 30% (153 cyclones) occurred during IOD events (both positive and negative). Fourth, it is remarkable that even in the absence of IOD at least 42% of these cyclones ($127 + 84 = 211$ out of $502 = 42\%$) occurred during alternating phases of El Niño and La Niña, again confirming teleconnection. Finally, it is equally remarkable that only about one-quarter of the tropical cyclones (138 out of $502 = 27\%$) occurred in the absence of both ENSO and IOD effects. This confirms the assumption that SST anomalies (ENSO and IOD) play a significant role in producing nearly three-quarters of tropical cyclones in the Bay of Bengal.

CYCLONE HAZARDS IN BANGLADESH

Nature of the Hazards

Among all climatic hazards in Bangladesh, undoubtedly tropical cyclones rank at the top for causing extensive property damage and massive loss of lives. Most of the impoverished people of coastal Bangladesh live in flimsy houses made of earthen floor, bamboo or wooden frames, and thatched or tin roofs and sidings (Fig. 2.1).

Figure 2.1 A typical house in coastal Bangladesh, illustrating flimsy structures made of thatched materials. *Photograph by M. Khaledur Rahman, reproduced with permission.*

These houses are classified as *kutcha* (literally meaning weaker houses) by the Government of Bangladesh Bureau of Statistics (BBS). About 78% of all rural houses of Bangladesh in 2007–2008 were listed by BBS as *kutcha*, whereas 13% were characterized as *semi-pucca* homes (medium strong homes), made of mostly earthen floor (sometimes cement floor), wooden frames, and corrugated tin roofs and sidings (www.bbs.gov.bd/). Like *kutcha* houses, most of the *semi-pucca* houses are not cyclone resistant. Very few well-to-do people (about 9% of rural population) live in *pucca* houses (brick buildings). Field evidence indicates that most of the brick buildings are resistant to cyclone damage. For example, a post-*Gorky* (1991) survey among a sample of 1123 cyclone survivors showed that nearly one-quarter who did not take shelter in a concrete or brick structure died, whereas those who sought refuge in such structures survived (Bern et al., 1993). Another survey by Haque and Blair (1992) confirmed that not only most of the wooden-frame or bamboo-frame structures were totally destroyed or badly damaged but also some of the brick buildings could not withstand the impacts of strong winds and storm surges associated with category 5 Cyclone Gorky (also called Cyclone 2B) (see Figs. 2.2–2.4).

Figure 2.2 Remnants of a bamboo-frame house after *Cyclone Gorky* had blown away almost its entire structure, including its roofs. A cyclone victim made an attempt to build a temporary shelter using leftover materials. A cruise ship grounded by the cyclone is visible in the background. *Photograph by Emdad Haque, reproduced with permission.*

Figure 2.3 *Cyclone Gorky* damage survey (May 1, 1991): During a post-*Gorky* survey by Emdad Haque (University of Manitoba) cyclone victims are seen here narrating their traumatic experience with flying sharp corrugated iron sheets, which were blown off from roofs injuring many victims. *Photo-courtesy: Emdad Haque; reproduced with permission.*

Figure 2.4 Even brick buildings, such as this retail store along the Chittagong coast, could not withstand the awesome power of category 5 *Cyclone Gorky*, which struck the Chittagong coast on April 29, 1991. In this picture (taken on April 30 1991), the masonry roofs of the building were blown away leaving behind the bare walls standing. *Photograph by Emdad Haque, reproduced with permission.*

Threat to human lives poses by far the greatest cyclone hazard in coastal Bangladesh, as nearly 1 million people had perished in the past 500 years under the crushing impacts of storm surges (Alam and Dominey-Howes, 2014; Alexander, 1993). The livelihoods of coastal residents are also at risk from rising sea levels and cyclone impacts. Saltwater intrusion due to rising sea levels and storm surges has been polluting their rice crops, the main subsistence crop of Bangladesh. Fishing in the open sea in small indigenous boats, which is another component of local subsistence economy for many coastal residents, is a hazardous undertaking, as many in the past were caught off-guarded and had lost their lives by rapidly advancing storms, irrespective of whether they had listened to cyclone warnings or not (Fig 2.5).

Cyclone Deaths in Bangladesh

Since magnitudes of cyclones striking the coast of Bangladesh have increased recently with a simultaneous rise in sea levels, it is reasonable to assume that there may be a corresponding increase in cyclone death rates in coastal Bangladesh. To test this assumption, we have reviewed two sets of historical data on cyclone deaths in Bangladesh (Tables 2.10 and 2.11).

Figure 2.5 Indigenous boats venturing into the Bay of Bengal for fishing often perish under rapidly advancing storm waves. Too often fishers may not be aware of advance storm warnings or they may ignore such warnings. *Photograph by M. Khaledur Rahman, reproduced with permission.*

Data in Table 2.10, based principally on historical archives, spans over a period of more than 500 years (1484–2009). Although different types of information on past major cyclones for such a long period of time have been retrieved most diligently by Alam and Dominey-Howes (2014), the quality of data on cyclone deaths has been compromised by several interrelated methodological issues, such as the original sources of information and their consistency, accuracy, and reliability. The authors themselves have admitted shortcomings in their data collection throughout several places in their manuscript, but these can be summed up as follows, in authors' own words:

- "The limited records of tropical cyclones prior to AD 1900 for the Bay of Bengal imply that historical records of tropical cyclones are incomplete" (Alam and Dominey-Howes, 2014, p. 807).
- "Discrepancies in the reporting of deaths are observed within the data from the very earliest (i.e., 15th century) to the recent period (i.e., 20th century)" (Alam and Dominey-Howes, 2014, p. 828).

On a closer scrutiny of this catalog of data we found that original discrepancies in the reporting of deaths were so severe that some of the older

Table 2.10 Cyclone Deaths in Bangladesh: 1484–2009

Years (With Historical Records of Cyclone Deaths)	Numbers of Recorded Cyclones	Districts of Cyclone Fatality	Total Numbers of Deaths
1484–1760 (pre-industrial)	4	K (Khulna), B (Barisal), (C) (Chittagong)	650,000
1761–1900	11	K, B, N (Noakhali), C	934,500
1901–1910	2	B, CB (Cox's Bazar)	578
1911–1920	5	K, B, CB	200,932
1921–1930	2	CB	612
1941–1950	4	K, B, C, CB	45,700
1951–1960	3	B,N,C	16,819
1961–1970	12	K, B, C, CB	399,120
1971–1980	6	K, C, CB	2018
1981–1990	10	K, B, N, CB	22,801
1991–2000	8	K, C, CB	139,908
2001–2010	4	K, B, C	4446
Total	**71**		2,417,434

Prepared by the authors based on relevant data in Alam, E., Dominey-Howes, D., 2014. A new catalogue of tropical cyclones of the northern Bay of Bengal and the distribution and effects of selected landfalling events in Bangladesh. International Journal of Climatology 35 (6), 801–835. For some years data from multiple sources were averaged.

Table 2.11 Cyclone Disasters in Bangladesh: Cyclone/Storm Deaths, 1961–2014

Periods	Number of Cyclones/ Storms[a]	Number of Deaths	Deaths Per Cyclone/Storm
1961–1970	17	384,791 (84,791 without *Bhola Cyclone*)	22,635 (5299)
1971–1980	14	6570	469
1981–1990	18	18,554	1031
1991–2000	38	141,791 (2925 without *Cyclone Gorky*)	3731 (79)
2001–2010	30	5319	177
2011–2014	4	145	36
Total number of cyclones/storms (1961–2014)	**121**	**557,170**	**4605**

[a]Cyclones/storms include tropical cyclones (cat 1–5), tropical storms and convective storms.
Prepared by the authors based on original data from EM-DAT: The OFDA/CRED International Disaster Database, Universite Catholique de Louvain-Brussels, Belgium (www.emdat.be/advanced_search/index.html/). Data from EM-DAT, 2015.

archival data do not seem to be reliable. For example, the authors have cited a gray literature for its earliest data on fatality from the 1484 cyclone. A local publication (in Bengali) by the Sandweep Association, a community-based association in the subdistrict of Sandweep (in the district of Chittagong) has cited 200,000 deaths due to storm surge impacts by this historic cyclone (Hasan, 1999). The main problem with this data is that Alam and Dominey-Howes (2014) have not provided any further details about the original source of information for such a large number of deaths. If 200,000 people lost their lives in the district of Noakhali, including Sandweep Island, it begs a question what proportions of the total population of the district accounted for this loss? It will be difficult to answer this question because historical records indicate that there was no official census of Bengal in the late 15th century when the territory was under the Bengal Sultanate rule. Further, it begs another question if there was any authentic oral history on this event that prompted Hasan (1999) to cite such a figure. Similarly, Alam and Dominey-Howes (2014) have cited three other historic cyclones (in 1582, 1584, and 1767), each with 200,000 to 300,000 fatalities, respectively, again without indicating their primary sources.

Another major problem with this catalog is that for some of the years authors have listed multiple estimates (i.e., a range of data) on cyclone deaths for each of the events, whereas at one point in their manuscript they have indicated that "whenever significant differences [occurred] concerning the number of deaths reported in different sources ... an average number of deaths [was] calculated from the numbers reported in all sources" (Alam and Dominey-Howes, 2014, p. 807). This has been confusing for us because multiple data (i.e., ranges of data) have been provided for each of the 16 out of 56 cyclones between 1822 and 1995 (without any average). Further, in the absence of primary sources, only secondary sources of these data have been cited. The main problem with this approach is that the authors of most of these publications have not provided any source of their data; instead, they have just produced some of the unattributed tables (see, for example, Frank and Husain, 1971; Dunn, 1962).

Despite these shortcomings, the main contribution of historical archives in cyclone studies is that often they provide valuable qualitative information on past cyclones. Among all archival resources, district gazetteers, which were maintained for the Greater Bengal by the British colonial rulers, provide valuable information on past physical and socioeconomic characteristics of different districts of Bengal (Bangladesh constitutes a major part of former Greater Bengal). Some of these gazetteers provide detailed references to many catastrophic cyclone events, although they may or may not provide quantitative data on cyclone damage and fatality (see for example, Beveridge, 1876; Hunter, 1876). Among these, archival contributions by Henry Beveridge, the District Magistrate of Barisal, a Bengal Civil Service officer during the British rule in India, are particularly rich in text and data. In the *Bakarganj (Barisal) District Gazetteer*, he documented several cyclone events in 1822, 1867, and 1869. Perhaps, a more significant finding of archival research by Alam and Dominey-Howse (2014) is that historical archives provide evidence of the fact that high-magnitude tropical cyclones have been occurring in the Bay of Bengal for a long period of time, well before the beginning of industrial emissions of carbon dioxide.

To sum up main findings of the catalog prepared by Alam and Downey-Howes (2014), at least 193 major cyclones struck the coast of Bangladesh between 1484 and 2009 and 71 of them were responsible for significant loss of human lives. Most of these cyclone deaths occurred in five coastal districts, namely, Khulna, Barisal, Noakhali, Chittagong, and Cox's Bazar. As far as the details of the number of deaths in these districts, we will not review the remaining data in Table 2.10 because of the methodological issues noted above. Instead, we focus below on post-1960 data in Tables 2.11 and 2.12, which are based on the *EM-DAT: OFDA/CRED International Disaster Database.*[8] The justification for analyzing post-1960 data lies in the fact that systematic record keeping of cyclone deaths in Bangladesh started from 1960 (Dunn, 1962). The EM-DAT original database confirms that pre-1960 records on cyclone deaths in East Pakistan (now Bangladesh) and Bengal (during British rule) were incomplete, whereas post-1960 records seem to be comprehensive. The EM-DAT is perhaps one of the most consistent, reliable, and frequently cited global database for mass disasters. This database is compiled from various sources, "including UN agencies, government and non-government organizations, insurance companies, research institutes and press agencies", using a systematic and rigorous methodology (www.emdat.be/).

Coming back to our initial question, if cyclone death rates in coastal Bangladesh have increased recently, the findings are contrary to the expectation of current public discourse on climate change in Bangladesh. The EM-DAT data on cyclone deaths in Bangladesh, retrieved through its advanced search (www.emdat.be/advanced_search/index.html/), includes

[8] See footnote 6.

Table 2.12 Cyclone Disasters in Bangladesh: Personal Injuries, Population Displacements, Other Effects, and Damage Estimates, 1961–2014

Decade	People Injured[a] (Number of Years of Data)	Homeless[b] (Number of Years of Data)	People Affected by Storms[c] (Number of Years of Data)	Estimated Storm Damage (US Dollar)[d] (Number of Years of Data)
1961–1970	616,775 (4)	5,323,943 (3)	16,384,003 (7)	238,479,000 (8)
1971–1980	25,250 (4)	10,000 (1)	26,000 (2)	50,000,000 (1)
1981–1990	2750 (5)	2,645,000 (4)	14,493,730 (6)	50,000,000 (1)
1991–2000	194,549 (9)	1,760,500 (7)	20,882,278 (8)	2,707,500,000 (3)
2001–2010	67,610 (10)	15,125 (4)	13,250,717 (8)	2,570,000,000 (2)
2011–2014	260 (3)	4000 (1)	1,629,204 (3)	NA
Total	**907,377**	**9,758,587**	**66,665,963**	**5,615,979,013**

Note: All definitions below have been retrieved from: www.emdat.be/explanatory-notes/.
[a]People injured: "People suffering from physical injuries, trauma or illness requiring immediate medical assistance as a direct result of a disaster."
[b]Homeless: "Number of people whose house is destroyed or heavily damaged and therefore need shelter after an event."
[c]People affected: "People requiring immediate assistance during a period of emergency, i.e. requiring basic survival needs such as food, water, shelter, sanitation and immediate medical attention."
[d]Estimated damage: "The amount of damage to property, crops and livestock (in US dollar) … may include breakdown figures by sectors: social, infrastructure, production, environment and others."
Prepared by the authors based on original data from EM-DAT: The OFDA/CRED International Disaster Database, Universite Catholique de Louvain-Brussels, Belgium (www.emdat.be/advanced_search/index.html/). Data from EM-DAT, 2015.

fatalities from all storms, ranging from convective storms (such as the pre-monsoon Norwesters) to tropical storms and tropical cyclones of all categories. As data in Table 2.11 indicate, 121 storms were responsible for loss of at least 557,000 lives in Bangladesh between 1961 and 2014. The storm/cyclone death rates by successive decades indicate that there was a huge spike in cyclone fatality in the first three decades (1961—1991). This spike was contributed mainly by disproportionately large fatalities from two catastrophic cyclones, namely the Bhola cyclone (1970: +300,000 deaths) and Cyclone Gorky (1991: 138,866 deaths). Together, these two cyclones accounted for nearly 79% of the total storm/cyclone deaths between 1961 and 2014 (438,866 out of 557,170). None of the subsequent high-magnitude cyclones was responsible for significant loss of lives, certainly not at comparable scales. Even if we exclude these two historic cyclones, perhaps the most unexpected trend for the remaining 119 storms between 1961 and 2014 was that storm/cyclone death rates in Bangladesh have been declining gradually following the 1991 historic event, irrespective of recent intensification of cyclones. It may seem to be a paradox but this decline may be explained more by coping strategies than by cyclone magnitudes. Thus, the gradual decline in post-1991 cyclone death rates in Bangladesh may be attributed largely to recent expansion of cyclone shelters in coastal Bangladesh and partly to a significant improvement in disaster management policies (Haque et al., 2012; Paul 2009, 2010) (for further details of these factors see Chapters 7 and 8).

Cyclone Injuries

Our Survey Data

Besides the obvious threat of loss of lives, threat of personal injuries is another major cyclone hazard in coastal Bangladesh. Both storm surge impacts and wind damage may be responsible for personal injuries, but trauma and injuries due to wind impacts are more widespread in the coastal area as these may be experienced further inland from the storm surge impact areas. Data on cyclone injuries in Bangladesh are much more limited than that on cyclone deaths. We made an attempt to gain some insights on this issue by asking a limited number of questions on cyclone injuries as a part of one of our post-Sidr surveys in 2007 (Paul, 2009). The following are the highlights of cyclone injuries from our survey among a systematic sample of 1443 respondents from four districts that were severely impacted by Sidr, namely, Bagerhat, Barguna, Pirojpur, and Patuakhali (see Paul, 2009, Fig. 1 for locations of these districts).

Nature of injuries: The nature of injuries may range from minor, such as soft-tissue injuries (cuts, bruises and scrapes), to more serious, such as blunt trauma, laceration, and fracture/dislocation. Some of the most common causes of cyclone injuries in Bangladesh are described below.

Injuries due to collapse of housing structures: Housing structures may collapse under the impacts of storm surges and winds or they may be blown away by strong winds. Major injuries have been reported when residents inside their homes were trapped under the burden of collapsed structures.

Injuries by flying and fallen debris: Both flying and fallen debris could be dangerous (sometimes lethal) weapons for personal injuries. Among these, blown away corrugated tin roofs are particularly dangerous as they tend to be heavy and have sharp edges. Even falling wooden debris could be major hazards for the coastal residents struggling to take refuge either inside or outside their homes. Wooden debris may include uprooted tree stumps or broken branches or even flying parts from housing structures (such as wooden poles or parts of sidings). Field evidence shows that these injuries had occurred when some of the residents decided not to evacuate so that they could personally protect their homes by employing different protective devices, such as tying their homes to nearby trees and holding on to them. Injuries occurred when these protective devices failed and the house collapsed on them (see also next section on indigenous adjustments).

Injuries due to fall from trees: This may sound to be unexpected but in reality a large number of people climb on trees and tie themselves against the trunk or branches to protect them from storm surge and wind impacts (see also next section on indigenous adjustments). Past experience shows that accidents and injuries had happened when some of these people simply fell down from trees when their protective ties had failed. More commonly, many people sustained major injuries when they were thrown off from broken tree branches or uprooted trees.

Pre- and post-storm injuries: Cyclone-related injuries may also include injuries during evacuation prior to the landfall of a cyclone or during clean-up of debris in the cyclone's aftermath. Injuries during clean-up may include puncture wounds from stepping on nail, or muscle strains from lifting heavy items. Surprisingly, even emergency management volunteers (for example, the staff of the Cyclone Preparedness Program) also sustained some injuries during the evacuation process.

Treatment: Most of the injuries were treated onsite by first responders, such as members of mobile medical teams (MMTs). Besides the Government of Bangladesh, Pakistan and USA among others participated in MMTs to treat people with injuries. In particular, the United States

MMT operations reached remote villages by making timely use of its US Marine CH-46E Sea Knight Helicopters (Paul, 2009). In addition, an 18-member US army medical team operated a 250-bed hospital in Patuakhali. Most of the injured recovered from injuries fairly quickly. In more serious cases some of the injured received treatment at local government clinics or at private clinics. Only in rare cases more serious injuries led to post-storm deaths.

Age and gender of victims: We investigated the role of age and gender in cyclone injuries but found that neither of these variables conformed to our *priori* expectation or to data from previous studies on cyclone deaths in Bangladesh. As indicated above, data on cyclone injuries in Bangladesh are limited, but several studies on previous cyclone deaths have shown repeatedly that women and children accounted for significantly larger proportions of cyclone fatality than men (see, for example, Ikeda, 1995; Bern et al., 1993; Chowdhury et al., 1993; Sommer and Mosley, 1972). In contrast, our post-Sidr survey showed that children under 15 years accounted for only 4% of the injured (among our sample), compared to 11% and 14% of two older-age groups (i.e., 15–49 and 50+, respectively). There is a need for further survey data and additional research for explaining this success story. In the absence of a satisfactory explanation, we concluded somewhat euphemistically that "possibly *Sidr* survivors took extra precautions to save their children's lives" (Paul, 2009, p. 492). Similarly, we found that there was no significant difference in injury rates between women and men: 9.27% women versus 9.03% men (among our survey respondents). Even considering a slightly favorable ratio of male population of the study area, this difference was not statistically significant (Paul, 2009, Table 2).

EM-DAT Data

In the absence of any alternative and reliable long-term data on cyclone injuries in Bangladesh, EM-DAT database seems to be perhaps the most comprehensive historical data on this critical variable. The original annual data from which we have prepared Table 2.12 indicate that the OFDA/CRED (source of EM-DAT data) reported storm injury data for 35 years during a period of 54 years (1961–2014). There was no way to verify from the database whether data were missing for some years or no cyclone injuries were reported for some of the years. The 1970 Bhola cyclone is a case in point. Whereas the OFDA/CRED has provided data on loss of lives (300,000) from this catastrophic cyclone, it has not provided any data on cyclone injuries, presumable because no such data were available. Despite such incomplete record, an in-depth analysis of the EM-DAT record provides valuable insights on historical trends in cyclone injury rates. Clearly, storm injuries constitute a major cyclone hazard in Bangladesh, as at least 900,000 people sustained storm-related injuries during the period of record (1961–2014). However, the statistical distribution of the original data, expressed as the average rates of injuries per storm, is highly skewed. In some years, lower magnitude storms caused fewer deaths but disproportionately large numbers of injuries. In 1965, for example, nearly 49,000 people lost their lives from three lower magnitude cyclones, but the latter were responsible for about 600,000 reported injuries (original EM-DAT data). Similarly, in 2007 only about 4000 people lost their lives from Cyclone Sidr, but nearly 55,000 people were injured by this category 4 cyclone (Table 2.13). The aggregate data in Table 2.12 hide these details. Therefore, we have provided data in Table 2.13 on death rates versus injury rates for selected years to demonstrate how either the death rates or the injury rates could influence some of the inflated ratios.

Despite the nonrandom distribution of data, it is possible to divide the injury data into two distinct periods (1961–1990 and 1991–2014), based on the completeness of data (or lack of it) and rates of injuries (Table 2.12). During the first period (1961–1990) storm/cyclone injury data are available only for 13 years (out of 30 years). The total number of storm/cyclone injuries from these storms was 644,775, resulting in an average of nearly 50,000 injuries per storm. This abnormally high injury rate, despite the lack of 1970 data, was pulled by the abnormally high injury rates for the 1965 storms (Table 2.13). In contrast, data for the second period (1991–2014) averaged at 11,928 injuries per storm (262,419/22), including high injury rates in 2007 (Cyclone Sidr). Thus, not only cyclone death rates

Table 2.13 Cyclone Disasters in Bangladesh: Selected Data on Cyclone Deaths Versus Cyclone Injuries

Year	Number of Deaths	Number of Injuries	Death Versus Injury Ratios
1965	48,921 (3)	600,000	1: 12.26
1969	849 (1)	15,503	1: 18.29
1985	15,071 (3)[a]	300	1: 0.02
1991	138,866 (2)[b]	138,849	1: 1
2007	4275	55,282	1: 12.93

[a]One of them was a convective storm.

[b]One of them was a lower magnitude cyclone, besides category 4 *Cyclone Sidr*.

Prepared by the authors based on original data from EM-DAT: The OFDA/CRED International Disaster Database, Universite Catholique de Louvain-Brussels, Belgium (www.emdat.be/advanced_search/index.html/). Data from EM-DAT, 2015.

in Bangladesh have been declining since 1991, cyclone injury rates also seem to be decreasing significantly.

Other Impacts

Population Displacement

Displacement from home, or homelessness, due to cyclone impacts is another major cyclone hazard for the residents of coastal Bangladesh. The nature of displacement may include two broad categories: (1) temporary displacement prior to or during or immediately following the devastating impact of a cyclone and its storm surges, and (2) permanent displacement in the aftermath of a cyclone. Among these two categories, our field observations indicate that large proportions of people displaced by a cyclone return to their homes and undertake necessary repairs or rebuild new homes if their original structures were totally destroyed. Based on such observations, one may assume that only a small proportion of the displaced population might have abandoned their ancestral homes and moved elsewhere permanently in search of a new life. Unfortunately, there are no reliable data on permanent displacement ("climate refugees"), to back up such an assumption. This lack of information on climate change–induced population displacement from coastal Bangladesh can be traced back to an absence of a deliberate program for collecting such data. This may sound to be surprising in view of relentless demands by the Government of Bangladesh for climate change compensation from industrial nations. Most often climate refugees from coastal Bangladesh figure out prominently in such demands. There may be a lack of political will or ideas and methodologies for collecting such data, but methodological problems (briefly outlined below) pose a major challenge for data collection on the number of climate refugees.

Sea level rise is a relatively slow, insidious (invisible), and ongoing process. Therefore, it is difficult to attribute population displacement to such a slow and relatively small amount of sea level rise (at least up to now). In contrast, high-magnitude cyclones are episodic, that is, they occur once in few years and their destructive impacts, especially property damage and population displacement, are highly visible. News media rush to the scenes of cyclone damage and publish reports on deaths and destructions caused by a given storm, often accompanied by different types of pictures as evidence of storm damage, such as damaged or destroyed homes and scenes of homelessness, including make-shift camps for temporary shelter. Often some of these reports publish estimates of property damage and population displacement. One of the problems with such news media data is that these may be based on estimates by different sources, including government and nongovernment sources. Consequently, such data may not be consistent in format or reliable in content. Thus, there is no shortage of anecdotal evidence of population displacement from coastal Bangladesh due to cyclone impacts and/or sea level rise but there is no reliable source of information on how many people have been displaced permanently following each of the successive high-magnitude cyclones (see, for example, Shamsuddoha, 2015).

In the absence of any alternative data source, the EM-DAT provides a systematic record of cyclone-induced population displacement for a period of 54 years (Table 2.12, column 3). Nearly 10 million people were displaced from their homes reportedly by 20 cyclones. As in the case of cyclone injuries, it is not clear if these data provided a complete record of population displacement or if data were missing for a number of years. Again, as in the case of cyclone injuries, the statistical distribution of the original data is highly skewed by several events of large numbers of displacements. For example, three lower-magnitude cyclones in 1965 were responsible for displacing nearly 5 million people (EM-DAT original data). Similarly, 2 million people were displaced by four lower-magnitude cyclones in 1988, whereas 300,000 became homeless during Cyclone Gorky in 1991. Following 1991, population displacement by cyclones has been declining gradually, especially during the last two decades, contrary to popular assumptions.

People Affected by Storms/Cyclones

Besides cyclone deaths, injuries, and homelessness, EM-DAT has provided a broader category of data on "people affected" by combining data on injuries and homelessness with other cyclone emergencies, such as "people requiring immediate assistance during a period of emergency, i.e., requiring basic survival needs such as food, water, shelter, sanitation and immediate medical attention." The total number of people affected thus by cyclone disasters amounted to a staggering 66.6 million during 34 cyclones (Table 2.12, column 4). Among these storms, Cyclone Gorky (also called Cyclone 2B) (1991) ranks at the top for its adverse effects, as about 15 million people of Bangladesh (14% of 1991 population) were affected by this historic cyclone. Next, Cyclone Sidr (2007) ranks second to *Gorky* for its adverse impacts, as it affected at least 8 million people. Among other catastrophic cyclones, the Bhola cyclone (1970) was not only "the killer storm of the century," it also affected adversely nearly 3 million people. Very few people (less than 400) died from Cyclone Aila in 2009, but it was notorious for adversely affecting at least 3 million people of coastal Bangladesh. Besides

Table 2.14 Five Costliest Cyclones in Bangladesh, 1961–2014

Year (Cyclone Name)	Damage Estimate in 2015 US Dollar (Accounting for Average Inflation Rates)	Original Damage Estimate in US Dollar (in the Year of Event)
1970 (*Bhola Cyclone*)	$530 million	$86.4 million (1970)
1991 (*Gorky*)	$3.11 billion	$1.78 billion (1991)
1995 (two unnamed cyclones)	$1.25 billion	$800 million (1965)
2007 (*Sidr*)	$2.64 billion	$2.3 billion (2007)
2009 (*Aila*)	$300 million	$270 million

Prepared by the authors based on original damage estimates (for respective years of each event) from EM-DAT data (see 2.12 for full reference). Inflation rates are based on a table in: www.dollartimes.com/inflation/.

these top ranking cyclones, three other unnamed cyclones in 1965 affected about 10 million people, whereas at least 10 other cyclones that occurred between 1966 and 2013 (specifically in 1966, 1981, 1988, 1995, and 2013) affected adversely between 1 and 2 million people each.

Estimated Storm Damage

Expressed in US dollar, estimated cyclone damage in Bangladesh amounted to a staggering total of $5.6 billion in 15 years (Table 2.12, column 5). Five costliest cyclones in Bangladesh are listed in Table 2.14. Among them, Cyclone Gorky in 1991 was not only one of the deadliest cyclones in Bangladesh (responsible for 138,866 deaths), but was also the costliest: estimated damage was about $3.12 billion (in 2015 dollars) (The original estimate in 1991 was US $1.78 billion). The second costliest was Cyclone Sidr in 2007 with an estimated damage of $2.64 billion ($2.3 billion in 2007), although less than 4000 people died from its impact. The 1970 Bhola cyclone, the "killer storm of the century" (responsible for at least 300,000 deaths), was also a costly cyclone with an estimated damage of $530 million dollars ($86.4 million in 1970). In 2009, Cyclone Aila had cost about US 300 million ($270 million dollars in 2009). Two other unnamed cyclones in 1995 caused extensive property damage and crop losses amounting to a total of US 1.25 billion (US $800 million in 1995).

INDIGENOUS ADJUSTMENTS TO CYCLONE HAZARDS

Reasons for Living With Cyclone Hazards

Despite massive numbers of cyclone deaths, injuries, population displacements, and other cyclone-related miseries, why do so many people continue to live in cyclone-vulnerable coastal Bangladesh? For the residents of coastal Bangladesh, perhaps the answer lies in what Gilbert White (1974, p. 4) has characterized as "lack of satisfying alternative opportunities." Because of high population densities and scarcity of land in Bangladesh most of the coastal residents have limited opportunities of living anywhere else. As indicated earlier, most of the residents of coastal Bangladesh are impoverished farmers and fishers who eke out a living out of the low-lying land and hazardous sea. Poverty figures out prominently in their vulnerability to cyclone hazards as it forces many people to live in hazard-vulnerable low-lying areas (Rashid, 2011; Alam and Collins, 2010; Paul et al., 2010; Haque, 1997; Haque and Blair, 1992). A *New York Times* reporter summed up this problem succinctly in the aftermath of *Cyclone Gorky* (1991): "In Bangladesh's storms, poverty more than weather is the killer" (Kristoff, 1991). Another reporter of the UK-based *Guardian Weekly* probed this connection in more detail:

> *But why do so many people die when the disaster is more or less predictable? The answer, according to Allister McGregor, lecturer in development studies at Bath University, is poverty. A large proportion of the population is landless and has gravitated for generations to those places where most silt has been deposited and islands have formed. ... These people are the very poorest, and they take the biggest risk. It is like living next to a precipice*
>
> **Simmons (1991)**

Nature of Indigenous Adjustments

Large numbers of cyclone deaths, cyclone injuries, and cyclone-induced displacements from homes in coastal Bangladesh can be attributed directly to poor-quality housing structures. The impoverished coastal residents cannot afford to build stronger, cyclone-resistant structures or structures incorporating cyclone-proofing measures. Recent expansion of cyclone shelters has contributed to gradual reductions in cyclone deaths following *Cyclone Gorky* (1991). Barring this modern structural intervention, for at least 500 years (based on archival evidence of cyclones in the Bay of Bengal coast), residents of coastal Bangladesh have been living with cyclone hazards by adopting a variety of indigenous adjustments, which have proven to be extremely ineffective against the awesome power of devastating cyclones. The term *indigenous adjustment* refers to "folk, or preindustrial, adjustments which are ... low in capital requirements, require action only by individuals or small groups" (White, 1974, p. 5). The nature of indigenous adjustments

Table 2.15 Indigenous Adjustments to Cyclone Hazards: Cyclone-Proofing of Homes

Nature of Adjustment	Purpose
Houses built on posts/piers	To strengthen foundation
Houses anchored to deep-rooted pillars	To reinforce foundation against strong cyclone winds
Frames tied to trees by strong ropes	To secure houses from wind damage
Raised homesteads and plinth levels to protect houses from storm surge flooding above surrounding lands	To protect houses from storm surge flooding
Special types of roofs	To streamline roofs with winds. Also, use roofs as shelters from storm surges
Brick-built houses	To protect homes from strong cyclone winds. Also to provide shelters from storm surges
Planting trees around homesteads and individual houses	Windbreak against strong cyclone winds
Embankments around property	To protect homes from tidal flooding

Prepared by the authors based on field evidence.

to cyclone hazards in coastal Bangladesh can be classified broadly into two categories: (1) elementary cyclone-proofing of homes and (2) climbing trees to save lives from onslaughts of cyclones and storm surges.

As a first line of defense against the onslaught of cyclonic winds and storm surges, coastal residents have incorporated several measures to cyclone-proof their homes (Table 2.15). One of the common measures, designed to provide protection from storm surge flooding, is the practice of raising the level of common yards of a homestead above the surrounding lands by earth-fills borrowed from the common property lands (Paul et al., 2010; Rasid and Paul, 1987; Khan, 1973; Islam, 1971).[9] This measure is almost identical to the practice of raising homesteads above flood levels throughout major floodplains in Bangladesh. Surveys from coastal areas indicate that nearly three-quarters of the respondents reported this practice at varied intervals: at least half of them at intervals of 2–5 years and 10–15% at intervals of greater than 5 years, whereas regular annual maintenance of

[9] Typical rural landscape in Bangladesh consists of clusters of houses built on elevated tracts of lands, called *bari* in Bengali. A *bari* has a common yard shared by several houses. Typically, a village has several *paras* (neighborhoods) consisting of clusters of *baris* (see also endnote 1 in Chapter 8).

the yards has been reported by about 10%, mostly from low-lying areas (Paul et al., 2010). In addition to elevating the homesteads above the surrounding lands, this practice also includes occasional reinforcing of plinth levels of houses above yard levels.

Most of the indigenous houses are not adequately cyclone-resistant principally because of their poor quality. As indicated earlier, most of the houses consist of thatched or mud walls and thatched roofs (typified by Fig. 2.1). Woven split bamboo panels on a wood frame are also extensively used, either with iron sheets or with a shingle roof (Islam, 1974). Although most of these flimsy indigenous houses are not adequately designed as cyclone-resistant structures, often residents employ different types of crude reinforcing devices to provide additional protection to their homes from the impacts of strong cyclone winds and storm surges. These include different types of anchorage devices, such as deep-rooted additional pillars to reinforce the foundation, or use of strong ropes to tie the main frame of the house to nearby trees (Ahmed, 2008). A limited number of houses are also built on posts (piers) to strengthen the foundation and to provide protection from storm surge flooding. To minimize wind impacts, roofs of some of the houses are streamlined with the prevailing direction of storms (i.e., the direction from which most of the past cyclones have struck). Often roofs are also used for temporary shelter from storm surge flooding. In contrast to these relatively weak structures, most of the brick buildings are cyclone-resistant, as indicated earlier. Also, many coastal homes are protected from tidal flooding and low storm surges by private embankments, but their elevation (most often no more than 3 ft or 1 m high) is no match for most of the high-magnitude surges. For example, storm surges reached up to 30–35 ft (9–10 m) during the 1970 Bhola cyclone and up to 20 feet (6 m) during *Gorky*. Additional storm protection is also provided by natural tropical vegetation and planted trees surrounding clusters of houses (as seen in Fig. 2.1).

During high-magnitude storms when warnings are issued to evacuate their homes, often many coastal residents, unable to reach cyclone shelters for a variety of reasons (see Chapter 7), try to save their lives by climbing trees. Many cyclone victims, indeed, saved their lives by climbing trees during the 1970 Bhola cyclone (Islam, 1974). However, during Cyclone Gorky (1991) there were extensive reports of this practice with mixed results, some with heart-breaking tragic outcome (Table 2.16). The latter relates to the decision by some of the household heads to tie their family members to tree trunks and branches to withstand strong winds and storm surges. These

Table 2.16 Indigenous Adjustments to Cyclone Hazards: Mixed Results of Climbing Trees During *Cyclone Gorky* (1991)

1. ***Successful efforts to save lives***
   - **a.** Some survived just by climbing trees.
   - **b.** Some survived by tying themselves to treetops.
   - **c.** One survivor clung to a coconut tree for 9 h.
   - **d.** A man climbed a tree, tied his wife and children to the strongest branches, and wrapped his arms around the trunks.
   - **e.** An individual survived by clinging to a banana tree as it floated 60 km down a raging river.
   - **f.** A three-month-old baby was found alive in the upper fronds of a palm tree.
2. ***Failures to save lives***
   - **a.** Some fled to forests but did not climb trees and was drowned in surge water inside forests.
   - **b.** Dead children were found still tied to the branches of uprooted trees.
   - **c.** Dead babies were found tied to tree branches that had broken from their trunks.

Courier-Mail [Melbourne] (May 7, 1991); *New Zealand Herald Sun* (May 6 and 11, 1991); *St. Petersburg Times* [Florida] (May 11, 1991); *Toronto Star* (May 6 and 7, 1991); *USA Today* (May 16, 1991); *Washington Post* (May 7, 1991).

measures worked for some survivors who lived to tell their stories, but proved deadly for a large number of children, whose dead bodies were found tied to uprooted trees or broken tree branches (Table 2.16). Most of their parents perished in the storm surges (Rashid, 2011).

CONCLUSION

In public discourse dealing with climate change impacts on tropical cyclones in Bangladesh it has often been claimed erroneously that frequencies of cyclones in the Bay of Bengal have increased recently. Contrary to this assumption, our analyses of historical data indicate that frequencies of lower-magnitude cyclones have declined in the Bay of Bengal at least for the last century. However, frequencies of landfall of higher-magnitude cyclones along the Bangladesh coast have increased significantly since the 1960s. In particular, at least five super-cyclones (with wind speeds exceeding 138 mph, 222 km/h) occurred during this relatively short period, providing evidence of recent intensification of tropical cyclones over northern Bay of Bengal. More significantly, Emanuel (2005) has developed an index of "total dissipation of power" (PD) of hurricanes/tropical cyclones (by integrating the effects of relevant meteorological parameters) to demonstrate that this index has increased significantly since the 1970s, corresponding to the period of significant global warming and increases in SSTs. The main implication of the increasing values of the PD index for coastal Bangladesh is that its formulation indicates that the total power of dissipation rises roughly as the cube of the wind speed (Emanuel, 2005). Since super-cyclones have exceedingly high wind speeds their destructive power increases at the rate of the cube of the wind speed. Cyclone fatalities of epic proportions, widespread property damage, personal injuries, and displacements from homes resulting from destructive powers of super-cyclones in coastal Bangladesh can now be firmly attributed to global warming, increases in SSTs, and ongoing sea level rise.

For generations, people of coastal Bangladesh have lived with cyclone hazards by adopting some of the most ineffective indigenous adjustments to cyclone hazards. These include elementary cyclone-proofing of housing structures using some of the crude methods and/or climbing on trees for protection from storm surges and strong winds. During *Bhola cyclone* (1970) and *Cyclone Gorky* (1991) most of these measures had mixed results, that is, often they worked out but most of the time they resulted in tragic loss of lives. Recent introduction of cyclone shelters and significant improvements in cyclone preparedness, evacuation, and other disaster management initiatives have reduced cyclone deaths significantly. Still the scale of property damage, personal injuries, and post-storm homelessness continue to dominate adverse impacts of tropical cyclones. Cyclone shelters provide evidence that strong structures can save lives. Thus, longer-term solutions to the problems of loss of lives, property damage, and displacements from homes seem to indicate that there is an urgent need for building cyclone resistant stronger homes. Several prototypes have already been tested in the coastal area. Since industrial societies have accepted some responsibilities for polluting the global atmosphere, which is now implicated in the intensification of tropical cyclones, one of the most productive ways they could help the people of coastal Bangladesh is through a comprehensive plan for building not only additional cyclone shelters but also a network of cyclone resistant homes. Besides accommodating the household members, these should be designed for temporary shelters for some of the neighbors during cyclone emergencies. In Chapters 7 and 8, we expand on this and other structural and nonstructural measures for confronting adverse impacts of climate change and sea level rise in coastal Bangladesh.

REFERENCES

Ahmed, A.U., 2008. Assessment of Vulnerability to Climate Change and Adaptation Options for the Coastal People of Bangladesh. Practical Action, Dhaka, Bangladesh.

Akter, N., Tsuboki, K., 2012. Numerical simulation of cyclone Sidr using a cloud-resolving model: characteristics and formation process of an outer rainband. Monthly Weather Review 140, 780–810.

Alam, E., Collins, A.E., 2010. Cyclone disaster vulnerability and response experiences in coastal Bangladesh. Disasters 34 (4), 931–954.

Alam, E., Dominey-Howes, D., 2014. A new catalogue of tropical cyclones of the northern Bay of Bengal and the distribution and effects of selected landfalling events in Bangladesh. International Journal of Climatology 35 (6), 801–835.

Alexander, D., 1993. Natural Disasters. Chapman & Hall, New York.

Azam, M.H., Samad, M.A., Kabir, M.U., 2004. Effect of cyclone track and landfall angle on the magnitude of storm surges along the coast of Bangladesh in the northern Bay of Bengal. Coastal Engineering Journal 46 (3), 269–290.

Barry, R.G., Chorley, R.J., 2003. Atmosphere, Weather and Climate, eighth ed. Routledge, London.

Battan, L.J., 1961. The Nature of Violent Storms. Doubleday Anchor, Garden City, NY.

Bern, C., Sniezek, J., Mathbor, G.M., Siddiqi, M.S., Ronsmans, C., Chowdhury, A.M., Choudhury, E.A., Islam, K., Bennish, M., Noji, E., 1993. Risk factors of mortality in the Bangladesh cyclone of 1991. Bulletin of World Health Organization 71 (1), 73–78.

Beveridge, H., 1876. The District of Bakarganj: Its History and Statistics. Trübner & Co, London. https://archive.org/details/districtofbkar00bve.

Buis, A., 2007. Dangerous Tropical Cyclone Sidr Heads North through Bay of Bengal. NASA: Goddard Space Flight Center/Jet Propulsion Laboratory (web document). www.nasa.gov/mission_pages/hurricanes/archives/2007/h2007_sidr.html.

Chowdhury, A.M.R., Bhuyia, A.U., Choudhury, A.Y., Sen, R., 1993. The Bangladesh cyclone of 1991: why so many people died? Disasters 17 (4), 291–304.

Davis, C.A., 2015. The formation of moist vortices and tropical cyclones in idealized simulation. Journal of the Atmospheric Sciences 72, 3499–3516.

Debsarma, S.K., 2009. Simulations of storm surges in the Bay of Bengal. Marine Geodesy 32, 178–198.

Dunn, G.E., 1962. The tropical cyclone problem in east Pakistan. Monthly Weather Review 90, 83–86.

Emanuel, K., 2003. Tropical cyclones. Annual Review of Earth and Planetary Sciences 31, 75–104.

Emanuel, K., 2005. Increasing destructiveness of tropical cyclones over the past 30 years. Nature 436 (7051), 686–688.

Frank, N.L., Husain, S.A., 1971. The deadliest tropical cyclone in history. Bulletin of the American Meteorological Society 52 (6), 438–445.

Gutro, R., Olsen, E., 2007. Powerful Tropical Cyclone Sidr Makes Landfall in Bangladesh. NASA: Goddard Space Flight Center/Jet Propulsion Laboratory (web document). www.nasa.gov/mission_pages/hurricanes/archives/2007/h2007_sidr.html.

Haque, C.E., Blair, D., 1992. Vulnerability to tropical cyclones: evidence from the April 1991 cyclone in coastal Bangladesh. Disasters 16 (3), 217–229.

Haque, U., Hashizune, M., Kolivras, K.N., Overgaard, H.J., Das, B., Yamamoto, T., 2012. Reduced death rates from cyclones in Bangladesh: what more needs to be done? Bulletin of the World Health Organization 90 (2), 150–156.

Haque, C.E., 1997. Atmospheric hazards preparedness in Bangladesh: a study of warning, adjustments and recovery from the April 1991 cyclone. Natural Hazards 16 (2), 181–202.

Hasan, F., 1999. Natural calamities and the life struggles of sandwipees (in Bengali). In: Mohammad, H. (Ed.), Sandweep Studies. Sandweep Association, Chittagong, Bangladesh, pp. 175–180 (cited in Alam and Dominey-Howes, 2014).

Hasegawa, K., 2008. Features of super cyclone Sidr to hit Bangladesh in Nov 07 and measures for disaster from results of JSCE investigation. In: Proceedings of the WFEO-JFES-JSCE Joint International Symposium on Disaster Risk Management. Science Council of Japan, Sendai, Japan, pp. 51–58.

Hendricks, E.A., Montgomery, M.T., Davis, C.A., 2004. The role of "vortical" hot towers in the formation of tropical cyclone Diana (1984). Journal of the Atmospheric Sciences 61 (11), 1209–1232.

Houze Jr., R.A., February 2010. Clouds in tropical cyclones. Monthly Weather Review 138, 293–344.

Houze, R.A., 2014. Clouds and precipitation in tropical cyclones. International Geophysics 104, 287–327 (Chapter 10 in special volume on *Cloud Dynamics*, edited by R.A. Houze).

Hunter, W.W., 1876. A Statistical Account of Bengal, Chittagong Hill Tracts, and Districts of Chittagong, Noakhali, Tripperah, Hill Tripperah, Vol. 6. Trübner & Co., London (cited in Alam and Dominey-Howes, 2014).

Ikeda, K., 1995. Gender differences in human loss and vulnerability in natural disasters: a case study from Bangladesh. Indian Journal of Gender Studies 2 (2), 171–193.

Islam, T., Peterson, R.E., 2009. Climatology of landfalling tropical cyclones in Bangladesh 1877–2003. Natural Hazards 48 (1), 115–135.

Islam, M.A., 1971. Cyclone hazard and the strategy of human occupance in the coastal areas of Bangladesh. Oriental Geographer 15, 37–45.

Islam, M.A., 1974. Tropical cyclones: coastal Bangladesh. In: White, G.F. (Ed.), Natural Hazards: Local, National, Global. Oxford University Press, New York, pp. 19–24.

Kastendeuch, P.P., 2007. Pressure gradient force, atmospheric circulation and climate in western Europe (1899–2002). International Journal of Climatology 27, 2055–2067.

Khan, A.A., 1973. Hazard Perception and Choice of Adjustments in the Chittagong Coastal Area (MA thesis). University of Dhaka, Department of Geography, Dhaka, Bangladesh.

Kristof, N.D., May 11, 1991. In: Bangladesh's Storms, Poverty More than Weather Is the Killer. The New York Times.

Mahala, B.K., Nayak, B.K., Mohanty, P.K., 2015. Impacts of ENSO and IOD on tropical cyclone activity in the Bay of Bengal. Natural Hazards 75, 1105–1125.

Murty, T.S., Flather, R.A., 1994. Impact of storm surges in the Bay of Bengal (Chapter 10). In: Finkl Jr., C.W. (Ed.), Coastal Hazards: Perception, Susceptibility and Mitigation. Coastal Education & Research Foundation, Inc, pp. 149–161. Journal of Coastal Research Special Issue No. 12.

Paul, B.K., Rashid, H., Islam, M.S., Hunt, L.M., 2010. Cyclone evacuation in Bangladesh: tropical cyclones *Gorky* (1991) vs. *Sidr* (2007). Environmental Hazards Human and Policy Dimensions 9, 89–101.

Paul, B.K., 2009. Why relatively fewer people died?: the case of Bangladesh's cyclone Sidr. Natural Hazards 50, 289–304.

Paul, B.K., 2010. Human injuries caused by Bangladesh's cyclone Sidr: an Empirical study. Natural Hazards 54 (2), 483–495.

Rashid, H., 2011. Interpreting cyclone disasters in Bangladesh and Myanmar from web-based newspaper discourse: media framing of cyclone vulnerability on the Bay of Bengal coast. The Arab World Geographer 14 (1), 1–32.

Rasid, H., Paul, B.K., 1987. Flood problems in Bangladesh: is there an indigenous solution? Environmental Management 11 (2), 155–173.

Riehl, H., 1954. Tropical Meteorology. McGraw-Hill Book Co, New York.

Shamsuddoha, M., 2015. Climate-induced Displacement and Migration Policy Gaps and Policy Alternative: A Likely Legal Instrument for a Rights-based Political Solutions. Center for Participatory Research Development, and Berlin, Germany: Bread For the World, Dhaka, Bangladesh. cprdbd.org/papers/BriefingPaperClimateInduced Displacement.

Siddiqui, Z.A., 2009. Storm surge forecasting for the Arabian sea. Marine Geodesy 32 (2), 199–217.

Simmons, M., May 12, 1991. Poverty Is the Real Tragedy. Manchester Guardian Weekly.

Sommer, A., Mosley, W.H., 1972. East Bengal cyclone of November, 1970: epidemiological approach to disaster assessment. Lancet 1 (7759), 1029–1036.

Wang, S.-Y., Buckley, B.M., Yoon, J.-H., Fosu, B., 2013. Intensification of pre-monsoon cyclones in the Bay of Bengal and its impacts on Myanmar. Journal of Geophysical Research 119 (10), 4373–4384.

White, G.F. (Ed.), 1974. Natural Hazards: Local, National, Global. Oxford University Press, New York and Toronto.

Zehnder, J.A., not dated. Tropical Cyclone. Encyclopedia Britannica (section written by Zehnder). www.britannica.com/science/tropical-cyclone.

CHAPTER THREE

Climate Change and Sea Level Rise in Bangladesh

INTRODUCTION

We begin this chapter with a review of complexities of estimating global sea level rise (SLR). Using global mean sea level (GMSL) data, based on tide gauge records for nearly 140 years (1880–2009) and more recent satellite altimetry data (1993–2007), we then review complexities and uncertainties associated with estimating GMSL. This provides a context for interpreting even greater uncertainties with estimating SLR in coastal Bangladesh, since tide gauge data for Bangladesh are available only for a limited number of stations and for a relatively short period of time (1979–2003). In the next section the focus is on SLR as a climatic hazard for Bangladesh resulting from potential inundation of large areas of coastal Bangladesh. Sea level fluctuations may result from a variety of geophysical factors but, in the current context, SLR may be considered as a climatic hazard since it is fueled largely by global warming–induced warming and expansion of sea water and melting of mountain/land glacier. Global warming is also likely to exacerbate storm surges associated with tropical cyclones since higher sea levels are likely to extend storm surge impacts further inland beyond the coastline. We have characterized these combined impacts of SLR and storm surges as SLR risks. Further, using topographic contours and past storm surge records, we have made an attempt to delineate SLR risk areas of each of the coastal districts of Bangladesh. Using data on percentages of district areas likely to be impacted by these combined hazards, SLR risks of each of the districts has been classified into one of the following three nominal categories: (a) major, (b) moderate, and (c) minor. We conclude this chapter emphasizing the need for monitoring SLR closely in the context of river sedimentation, which seems to provide conflicting field evidence of land gains in coastal Bangladesh, minimizing the impact of SLR.

Climatic Hazards in Coastal Bangladesh
ISBN 978-0-12-805276-1
http://dx.doi.org/10.1016/B978-0-12-805276-1.00003-X

COMPLEXITIES OF SEA LEVEL RISE

Bangladesh as "Most Vulnerable Country" for Climate Change Impacts

Bangladesh is rated by Maplecroft (2010) as the "most vulnerable country" for climate change impacts.[1] This rating is based on an assumption of 1–2 m (3–6.5 ft) of SLR along the Bangladesh coast by the end of this century. Bangladesh is particularly vulnerable to rising sea levels as its agrarian settlements in coastal areas (henceforth coastal Bangladesh), comprising nearly one-fifth of the total population of the country, face the grim prospect of large-scale inundation and permanent displacement in not so distant future. Stated in this manner, it seems to be a *prima facie* logical assumption. Thus, drawing information from IPCC projections and other estimates presented in non-IPCC technical reports and peer-reviewed scholarly articles, the media and public discourse tend to frame impacts of projected SLR of such magnitudes (1–2 m) as an impending disaster for millions of people of Bangladesh. The reality of SLR is much more complex than a simplistic assumption of contour-by-contour inundation of coastal Bangladesh due to a linear rise in sea level (Brammer, 2014a,b). The complexities may be attributed to both methodological procedures for estimating the rates of sea level changes as well as to several geophysical and anthropogenic processes that may either increase or decrease such changes (Table 3.1).

Sea Level Fluctuations at the Geological Time Scale

At the geological time scale, sea level fluctuation is a familiar concept as it is associated with the impacts of ice ages. During past ice ages, huge quantities of freshwater from precipitation (mostly snowfall) were landlocked in glaciers, such as the Pleistocene glaciers in North America, Europe, and Siberia (approximately 3 million years before present [BP] to 10,000 years BP). As the oceans were deprived of runoff from continental surfaces, sea levels dropped, sometimes by hundreds of meters/feet. Drop in sea levels were further accompanied by land subsidence due to the overburden of glaciers. For example, land masses in central Canada adjacent to Hudson Bay depressed significantly due to the weight of the Laurentide ice sheets (part of the Pleistocene glacier), which reached maximum thickness of up to 3.4 km (2.2 miles) (Simon et al., 2014; Bryson et al., 1969). By the end of the Pleistocene epoch, i.e., beginning of the warmer Holocene epoch,

Table 3.1 Complexities of Estimating Sea Level Rise

Time-Scales	Nature of Change	Implications
Geological time scale (thousands to millions of years)	Major fluctuations of sea levels (hundreds of meters/feet) between ice ages and warmer interglacial/postglacial periods	Much higher amplitudes of sea level fluctuations than more recent global warming induced SLR
Historical time scale (postindustrial period)	Gradual increases in sea level at rates of several mm/inches/year	Sea level fluctuations vary regionally (from one ocean basin to another) and from one year to another (besides seasonal variations)
Geological time scale	Isostatic rebound (uplifting of land, also called GIA: glacial isostatic adjustment)	Tends to cancel out SLR
Geological time scale	Plate tectonics (either uplifting or subsidence of land)	Either decrease or increase in SLR
Historical + geological time scale	River sedimentation (either land building or subsidence)	Either decrease or increase in SLR
Historical time scale (mostly past several decades)	Groundwater withdrawal (resulting in subsidence)	Increases SLR
Historical time scale (mostly past several decades)	Storage of freshwater in stream reservoirs (withholding surface runoff to sea)	Lowers sea level
Historical time scale (mostly past several decades)	Riverbank levees and polders in coastal areas (such as along Bangladesh coast that tend to confine flow near the river mouth)	Raises local tide gauge levels
Historical time scale (postindustrial period)	Limited tide gauge record	Methodological problems
Historical time scale (postindustrial period)	Differences between tide gauge and satellite altimetry data	Methodological problems

Source: Prepared by the authors based on various sources.

approximately 10,000 years BP, global sea levels were about 100 m (300 feet) lower than the current level (Houghton, 2006, p. 545). Sea levels started rising gradually with melting of glaciers until "the final retreat of the continental ice sheets from Europe and North America between 10,000 and 7000 years ago" (Barry and Chorley, 2003, p. 361). The sea levels stabilized to the preindustrial levels within the last 3000 to 2000 years, until global warming–induced SLR set in motion for at least the last 200 years (Cazenave and Llovel, 2010). The rates of post-Pleistocene sea level fluctuations have been somewhat moderated by isostatic rebound, also called glacial isostatic adjustment (GIA), which refers to uplifting (raising) of continental surfaces due to the release of elastic pressure following removal of the glacial overburden (Cazenave and Llovel, 2010).

Globally, GIA differs significantly from one coast to another. Therefore, regional variation in isostasy affects the rates of sea level changes at a given coastline as it tends to cancel out SLR by certain amounts. This makes it harder to estimate actual sea level changes at a given coastline, especially because GIA data may not be available for all regional coastlines. Among other geological processes that may complicate estimates of sea level changes in coastal Bangladesh plate tectonics may contribute to significant changes in its unstable landscapes. Major landscape shifts had already occurred along the Brahmaputra Valley due to earthquakes during the last 120 years, such as in 1897 (Johnson, 1982, p. 20). Similar changes in coastal streams, especially in the tectonically unstable estuary of the Ganges-Brahmaputra-Meghna (GBM) Rivers, cannot be ruled out because the entire Bengal delta (Bangladesh and West Bengal) is tectonically active. The 2015 Nepal earthquake has reinforced this assumption as it affected northern parts of both Bangladesh and West Bengal (India).

Global Warming–Induced Sea Level Rise

There are two components of global warming–induced SLR: (a) model-based projections for SLR by the end of the 21st century and (b) historical trends of sea level fluctuations during the 20th century and the first decade of the 21st century, based on tide gauge records and recent satellite-altimetry data. We will deal briefly with model-based projections in a later section in the context of SLR as a climatic hazard for coastal Bangladesh. Here we will focus on historical trends using tide gauge and satellite altimetry data. The largest tide gauge database of monthly and annual mean sea level records, based on 2000 sites in 200 nations, is maintained by the Permanent Service for Mean Sea Levels (PSMSL). These data are readily available at the PSMSL Website: http://www.pol.ac.uk/psmsl. Tide gauges measure sea level in relation to the land surface. Typically, gauges are placed on piers and they measure the sea level inside a stilling well (similar to a stream gauge stilling well) relative to a nearby geodetic benchmark (i.e., elevation record on a permanent monument) (Colorado University Sea Level Research Group Website: sealevel.Colorado.edu/content/tide-gauge-sea-level). The gauges may be subjected to uplift due to isostasy (GIA) and uplift or subsidence due to plate tectonics, sedimentation, and anthropogenic processes (see Table 3.1).

"Sea level at any location contains the influence of local and meteorological effects (including storm surges), modes of climate variability (for example the El Niño Southern Oscillation) and long-term trends (from both the ocean surface and land movements), including the impact of anthropogenic climate change" (Church and White, 2011). Because of the influences of these variables sea level elevations demonstrate strong regional variability at different spatial scales:

Macroscale Variations

- From one ocean to another
- From one ocean basin to another basin within a given ocean
- From low latitudes to high latitudes (reflecting differential heating of the atmosphere and ocean water)

Mesoscale Variations

- Within relatively small ocean basins, such as the Bay of Bengal
- Along a coastline, such as the Bangladesh coast

Microscale Variations

- Along large river estuaries (such as the GBM estuary in Bangladesh, subject to higher sea levels due to backwater and funneling effect)
- Along coastal zones with dense vegetation, such as the mangrove forests in southwestern coast of Bangladesh (which plays a role in sediment accretion and land building)
- Along coastal zones with anthropogenic interventions (such as embankment and polders in southwestern coast of Bangladesh inducing higher local sea levels)

Whereas mesoscale and microscale variations in sea levels are relevant for regional studies, at the global scale commonly global mean sea level (GMSL) data are used for detecting historical (temporal) changes in sea levels. The GMSL data may be computed crudely by using arithmetic averages of all tide gauge records or, more appropriately, by employing some of the sophisticated mathematical techniques, such as the "area-weighted mean of all sea surface elevations." Based on the latter technique and tide gauge data from 1880 to 2009 and satellite altimeter data from 1993 to 2009, Church and White (2011) and Cazenaive and Llovel (2010) have analyzed long-term historical trends of GMSL for the last 130 years. We have reproduced selected data from these studies in a summary table (Table 3.2).

The data in Table 3.2 demonstrate strong temporal variability during the period of record. Perhaps one of the important findings from these data is that the GMSL has risen by 210 mm (8.26 inches) in 130 years, i.e., at an average rate of 1.5 mm/year (0.059 inches/year). This may seem to be a small change but the long-term average masks several periods of above-average and below-average rates of changes. Whereas the first 55 years (1880–1935) recorded annual average rise of only 1.1 ± 0.7 mm/year (0.043 ± 0.028 inches/year), SLR accelerated from 1936 to 2009 (1.8 ± 0.3 mm/year or 0.071 ± 0.012 inches/year). During the 20th century there were two major periods of acceleration: 1967–1982 (2.4 mm/year or 0.094 inches/year) and 1993–2009 (2.8 ± 0.8 mm/year or 0.11 ± 0.031 inches/year). The latter period includes recent acceleration in SLR that started in the last decade of the 20th century and continued throughout the first decade of the 21st century. Based on improved methods

Table 3.2 Historical Trends of Global Mean Sea Level (GMSL), 1880–2009

Tide Gauge Period	GMSL Rise in mm/Year (±Indicating Standard Deviation) (Inches/Year With Standard Deviation)
1880–2009	1.5 ± 0.1 (0.059 ± 004)
1900–2009	1.7 ± 0.2 (0.067 ± 0.004)
1880–1935	1.1 ± 0.7 (0.043 ± 0.028)
1936–2009	1.8 ± 0.3 (0.071 ± 0.012)
1967–1982	2.4 (0.094) ± ?
1961–2009	1.9 ± 0.4 (0.075 ± 0.016)
Altimeter period	
1993–2009[a]	2.8 ± 0.8 (0.11 ± 0.031)
1993–2009[b]	3.3 ± 0.4 (0.13 ± 0.016)

[a]Church and White (2011).
[b]Cazenave and Llovel (2010).
Source: Prepared by the authors, based on Church, J.A., White, N.J., 2011. Sea level rise from the late 19th to the early 21st century. Surveys in Geophysics 32 (4), 585–602 and Cazenave, A., Llovel, W., 2010. Contemporary sea level rise. Annual Review of Marine Science 2 (1), 145–173.

of estimates by satellite altimetry, Cazenave and Llovel (2010) have further updated these rates of acceleration to 3.3 ± 0.4 mm/year (0.13 ± 0.016 inches/year) (Table 3.2).

Acceleration of SLR during the post-industrial period, especially during the latter two-thirds of the 20th century and its continuation into the first decade of the 21st century, has been attributed primarily to two factors related to climate change: (a) thermal expansion of seawater due to absorption of increasing amounts of energy from the atmosphere and (b) melting of land ice, i.e., glaciers and ice caps (which include the small glaciers and ice caps fringing the major ice sheets) and the major ice sheets of Antarctica and Greenland (Church and White, 2006). Estimates of the relative contributions of thermal expansion and ice melting vary widely, depending on the periods of observations and climate models employed to assess such contributions. Some of the earlier coupled atmosphere-ocean general circulation models (AOGCMs) estimated about 60% of the rise due to thermal expansion and only about 25% due to land ice changes (Houghton, 2006, p. 147). More recent changes in sea levels and improved methods of satellite altimetry have reversed such estimates. For example, estimates by Cazenave and Llovel (2010) indicated that only about 30% of the SLR during 1993–2009 was attributable to thermal expansion compared to 55% due to land ice melt. Perhaps the most striking finding of this study was that the recent acceleration in glacier melting and ice mass loss from ice sheets increased the land ice contribution up to 80% of SLR for a period of five years (2004–2009) (Cazenave and Llovel, 2010).

Estimation of the relative contribution of seawater expansion to SLR is extremely complex because it depends critically on the water temperature. For cold water the expansion for a given change of temperature is small (Houghton, 2006, p. 146). The rates of ocean water circulation and mixing also vary widely within a given ocean basin, depending on its configuration and latitudinal position (that determines air and water temperatures). For these reasons geographical variations in sea levels are substantial even within a single ocean. For example, the rate of SLR in the 20th century was twice as large at New York than at Buenos Aires (Cazenave and Llovel, 2010). In some regions, such as the western Pacific, North Atlantic around Greenland, and southeast Indian Oceans, comparable rates were up to three times faster than the global mean (e.g., sea level was higher in these regions by ~15 cm [6 inches] compared to 16 years ago), while the eastern Pacific and west Indian oceans exhibited a lower rate (Cazenave and Llovel, 2010). Thus, all oceans of the world may be interconnected but the rates of sea level changes vary greatly from one ocean basin to another. Even within a relatively small basin, such as the Bay of Bengal and its coastline in Bangladesh, large variations in sea levels have been observed (Singh, 2001).

Besides thermal expansion and melting of land ice, other variables attributable to certain amounts of sea level fluctuations include (a) changes in surface runoff contributions to oceans due to climate change-induced perturbations of the hydrologic cycle and (b) anthropogenic interventions, such as groundwater withdrawal (resulting in land subsidence, hence increased SLR), and reductions of surface runoff contributions to oceans due to reservoir storage. Most of the anthropogenic factors have regional and local implications but at least one study suggests that water impoundment by nearly 30,000 stream reservoirs built during the 20th century contributed to lowering of the global sea levels by about 0.55 mm/year (0.022 inches/year) (Chao et al., 2008).

SEA LEVEL RISE IN BANGLADESH

Data Source

Tide gauge data for all of the stations on the Bangladesh coast are constrained by their short periods of records, missing data (some years), and station instability. The following three provide data for about two decades without major interruptions: (a) Hiron Point in the Sundarban mangrove forest area (southwestern Bangladesh), (b) Char Changa in the Meghna estuary (central Bangladesh), and (c) Cox's Bazar on the southeast coast.[2] Data for these stations are available on the PSMSL Website (http://www.pol.ac.uk/psmsl). The original tide gauge data are monitored and collected by the Government of Bangladesh Inland Waterways and Transportation Authority (BIWTA). One of the advantages of the PSMSL database (compared to the original BIWTA data) is that all of its sea level data are standardized by using the "revised local reference" (RLR), which is defined as a reference datum starting from 7000 mm below the mean sea level (i.e., starting from minus 7000 mm below 0 MSL). The PSMSL justifies this methodology to avoid negative values for fluctuating sea levels. Another useful characteristic of its database is that each of its datasets is accompanied by footnotes, some of which may include notes on tide gauge monitoring issues. In the case of the selected Bangladesh stations, for example, the PSMSL noted that some of the original data for these stations (supplied by BIWTA) had some inconsistencies and errors. Therefore, the PSMSL had to apply some of its standard procedures and formulas for correcting and adjusting these data in order to standardize them.

Nature of Sea Level Fluctuations

Table 3.3 summarizes the extent of geographical and temporal variations in observed rates of sea level fluctuations along the indented coastline of Bangladesh that stretches for about 500 miles (800 km of indented coastline). The coastline of Bangladesh consists of several segments of highly dynamic terrestrial and marine environments. This is reflected in some of the characteristics of the PSMSL data. One of the basic characteristics of the data was that there were significant variations in sea surface elevation, both from one station to another and also from one year to another. The differences between the maximum and the minimum levels ranged from 286 mm (11.26 inches) at Hiron Point and 386 mm (15.2 inches) at Char Changa to 487 mm (19.17 inches) at Cox's Bazaar (Table 3.3). Thus, the ranges of sea level fluctuations increased from southwest to southeast. The trend was reversed by the average values of sea levels for the period of record, with a declining trend from southwest to southeast: 7058.52 mm at Hiron Point, 6924.95 mm at Char Changa, and 6941.65 mm at Cox's Bazaar. There are several competing processes that may explain such station-to-station variations in sea levels.

Using limited data on subsidence and isostatic rebound (GIA), Pethic and Orford (2013) have attributed higher sea levels at Hiron Point to "a combination of deltaic subsidence, including sediment compaction, and eustatic sea level rise, but principally [to] increased tidal range in estuary channels recently constricted by embankments." Char Changa station is situated on Hatia Island, one of the unstable islands of the GBM estuary, which is characterized by extensive erosion and sedimentation processes. Consequently, many smaller islands disappear and reemerge from time to time. Coastlines (outlines) of some of the larger islands, particularly Bhola, Hatia, and Kutubdia, have changed significantly within a relatively short period of time (Brammer, 2014a,b). In such an environment reliable data on sedimentation and subsidence are not readily available (see "Reassessing sea level rise in Bangladesh" under conclusion for an expanded explanation on river sedimentation). Cox's Bazaar, situated on the southeast coast of Bangladesh, on the other hand, lies in a tectonically active zone. Tectonic data (either uplift or subsidence) for this area are not readily available despite recent advances in GPS technology. In short, interpreting temporal variations

Table 3.3 Geographical Variations in Recent Sea Level Changes Along the Bangladesh Coast: 1979–2003

Sea Level Data	Hiron Point 21.78°N/ 89.46°E (Southwestern Bangladesh)	Char Changa 22.21°N/ 91.05°E (Central Bangladesh)	Cox's Bazar 21.45°N/ 91.83°E (Southeastern Bangladesh)
Years of record	1983–2003	1979–2000 (Missing data: 1986, 1993)	1979–2000 (Missing data: 1986, 1996)
Sea Surface Elevation (mm)			
Minimum (mm)[a]	6941 (1997)	6757 (1982)	6672 (1994)
Maximum (mm)	7227 (2001)	7143 (1999)	7159 (1995)
Range in mm (inches)	286 (11.26)	386 (15.2)	487 (19.17)
Average sea surface elevation (mm)	7058.52 (1983–2003)	6924.95 (1979–2000)	6941.65 (1979–2000)
Average sea surface elevation during the first half of the record (mm)	7044.36 (1983–1993)	6898.3 (1979–1989)	6915.6 (1979–1989)
Difference between average and first half average: mm (inches)	−14.16 (0.56)	−26.65 (1.05)	−26.05 (1.03)
Average sea surface elevation during the second half of the record (mm)	7074.1 (1994–2003)	6951.6 (1990–2000)	6967.7 (1990–2000)
Difference between average and second half average: mm (inches)	+15.58 (0.61)	+26.65 (1.05)	+26.05 (1.03)
Net SLR trends (1980–2000)[b]	3.39 mm/year (0.13 inch/year)	8.4 mm/year (0.33 inch/year)	2.62 mm/year (0.1 inch/year)

[a]Sea levels in PSMSL database are above the "Revised Local Reference" (RLR) datum, defined as "approximately 7000 mm below mean sea level." Normally sea level data are expressed in mm. To convert mm into inches, divide mm data by 25.4.

[b]Based on a least-square regression analysis by the authors.

Source: Prepared by the authors based on original sea level data from Permanent Service for Mean Sea Levels (PSMSL): http://www.pol.ac.uk/psmsl.

(i.e., year-to-year variations) in sea levels along the Bangladesh coast is a difficult proposition. In a later section, we have reviewed some of the relevant publications that make an attempt to explain geographical variations in longer-term SLR trends along the Bangladesh coast.

Another characteristic that was helpful for interpreting year-to-year fluctuations in sea levels was that the second half of the record coincided with recent accelerations in global SLR, providing a basis for exploring if the changes in sea levels along the Bangladesh coast demonstrated an upward trend. Consistent with the global trend, the second half of the data showed that, indeed, the average sea level along the Bangladesh coast rose by 26 mm (about 1 inch) above the 20-year average at Char Changa and Cox's Bazar, and by 15.58 mm (0.61 inch) at Hiron Point (Table 3.3). Spread over the entire 20-year period, the average sea level between the first and the second half of the record rose by 29.74 mm (1.17 inches) at Hiron Point, 53.3 mm (2.1 inches) at Char Changa, and by 52.1 mm (2 inches) at Cox's Bazar. In short, the overall distribution of the data seemed to indicate an upward trend for the sea levels along the Bangladesh coast.

Sea Level Trends Along the Bangladesh Coast

To determine if the upward sea level trend evident from the 20-year tide gauge record was statistically significant, we performed a least-square regression analysis between the PSMSL annual sea level data and the years of record for each of the three stations. The results of the regression analysis are summarized below:

$$\text{Hiron Point: } Y = 7058.52 + 3.56X$$

$$\text{Char Changa: } Y = 6826.86 + 9.34X$$

$$\text{Cox's Bazar: } Y = 6911.15 + 2.91X$$

where Y = mean annual sea level in mm; X = years of observation (for example, for Hiron Point, 1983 = 1, 1984 = 2 ... 2000 = 18).

All of the regression coefficients are significant at 95% level. Further, using the regression equations, we have estimated the projected sea levels (trends) in 1985, 1990, 1995, and 2000 (Table 3.4). Based on the trend-line data for 1980 and 2000 (as reported in Table 3.4), we then estimated the following SLR trends for 20 years (1980–2000) for each of the stations: 3.39 mm/year (0.13 inch/year) at Hiron Point, 8.4 mm/year (0.33 inch/year) at Char Changa, and 2.62 mm/year (0.1 inch/year) at Cox's Bazaar.

Table 3.4 Sea Level Rise Trends for Three Coastal Stations in Bangladesh

	SLR Trend for Each Station: Projected Average Sea Level (mm)[a]		
Projection Years	**Hiron Point**	**Char Changa**	**Cox's Bazar**
1980	7062.08 (1983 data)	6845.54	6916.97
1985	7069.2	6892.24	6931.52
1990	7087	6929.6	6943.16
1995	1995	6966.96	6957.71
2000	7122.6	7013.66	6969.35
1980–2000 SLR trend: mm/year (inches/year)	3.39 (0.13) (1983–2003 data)	8.4 (0.33)	2.62 (0.1)

[a]All data on projected sea levels are based on the least-square regression trend line.
Source: Prepared by the authors based on original PSMSL data and a least-square regression.

How do these rates compare with data from some of the other studies on SLR? As a part of a broader study on global SLR, Church et al. (2004) provided an authoritative interpretation of sea level variability in the North Indian Ocean, including the Bay of Bengal. Based on a "reconstruction method" employing "empirical orthogonal functions" of satellite altimetry data and combining them with monthly tide gauge data for 51 years (1950–2000), Church et al. (2004, p. 2622) estimated an average SLR of 2 mm/year (0.08 inch/year) in the North Indian Ocean, with the exception of the northeastern part of the Bay of Bengal where they found values of more than 4 mm/year (0.16 inch/year). Our data for both Hiron Point and Cox's Bazaar seem to fall within this range. On a closer scrutiny, another study by Unnikrishnan and Shankar (2007) provides a more direct regional context for our data when these are compared with the general trends of SLR along the North Indian Ocean coasts. In their study, Unnikrishnan and Shankar (2007) used a linear correlation technique for comparing the rates of SLR at several tide gauge stations on the North Indian Ocean coast with two "control stations": Mumbai on the Arabian Sea coast and Vishakhapatnam on the Bay of Bengal coast, respectively. The net SLR for Vishakhapatnam, based on initial calculations of tide gauge data and GIA corrections (−0.39 mm/year or −0.015 inch/year), amounted to 1.09 mm/year (0.04 inch/year). This was the lower end of the estimate for Vishakhapatnam as the authors also reported a higher rate of 2.42 mm/year (0.095 inch/year) for this station, by using Church et al.'s (2004) reconstruction estimates. The correlation results showed that data for both Hiron Point and Cox's Bazaar correlated well with that for Vishakhapatnam (linear correlation coefficients: 0.75 for Hiron Point, significant at 99.9%, and 0.43 for Cox's Bazaar, significant at 90%) (Unnikrishnan and Shankar, 2007, p. 302). Thus, it appears that a triangulation of these two studies provides a basis for relating SLR along the Bangladesh coast to regional rates, by using the following steps:

- Unnikrishnan and Shankar's (2007) linear correlation of rates of SLR at Hiron Point and Cox's Bazaar with that of Vishakhapatnam (control station)
- Church et al.'s (2004) reconstruction estimates for SLR at Vishakhapatnam: 2.42 mm/year (0.095 inch/year)
- Church et al.'s (2004) reconstruction estimates for the northeast coast of Bay of Bengal: +4 mm/year (0.16 inch/year)

Whereas the trends for SLR at both Hiron Point (3.39 mm/year or 0.13 inch/year) and Cox's Bazaar (2.62 mm/year or 0.1 inch/year) fall within the range of the preceding "reconstruction rates," in this triangulation the data for Char Changa (8.4 mm/year or 0.33 inch/year) is well beyond this range. Probably for this reason both Church et al. (2004) and Unnikrishan and Shankar (2007) did not include this station in their respective studies.

Another study by Singh (2001), exploring "cause-effect relationships between sea surface temperature, precipitation and sea level along the Bangladesh coast," estimated the following rates of SLR implying a west-to-east increasing linear trend along the Bay of Bengal coast:

- Hiron Point: 4 mm/year (0.16 inch/year)
- Char Changa: 6 mm/year (0.24 inch/year)
- Cox's Bazaar: 7.8 mm/year (0.3 inch/year)

Since there are very few publications on SLR along the Bangladesh coast, Singh's (2001) study has been cited extensively. However, the estimated rates seem to be significantly higher than the global and regional rates (such as those reported in Church et al., 2004; in Unnikrishnan and Shankar, 2007). Singh (2001, p. 238) has provided a limited interpretation of the trend line for the mean annual tidal level by stating that the "upward trends in annual mean tidal level (i.e. from January to December) along the Bangladesh coast are also increasing but the magnitudes are much less as compared to the monsoonal trends in the eastern and central regions". The latter, i.e., the monsoonal trends in sea level variability, seems to be his main contribution as he has provided an in-depth analysis of the sea surface dynamics of the monsoon season, using monthly data on sea surface temperatures (SSTs) for 14 years (1985–1998), BIWTA tide gauge data for 20 years (1977–1998), and monsoon rainfall data for 31 years (1961–1991). Based on linear regression trend lines, Singh's (2001) study found that the monsoon season (June to September) mean tidal levels along the Bangladesh coast showed an upward trend both spatially (from west to east) and temporally (from year to year within the period of analysis: 1977–1998). Among the three comparable stations, the highest rates were evident at Char Changa, especially during the peak of the monsoon season (July, August, and September) with the following rates (Singh, 2001, p. 237):

- July: 15.4 mm/year (0.6 inch/year)
- August: 15.5 mm/year (0.61 inch/year)
- September: 17.5 mm/year (0.69 inch/year)

As the monsoon just sets in the month of June, sea levels in this month rose by only 4.2 mm/year (0.16 inch/year). For Cox's Bazaar, trends for June, August, and September ranged between 4.7 and 9.2 mm/year (0.18 and 0.36 inches/year); only the July trend was comparable to Char Changa's (13.4 mm/year or 0.52 inch/year). In contrast, sea levels at Hiron Point rose between 1.1 and 3.9 mm/year (0.04 and 0.15 inch/year) for the entire monsoon season. Singh (2001) attributed both temporal (year-to-year) and spatial (west to east) increases in monsoon season sea levels in the Bay of Bengal to several interrelated variables that are supported by existing literature and climatic data. For example, the literature and climatic data indicate that the magnitudes of monsoon rainfall over the GBM drainage basin have been increasing for at least three decades (Singh, 2001). He postulated that the upward trends of monsoon season increases in sea levels along the Bay of Bengal coast could be attributed partly to these increases in summer monsoon rainfall with corresponding increases in freshwater contributions to the Bay of Bengal from the GBM drainage basin.

For explaining west-to-east increases in monsoon season sea levels, again he attributed such spatial variations in sea levels partly to the differential effect of increased freshwater flow into the GBM estuary. The combined flow of the Ganges, the Brahmaputra and the Meghna (GBM) rivers exceeds 160,000 m^3/s (5.6 million ft^3/s), nearly three times the amount borne by the Mississippi River and second only to the amount carried by the lower Amazon (Rashid and Paul, 2014; Coleman, 1969). The bulk of this flow drains into the estuary mostly during the monsoon season. Large amounts of discharges into the estuary during a relatively short period of time raise the sea level locally within the estuary. Further, there is field evidence that excessive monsoon season flow into the estuary creates temporary backwater effect raising the local water levels (Ali, 1995). Char Changa, situated in the middle of the estuary almost along the centerline of the mainstream flow, is likely to experience above-average sea levels due to such hydraulic effects.

Regarding the mangrove forest's potential effects on adjacent sea levels, Singh (2001) postulated that it seemed that the mangrove forest "inhibit the spreading of freshwater discharge of the GBM river systems into the Sundarban (mangrove) area during the monsoon season. As a result, the excess monsoon discharge is driven toward the eastern region from central region" (Singh, 2001, p. 237). Perhaps, his most convincing explanation of west-to-east increases in monsoon season sea levels in the Bay of Bengal lies in his assumption of the effect of southwesterly monsoon winds over the Bay of Bengal. The prevailing wind over the Bay of Bengal is northeasterly (northeast trade winds). During the monsoon season the dominant wind direction is reversed into southwesterly. Such seasonal reversal of the dominant wind system is literally translated as mausim or mausum (meaning seasonal weather or wind, in both Hindi and Bengali). The term monsoon has been anglicized from such vernacular names. Nearly constant flow of strong rain-bearing southwesterly winds toward the northeastern part of the Bay of Bengal creates a strong frictional/drag effect on sea surface (also called Eikman effect) resulting in raising of the sea levels along a west to east gradient, possibly explaining higher monsoon season sea levels at Char Changa and Cox's Bazaar. Again, Char Changa, situated at the center of the GBM estuary, perhaps experiences the full force of the southwesterly and southerly winds, explaining higher sea levels at this station than at the remaining two stations. However, we are not certain if this process was responsible (in our study) for the mean annual SLR of 8 mm/year for Char Changa, which was an outlier from our trend line. Similarly, Singh (2001) has not provided an explanation for the abnormally high rate for SLR at Cox's Bazaar (7.8 mm/year or 0.3 inch/year), contrary to our PSMSL-based trend line.

SEA LEVEL RISE AS A CLIMATIC HAZARD FOR COASTAL BANGLADESH

Why Sea Level Rise May Be Considered as a Climatic Hazard?

In our earlier work we have characterized potential impacts of climate change on Bangladesh as an "impending disaster" principally because of likely displacement of millions of people from coastal Bangladesh due to inundation by SLR (Rashid and Paul, 2014). For generations, residents of coastal Bangladesh have lived in this hazard-prone land by adjusting their homesteads and agricultural practices to different types of natural hazards, such as recurring river floods, riverbank erosion, coastal erosion, and tropical cyclones with their devastating storm surges. Recent SLR has added a new two-pronged hazard, firstly, because of its potential for exacerbating storm surge impacts and, secondly, because of its ultimate outcome of displacing millions from their homes due to inundation. There is some field evidence suggesting that the process has already started. For example, storm surge water stagnated for more than one year in low-lying coastal areas following cyclone Aila (2009) due to higher than normal sea levels, which inhibited timely drainage of stagnant water following the storm. There is also some field evidence that many people have already been displaced by some of the hazards associated with rising sea levels, although there is no systematic or reliable database on this process (Rashid and Paul, 2014).

In our previous study, we have characterized all people displaced by different types of climatic hazards (such as floods, riverbank erosion, droughts, and cyclones) as climate refugees (Rashid and Paul, 2014). Our focus in this chapter is on millions of future climate refugees who are projected to be displaced by a 1 m (3.281 ft) rise in sea level along the Bangladesh coast. We consider SLR as an impending climatic disaster for Bangladesh because of its both practical and epistemological implications. Epistemologically, hazards associated with rising sea levels may be linked to the existing literature, which suggests that every disaster starts with a hazard (Thywissen, 2006). A hazard may turn into a disaster "principally because the human population has placed itself in a situation of vulnerability" (Alexander, 1993, p. 7). Addressing vulnerability, broadly defined as the potential for loss, is an essential component of hazard mitigation strategies (Cutter, 1996). Clearly, coastal Bangladesh is one of the most vulnerable areas for SLR because millions of its residents, mostly impoverished subsistence farmers living in flimsy houses, have been exposed to increased hazard vulnerability due to global warming–induced SLR, which is entirely beyond their control. While SLR may not appear to be a conventional climatic hazard, such as floods and storms, we consider it as a climatic hazard because it has been induced by global warming, which is a distinct climatic phenomenon.

How Many Are Likely to be Displaced by Sea Level Rise?

The answer to this critical question is contingent upon the amounts of SLR projected by different climate change models. It is also related to population projections for Bangladesh. Following the IPCC First Assessment Report, initial estimates for SLR along the Bangladesh coast ranged from 1 m (3.281 ft) by 2050 to 2 m (6.56 ft) by 2100 (Houghton, 2006). According to these estimates, about one-fifth of the land area of Bangladesh would be lost due to an SLR of 1 m (3.281 ft) with a corresponding displacement of 15 million people. Subsequently, the IPCC revised its projections for global SLR several times. In its Fourth Assessment Report (2007) the projections for global SLR by the end of 2100 ranged between 28 cm (best estimate due to low emissions of greenhouse gases) and 43 cm (worst estimate due to high emission scenarios). Following recent acceleration in melt-water discharge from Greenland and Antarctica, the IPCC revised its projection for a maximum rise of 54 cm (1.77 ft) by the end of this century. It appears from our preceding review that the latest data on accelerated rates of global SLR for 1993–2009, i.e., 28 cm or 11 inches by the end of this century (2.8 mm/year or 0.11 inches/year) seem to coincide with IPCC's best estimates of global rates for the same period. The IPCC worst estimates (a global SLR of 54 cm or 1.77 ft by 2100, i.e., 5.4 mm/year or 0.21 inch/year) are only 19 cm (7.4 inches) higher than Cazenave and Llovel's (2010) revised estimates for 1993–2009 (3.3 ± 0.4 mm/year or 0.13 ± 0.016 inch/year). In its Fourth Assessment Report, the IPCC does not provide any specific projections for sea level along the Bangladesh coast. However, it has used a benchmark of 1 m (3.281 ft) increase in sea levels for postulating adverse impacts on several low-lying coastal areas, including Bangladesh. Further, the IPCC estimates of population displacement due to inundation by SLR have been relatively modest compared to some of the other non-IPCC estimates. For example, it makes the following statement in its Fourth Assessment Report: "Using a coarse digital terrain model and global population distribution data, it is estimated that more than one million people will be directly affected by sea-level rise in 2050 in each of the Ganges-Brahmaputra-Meghna delta in Bangladesh, the Mekong delta in Vietnam and the Nile delta in Egypt" (IPCC, 2007, Chapter 10.4.3.1).

Compared to the IPCC projections, some of the non-IPCC studies have produced wildly divergent estimates of population displacements due to unrealistically high SLR estimates. To address such discrepancies between IPCC and non-IPCC estimates, we have taken two distinct approaches in the subsequent sections of this chapter. First, we have reviewed data from some of the non-IPCC studies. Second, we have made our own contribution to this issue of potential population displacements in coastal Bangladesh due to SLR, by adopting an approach that relies on (a) 2013 census data for population distribution in each of the districts and subdistricts that are likely to be inundated by SLR and (b) potential areas of combined hazards of SLR and tropical cyclone storm surges associated with a 1 m (3.281 ft) rise in sea level. Subsequently, we have characterized such combined hazards as SLR risks.

Non-IPCC Projections for Sea Level Rise in Bangladesh

Table 3.5 compares two non-IPCC projections for SLR along the Bangladesh coast—one by World Bank and another by Greenpeace India. These are essentially macroscale analyses and lack some of the field-level data on areas of SLR risks. It should be stressed here that we have obtained data presented in Table 3.5 from original reports (available on websites), which do not provide any details of methodologies employed for these projections. The World Bank projections are much more conservative than those by Greenpeace India and seemed to be motivated by the bank's concerns for likely effects of climate change and SLR on Bangladesh's economy, particularly on the agricultural sector. To assess these impacts, the South Asian Rural Development Unit of the World Bank prepared a country report on Bangladesh Climate Change and Sustainable Development (World Bank, 2000). Consistent with the mandate of the bank, the emphasis of the report is on shorter-term climate change impacts on development activities in Bangladesh. Since population displacement due to SLR is a major concern for development activities, the report provides some scenarios of both short-term and long-term population displacement due to SLR within the ranges of 4–39 inches (10 cm–1 m) (Table 3.5).

The World Bank projections in Table 3.5 have at least two major planning implications. First, the rates of SLR, based on the IPCC projections, indicate a relatively slow linear rate of rise in the first half of this century (only about 10 inches or one-quarter of the total projected rise). This would allow significant adaptation to SLR within a planning time horizon of the next 30–40 years. The rates of rise would accelerate dramatically in the second half of this century. Second, estimates on population displacement are, at best, tentative because these are contingent upon population projections. Data in Table 3.5 are based on the 2013 census update, implying that significantly higher numbers are likely to be displaced depending upon population projections at given time horizons. The World Bank's own estimate for Bangladesh population in 2100 is 218 million (World Bank, 2000). Based on this projection, one could assume that about 44 million people (i.e., 20.4% of 218 million) might be displaced due to a 39-inch (1 m) rise in sea level (compared to 29.4 million listed in Table 3.5, based on the 2013 census). Similarly, if data for different population projections for

Table 3.5 Estimates of Population Displacement From Coastal Bangladesh, Based on Two Non-IPCC Projections

Nature of SLR Projections	World Bank Studies (2002, 2009)	Greenpeace India Study (Rajan, 2008)
10 Inches (25 cm) Rise By 2050		
Flooded area in mi^2 (km^2)	2432 (3300)	NA
Affected population in millions (% of 2013 population)	3.6 (2.5)	
39 Inches (1 m) Rise By 2050		
Flooded area in mi^2 (km^2)	NA	NA
Affected population in millions (% of 2013)		5.73 (NA)
39 Inches (1 m) Rise By 2100		
Flooded area in mi^2 (km^2)	10,174 (26,350)	NA
Affected population in millions (% of 2013 population)	29.4 (20.4)	41.61 (NA)
10 ft (3 m) Rise By 2100		
Flooded area in mi^2 (km^2)	NA	NA
Affected population in millions (% of 2013 population)	NA	58.25 (NA)
15 ft (5 m) Rise By 2100		
Flooded area in mi^2 (km^2)	NA	NA
Affected population in millions (% of 2013 population)	NA	75.9 (NA)

Source: Prepared by the authors based on data presented in Rajan, S.C., 2008. Climate Migrants in South Asia: Estimates and Solutions, a Report by Greenpeace. Greenpeace India, Bangalore, India; Rashid, H., Paul, B., 2014. Climate Change in Bangladesh: Confronting Impending Disasters. Lexington Books, Lanham, MD, Boulder, CO, New York, Toronto, Plymouth, UK; World Bank, 2009. World Development Report 2010: Development and Climate Change. World Bank, Washington, DC; and World Bank, 2000. Bangladesh: Climate Change & Sustainable Development, Report No. 21104. World Bank, Dhaka, Bangladesh.

2050 were used in the estimates, significantly more than 3.6 million (as listed in Table 3.5) would be displaced by a 10-inch rise in sea level.

Another study that has received significant media attention is a Green Peace document entitled Climate Migrants in South Asia: Estimates and Solutions—A Report by Greenpeace (Rajan, 2008). The document has been authored by Sudhir Chella Rajan, a Social Science faculty member at the Indian Institute of Technology, Madras (IIT, Chennai). This study has estimated wide ranges of population displacements due to three hypothetical scenarios of SLR (Table 3.5). These scenarios are contingent upon a series of physical chain effects starting with (a) unconstrained greenhouse gas emissions under a business-as-usual scenario, leading to greater than 750 ppm of atmospheric concentration of carbon dioxide, (b) resulting increase in global average temperature by 7°F to 9°F (4°C to 5°C) (Rahman et al., 2011; Rajan, 2008), and (c) a catastrophic acceleration in breakup and melting of land ice in Antarctica and Greenland. A significant feature of Rajan's (2008) estimates, based on some of these IPCC-generated scenarios, is that the SLR is likely to accelerate in a nonlinear rate following a threshold point in 2050, resulting in the bulk of the population displacement in the second half of this century (similar to the World Bank estimates). In the first half of this century, the estimates indicate a linear rate of SLR with relatively small displacement of population at the beginning of the century. For example, according to the first scenario of 39 inches (1 m) rise in sea level, a displacement of only 5.73 million is projected for 2050, whereas the estimate increases to a staggering 41.62 million by 2100. The projected displacements for the second half of this century (i.e., by 2100) increase to 58.25 million with an SLR of 10 ft (3 m) and a whopping 75.82 million with a 15 ft (5 m) rise scenario.

The preceding data have been reproduced by Mondal and Rajan (2009) in a conference proceeding and in other studies related to climate change in Bangladesh (for example, Figure 6.2 and Table 6.3 in Rahman et al., 2011). Without referring specifically to the Greenpeace study (Rajan, 2008), the Bangladesh Climate Change Strategies and Action Plans (BCCSAP) 2009 has postulated that six to eight million people could be displaced from coastal Bangladesh by 2050 due to SLR and would have to be resettled (GoB, 2009). From a planning perspective this time horizon (i.e., 2050) may be too distant. Most of the action plans of the BCCSAP are, therefore, based on the next 10-year cycle of relatively small changes in sea level, which are consistent with most of the recent observations and model studies.

Compared to these details, much greater attention has been given in the media and public discourse to Greenpeace Report's long-term projections of population displacement in Bangladesh by 2100: 42 to 76 million under scenarios of 3–15 ft (1–5 m) of SLR (Table 3.5). Some of the critics have questioned the reliability of such long-term projections based on a set of hypothetical scenarios but, as we have emphasized earlier, these projections are contingent upon not only the scenarios of SLR but also on population projections. If some of the projections materialize, Bangladesh population is expected to reach 250 million by 2100 (Streatfield and Karar, 2008). Using both World Bank's and our own estimates for 2100, at least one-fifth of the population of Bangladesh is likely to be displaced or destabilized by the end of this century due to a 1-m rise in sea level and its associated storm surges. This would amount to 50 million (i.e., 20% of 250 million). Thus, Rajan's (2008) data for the midcentury scenario (line 2 in Table 3.5) do not seem to be exaggerated. Other estimates under scenarios 2, 3, and 4 (lines 3, 4, and 5), with projected displacements of 42–76 million, have drawn extensive media attention. The Hindustan Times, a prominent Indian English daily newspaper, for example, published a sensational headline: "75 million Bangladeshis may inundate India" (Nag, 2008). The author himself (i.e., Rajan) has politicized his scientific study by a blanket assumption that the bulk of the 75 million climate refugees from such a scenario are likely to migrate to India: "Given the proximity of Bangladesh to India and the large land area that would be inundated, it is also likely that the bulk of these people will end up being migrants in India" (Rajan, 2008, p. 10). The central problem with such a sensational projection is that it is based on a hypothetical scenario of a 15 ft (5 m) rise in sea level, which is one of the worst IPCC scenarios (i.e., A1 SRES scenario) (Rajan, 2008). If such a projection materializes, not only a large part of Bangladesh would be inundated, but many coastal cities of South Asia, including Kolkata (Calcutta), Chennai (Madras), Mumbai (Bombay), and Karachi would be submerged. Indeed, it would constitute a global catastrophe as many low-lying coastal areas throughout the world would be at grave risks of inundation, resulting in massive population displacement.

Our Estimates of SLR-Risk Areas in Bangladesh

SLR along the Bangladesh coast is likely not only to displace large numbers of people due to direct inundation of coastal lands it is further likely to destabilize a second group of people within the striking distance of storm surges above the new coastline. Therefore, for estimating numbers of people likely to be impacted by a 1 m (3.281 ft) rise in sea level, we have defined SLR-risk areas as a broad geographical belt consisting of two spatial units: (a) a primary area likely to be inundated by SLR and (b) a secondary area impacted by the maximum extent of cyclone storm surges (Fig. 3.1). We have delineated these two areas in Fig. 3.1 by drawing two separate lines through the boundaries of the impacted districts. The lower line delineates areas likely to be inundated directly by a 1-m (3.281 ft) rise in sea level. The upper line

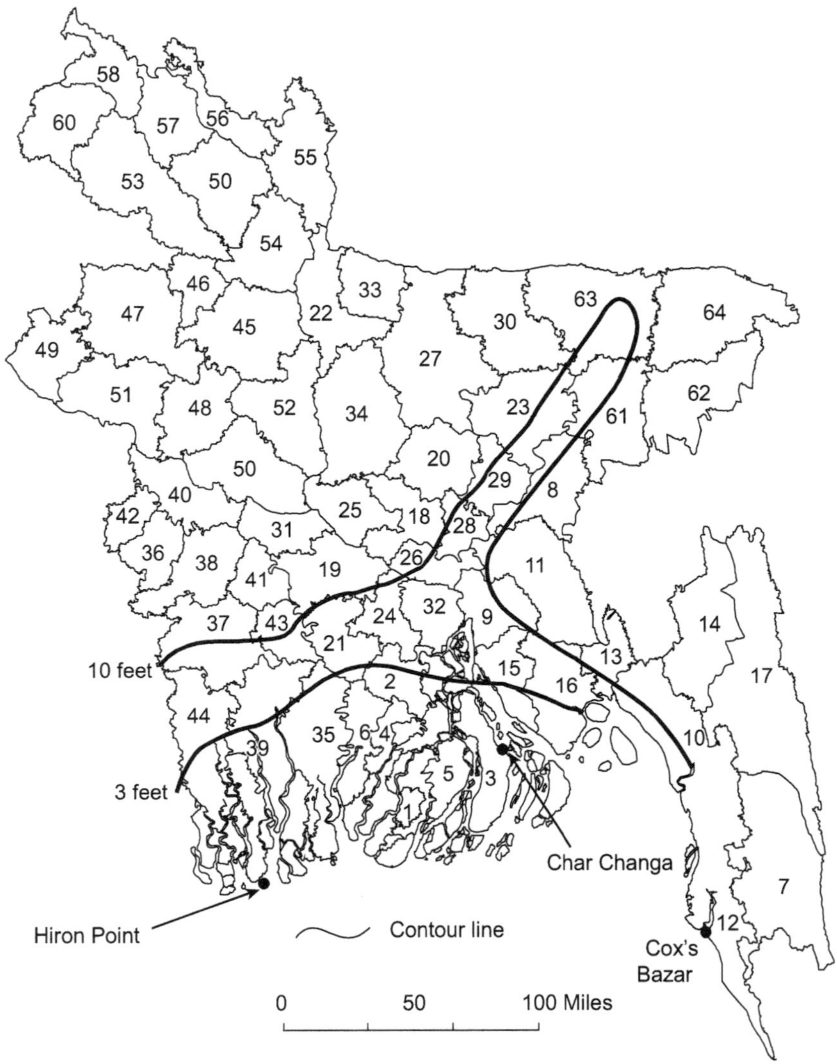

1 Barguna, 2 Barisal, 3 Bhola, 4 Jhalkati, 5 Patuakhali, 6 Pirojpur, 7 Bandarban, 8 Brahamanbaria, 9 Chandpur, 10 Chittagong, 11 Comilla, 12 Cox's Bazar, 13 Feni, 14 Khagrachhari, 15 Lakshmipur, 16 Noakhali, 17 Rangamati, 18 Dhaka, 19 Faridpur, 20 Gazipur, 21 Gopalganj, 22 Jamalpur, 23 Kishoreganj, 24 Madaripur, 25 Manikganj, 26 Munshiganj, 27 Mymensingh, 28 Narayanganj, 29 Narsingdi, 30 Netrakona, 31 Rajbari, 32 Shariatpur, 33 Sherpur, 34 Tangail, 35 Bagerhat, 36 Chuadanga, 37 Jessore, 38 Jhenaidah, 39 Khulna, 40 Kushtia, 41 Magura, 42 Meherpur, 43 Narail, 44 Satkhira, 45 Bogra, 46 Joypurhat, 47 Naogaon, 48 Natore, 49 Nawabganj, 50 Pabna, 51 Rajshahi, 52 Sirajganj, 53 Dinajpur, 54 Gaibandha, 55 Kurigram, 56 Lalmonirhat, 57 Nilphamari, 58 Panchagarh, 59 Rangpur, 60 Thakurgaon, 61 Habiganj, 62 Maulvibazar, 63 Sunamganj, 64 Sylhet

Figure 3.1 Areas likely to be inundated by two scenarios of SLR 1 m (3 ft) and 3 m (10 ft)—have been superimposed on district boundaries, identifying district areas at risk of inundation. *Prepared by authors based on information in Fig. 7.1 in Rashid, H., Paul, B., 2014. Climate Change in Bangladesh: Confronting Impending Disasters. Lexington Books, Lanham, MD, Boulder, CO, New York, Toronto, Plymouth, UK.*

coincides approximately with the 10 feet (3 m) contour. The latter is based on past observations that some of the high magnitude storm surges reached inland at least up to the 5 feet (1.5 m) contour. We have combined these two values (i.e., future surge level reaching up to the 5 feet contour +3.281 feet future sea level) and rounded the resulting odd number (8.281 ft) as 10 ft. Thus, we have delineated SLR-risk areas (up to the 10-ft contour) on a base map with outlines of district boundaries, by using the following set of data:

- District area (in mi^2, based on the census data)
- District population (based on the latest census update by the Bangladesh Bureau of Statistics, as of March 2013)
- Proportions of district areas under the 10-feet contour (estimated by visual interpretations of contour positions through district boundaries at map scales of 1 inch = 30–35 miles)
- Estimates of percentages of district population (based on the distribution of the smallest administrative units of Bangladesh, namely subdistricts and *thanas* on district maps)

Although we employed a visual interpretation method without using either a digital elevation model or a GIS analysis, we assume that our estimates have relatively small error margins because of relatively small district areas shown at large scales of district maps (1 inch = 30–35 miles). Further, our estimates of percentages of district population under a given contour are verifiable through census data for all subdistricts and *thanas* (under the contour line). Table 3.6 summarizes the results of our analysis.

Data in Table 3.6 indicate that, like the World Bank's (2009) estimates, approximately one-third of the land area of Bangladesh is likely to be impacted by a 1-m rise in sea level by 2100. Further, the data indicate significant spatial variations in likely impacts from one coastal zone to another, largely because of differences in their physical settings. Despite apparent similarities in topographic characteristics of the southwest and the central zones, storm surge impacts in the southwest zone are significantly reduced by the mangrove forest stands of the Sundarbans. There are no permanent settlements in the Sundarbans forest. Only about 5–20% of the population situated along the southern fringe of the settlement north of the forest is affected by storm surges. The central zone is by far the most storm surge vulnerable area of Bangladesh because it is exposed directly to advancing surges without any topographic or vegetation barriers. This zone has an extremely gentle topographic slope as the entire area has an elevation of less than 5 ft (1.5 m) above the sea level. Currently, the 5-ft (1.5 m) contour seems to

Table 3.6 SLR Risk Areas in Bangladesh Due to a 3-ft (1-m) Rise in Sea Level[a]

Geographical Area and Districts Impacted	Area Impacted in mi^2 (% of District)	Population Impacted in Millions (% of District)	SLR Risks[a] (Major, Moderate, Minor)
Southwest coast:			
Satkhira	1474 (100)	1.98 (100)	Major
Khulna	1697 (100)	2.31 (100)	Major
Bagerhat	1529 (100)	1.48 (100)	Major
Southwest inland:			
Jessore	302 (30)	0.69 (20)	Minor
Narail	187 (20)	0.29 (40)	Minor
Central coast:			
Barguna	707 (100)	0.89 (100)	Major
Patuakhali	1244 (100)	1.53 (100)	Major
Bhola	1314 (100)	1.78 (100)	Major
Noakhali	1.281 (90)	2.80 (90)	Major
Lakshmipur	501 (90)	1.30 (90)	Major
Pirojpur (coastal + inland)	493 (100)	1.11 (100)	Major
Central inland:			
Jhalokati	273 (100)	0.68 (100)	Major
Barisal	1075 (100)	2.32 (100)	Major
Gopalganj	567 (100)	1.17 (100)	Major
Madaripur	348 (80)	0.93 (80)	Major
Shariatpur	299 (66)	0.87 (75)	Major/ moderate
Faridpur	79 (10)	0.19 (10)	Minor
Eastern coast:			
Feni	792 (80)	1.29 (90)	Major
Cox's Bazaar	385 (40)	1.14 (50)	Moderate
Chittagong	204 (10)	3.81 (50)	Minor
Meghna Valley:			
Chandpur	381 (60)	1.69 (70)	Moderate
Kishoreganj	1344 (50)	1.46 (50)	Moderate
Brahmanbaria	291 (40)	1.14 (40)	Moderate
Comilla	243 (20)	1.08 (20)	Minor
Narsingdi	89 (20)	0.59 (25)	Minor
Sunamganj	289 (20)	0.49 (20)	Minor
Munshiganj	39 (10)	0.14 (10)	Minor
Narayanganj	26 (10)	0.29 (10)	Minor
Netrokona	108 (10)	0.11 (5)	Minor
Habiganj	102 (10)	0.21 (10)	Minor
Total (% of Bangladesh)	**17,662 (31)**	**36 (25)**	

[a]SLR risk areas have been classified into three categories by percentages of district areas impacted by SLR risks. Major risk (67–100% area impacted); moderate risk (34–66%); minor risk (<34%).
Primary data source: BBS 2013 census data on district area and population. *Secondary data source*: Jahan, S., 2009. Bangladesh and World in Map (in Bengali). Mela Book Publisher and Seller, Dhaka. Plate 11: district boundaries; plate 19: contour map; plate 20: surge area; plate 23: SLR map; and several other larger-scale district maps (at scales 1 inch = 30–35 miles): distribution of population.

provide a good approximation of storm surge hazard areas in the western part of Bangladesh. Along the east coast, as land rises toward the Chittagong Hill Tracts (parallel ranges of hills), recent storm surges had reached as high as 20–25 ft (6–8 m) in the districts of Chittagong and Cox's Bazaar and 15–20 ft (5–6 m) in the district of Feni. Because of the steep land slopes less than 50% of the eastern zone had experienced past storm surges, but 55% of the zone population is at risk since the zone includes two large metropolises, namely Chittagong and Cox's Bazaar.

The Meghna valley plays a special role in dissipating storm surge impacts from coastal areas deep into the interior of Bangladesh. In 5 of the 10 districts likely to be impacted most of the surge water would remain confined within the valley. Approximately, 10–20% of the land areas of these districts flanking both sides of the valley would experience deep flooding. With relatively gentle slopes along its left bank, about 40% of the land area of the district of Brahmanbaria had experienced past storm surge flooding. As the left-bank floodplain of the Meghna widens further downstream toward its confluence with the Ganges, approximately 60% of the land area of Chandpur had experienced past storm surges, but the southern part of this district, lying along the bank of the combined GBM channel, belongs more to the central coastal zone than exclusively to the Meghna valley.

Unlike inundation by SLR, which may be determined more precisely by topographic contours, areal extent of storm surge impacts is likely to be determined by the orientation and magnitude of a given tropical cyclone. All of the three coastal zones are vulnerable to varied magnitudes of storm surge impacts. The central zone is more directly exposed to surge impacts with potentials for maximum areal spread. The highest historical fatalities have been experienced in this zone during the 1970 *Bhola* cyclone, often called the killer storm of the 20th century. The east coast is also notorious for storm surge fatalities associated with the 1991 cyclone Gorky. However, surge impacts in this zone are confined to a narrow coastal zone because of relatively steep topographic slopes. The surge impacts in the Meghna valley are not notable for human casualties, but the valley helps in diffusing surges in an inland structural depression, called the Sylhet Trough (*Sylhet haor*). Surge-induced deep flooding in the valley affects settlements along the edges of the valley.

The varied geography of SLR in Bangladesh may be described as follows. First, slightly more than one-third of the current population (2013 census update) living in about one-third of the land area of Bangladesh is likely to be impacted. This amounts to 30 out of 64 districts of Bangladesh.

Second, out of the 30 districts, 10 are at the front-line of rising sea levels as their entire areas (100%) are likely to be impacted. Among them, the following are coastal districts (i.e., districts with the Bay of Bengal coastline): Satkhira, Khulna, Bagerhat, Barguna, Patuakhali, Bhola, Lakshmipur, and Noakhali. Consequently, large areas of these districts are liable to direct inundation with a 1-m rise in sea level. In addition, entire areas of some of the low-lying districts that are situated further inland of the coastal districts may still be vulnerable to SLR risks. These include: Pirojpur, Jhalokati, Barisal, and Gopalganj. Although Gopalganj is situated about 60 miles (100 km) inland from the coastline it occupies a structural depression, called Faridpur Trough (*Faridpur beel*), making it vulnerable to surge water accumulation. About 80% of Madaripur is also part of this structural depression. Similarly, about 80% of Shariatpur, another low-lying district, is liable to the SLR risks. Third, in the Meghna valley, only three districts (Chandpur, Kishoreganj, and Brahmanbaria) are at moderate risks, whereas the remaining seven are likely to experience minor surge impacts.

Regarding the number of people likely to be displaced by the SLR risks by 2100, our estimates may be modified (decreased) by prorating all data in Table 3.6 by a factor of 82.81%, since we rounded our areal estimates to a 10-feet contour from 8.281 ft. However, assuming a similar population growth rate in all of the districts, one can further extrapolate from our estimates that about one-fifth of the future population of Bangladesh is likely to be vulnerable to SLR risks. Even at the current population of 159 million (as of 2014), this would amount to at least 32 million people. These estimates seem to be very similar to those by the World Bank.

Where Would the Displaced People Resettle?

Field evidence suggests that vast majority of the current coastal residents of Bangladesh are willing to rebuild their homes following repeated cyclone disasters. In the short run with relatively small increases in sea levels, perhaps up to 2050, it may be possible to live with the rising sea levels by developing a variety of proactive mitigation and in situ adaptive measures (Lal and Aggarwal, 2000). Some of the indigenous measures (commonly considered as nonstructural adaptations) include elevating coastal embankments and polders, raising foundations of homes and redesigning housing architecture in coastal Bangladesh, significant changes in cropping practices, including development of salt-resistant crops, and many other innovative measures (see Chapter 7). In addition to these formal measures, the impoverished residents of coastal Bangladesh have lived for generations with cyclone and storm surge hazards by adopting some of the indigenous coping strategies, such as climbing trees or taking refuge in the nearby forests, often with catastrophic loss of lives (see Chapter 2; Rashid, 2011; Paul et al., 2010). More recently, a network of public cyclone shelters (perhaps the most successful structural adaptation) has been constructed to provide significant protection to coastal residents from storm surge impacts (Paul et al., 2010; Paul, 2009). An estimated 1.5 million people took refuge in these shelters during the category IV cyclone *Sidr*, which made landfall on November 15, 2007 (Paul, 2012). Field evidence from another cyclone (Cyclone Aila, May 25, 2009) suggested that many coastal residents, particularly the elderly, were reluctant to abandon their ancestral homes. Therefore, many cyclone victims had chosen to remain in the vicinity of their damaged or destroyed homes. As of late 2010, at least 100,000 of 3.9 million victims had been living on embankments and narrow roadsides in temporary shacks made of bamboo, mud bricks, and plastic sheeting (IOM, 2010; FAO, 2010). Among potential factors that discouraged outmigration from a hazardous landscape like coastal Bangladesh, the following have been identified in the literature: lack of financial resources to move to a new place; uncertainty and competition in the new place, leaving family behind; difficulty in adjusting to a new place; households with female heads; and inability to earn enough to send home and maintain two households (Paul, 2011; Penning Rowsell et al., 2011). Local communities often prefer in situ adjustment to migration. Thus, a premature focus on migration can block policies and programs that support these preferences. Other scholars have cautioned that large-scale outmigration can increase vulnerability among those remaining behind, reducing their capacity to implement in situ adaptations to environmental change because of distortions in markets, obsolescent political and administrative institutions, a hyperconcentration of social capital, and increased demands for labor (Barnett, 2012).

In the long run, especially after 2050, the displaced people (climate refugees) due to SLR will not have the option to settle in their immediate vicinity because of permanent inundation of coastal areas. Some of these displaced people may seek resettlement on higher grounds in rural areas further inland from their current location, presumably because of their familiarity with the adjacent rural settings. However, this option is not promising since agricultural land is extremely scarce in Bangladesh and opportunities for making a living in rural Bangladesh are also limited.

In the absence of opportunities for resettling in rural areas, it is widely assumed that large proportions of the displaced people would migrate to urban areas and most of them would end up in urban slums. According to a 2005 survey, nearly 5.4 million people lived in at least 9000 *bustees* (slums delineated by certain criteria) in six metropolitan centers of Bangladesh, accounting for more than one-third of its urban population (Islam et al., 2006). More than one-half of these slums (approximately 5000) are scattered throughout the metropolitan area of Dhaka. Besides Dhaka, other major cities of Bangladesh with significant slum population include Chittagong, Khulna, Rajshahi, Sylhet, and Barisal. What proportions of the slum residents constitute climate refugees? We do not know that as yet because most of the evidences of population displacement due to SLR are anecdotal. So far very few respondents of successive surveys had identified themselves as climate refugees from coastal areas. However, in the coming decades, climate refugees may no longer remain "invisible" as the rates of displacement from coastal Bangladesh are likely to accelerate with rising sea levels.

Large-scale migration of climate refugees from Bangladesh to advanced industrial countries is highly unlikely for at least two reasons. First, most of the displaced residents of coastal Bangladesh are likely to be impoverished and without significant formal education. It would be difficult for them to resettle in distant and unfamiliar cultural environments. Second, in the absence of a Bangladesh national policy specifying criteria for emigration, many are skeptical that the poor and largely illiterate climate refugees from coastal Bangladesh would never be able to migrate to advanced industrial countries. Some of the necessary requirements for long-distance and international migration include many challenging resources, such as finance, relevant information, social networks (both at home and abroad), and job skills. Those who are most vulnerable to SLR and other environmental disasters are unlikely to have access to these resources to undertake journeys of this sort (Walsham, 2010).

Compared to long-distance overseas migration, cross-border migration from Bangladesh to India is a much more plausible scenario largely because of their geographical contiguity. Moreover, there are examples of past mass migration during episodes of man-made calamities, such as the communal strife leading to the Partition of India in 1947 when millions crossed both sides of the borders and resettled in India and East Pakistan (now Bangladesh). Among other variables that have fueled speculation of mass migration to India is the geographical reality that, confronted with the prospect of large-scale loss of land due to SLE and lack of resettlement alternatives within the truncated territory of Bangladesh, millions of displaced people may have no other alternative than to consider resettlement in India, legally or illegally. India may be a vast country but it has also a vast population and, consequently, its resources are limited for absorbing such massive numbers of migrants. This has led to the birth of a (potential) geopolitical crisis. The New Delhi-based Institute for Defense Studies and Analyses estimates that already 10 to 20 million Bangladeshi immigrants are living in India illegally (Carney et al., 2011). Some security specialists contend that further increases in such illegal migration would surely lead to conflicts across international borders (Gaan, 2005). Futher, media coverage of Rajan's (2008) prediction that 75 million climate refugees from Bangladesh are likely to seek resettlement in India has reinforced such speculation. In response to the security concerns posed by cross-border migration, India has started building a 2500-mile (3730-km) security fence along the Bangladesh—India border. Media reports indicate that the Indian security guards have routinely shot people crossing illegally into India (Belt, 2011). While such incidences of illegal border crossing are common there is still a major problem with an assumption that vast numbers of climate refugees would like to resettle in India. Limited field data indicate that, on the contrary, most of the current climate refugees are not necessarily eager to move to India illegally. Their first preference is to resettle within the borders of Bangladesh, most likely in larger cities located inland, such as Dhaka and Khulna (Islam, 2009). Data on cross-border migration following some of the recent climatic disasters are largely nonexistent or at best anecdotal. For example, according to one report, among a sample of 300 survivors of cyclone Aila from an affected village most migrated to nearby towns, whereas a limited few crossed the border to India, "where they live in fear of being exposed as illegal immigrants and thrown into prison" (Aulakh, 2013, p. 2). The same report claims that "It is the same story in nearby villages, and in much of rural southwest Bangladesh."

CONCLUSION

Reassessing Sea Level Rise in Bangladesh

Our entire analysis presented above is based on an assumption of a 1-m rise in sea level by the end of this century. How realistic is this assumption? Based on "sea-level changes under increasing CO_2 in a transient coupled ocean-atmosphere experiment" (Gregory, 1993; Broadus,

1993), Houghton (2006, p. 152) reported that the sea level along the Bangladesh coast was expected to rise by 2 m by 2100. This estimate amounted to a net increase in sea level resulting from a combination of both SLR and land subsidence. Unlike tide gauge-based sea level data, earlier estimates of land subsidence along the Bangladesh coast were based on very limited data. More recently, satellite-based altimetry and advanced GPS surveys have added significant insights into SLR versus land subsidence along the Bangladesh coast. A recent study by Auerbach et al. (2015), based on a "fast-static GPS and theodolite survey of land-surface elevation" in a poldered island in southwestern Bangladesh, including measurements from the adjacent Sundarbans forest (natural SLR) found that the embanked island, deprived of natural sedimentation, experienced a subsidence rate of 20 mm/year (0.78 inches/year) during a period of 50 years since polder construction in the 1960s. In contrast, the areas outside the embankment "received tens of centimeters of tidally deposited sediment, equivalent to decades' worth of normal sedimentation." The rates of sedimentation outside the polder "have kept pace even with the enhanced rates of effective sea-level rise associated with tidal amplification" (Auerbach et al., 2015). This is a very significant finding that backs up some of the earlier assessments of the role of river sedimentation in SLR along the Bangladesh coast by several researchers, including Brammer (2014a,b), CEGIS (2010), Rob (2008), and Allison (1998).

Whereas SLR poses an "impending disaster" for Bangladesh, Brammer (2014a) has pointed out that the ongoing change of greatest significance in Bangladesh's coastal area is the formation of new land in the Meghna estuary. Based on a comparison of satellite images taken in 1984 and 2007, Bangladesh had a net land gain of 174 mi^2 (451 km^2) in the Meghna mouth (Brammer, 2014a, p. 151). This represented an average growth rate of 7.3 sq mi/year (19.6 km^2/year) during a relatively short period of 24 years (1984–2007), which was not much higher than two other longer-term estimates: an average rate of 5.7 mi^2/year (14.8 km^2/year) (1792–1840) (Allison, 1998) and 6.5 mi^2/year (17.07 km^2/year) (1793–2008) (CEGIS, 2010). Another study, using the technique of superimposing historical coastlines of Bangladesh, indicate that the southeastern land mass of the active delta has migrated 50 miles (80 km) seaward during a period of 200 years (1780–1980) (Rob, 2008). Such land growth has been sustained by massive amounts of sediment load—estimated at about 1.5–2 billion metric tons per year—which are carried by numerous (at least 230) tributaries of the GBM rivers (Ahmad, 2011; Ahsan et al., 2011; Islam et al., 2001, 1999; Rasid, 2000; Coleman, 1969). Most of the literature suggests that about one-half of the suspended sediment loads of these rivers are deposited on the floodplain and the remaining half is delivered to the coastal area (Islam et al., 1999). About 15% of the latter is lost to the sea in a deep submarine canyon and the remaining 35% is deposited on the subaqueous delta in the nearshore zone, contributing to delta building (Khalequzzaman, 2012).

Sedimentation in the deltaic environment is a three-dimensional process. Successive accumulation of sediments on river floodplains and coastal delta plains raises land levels in these settings. Contrary to a widely held expectation, a "rising sea level would not gradually overwhelm the country, contour by contour ... More likely in the active Ganges-Brahmaputra-Meghna delta is that sedimentation on tidal and young estuarine floodplains and on river levees would raise these areas simultaneously with sea-level rise" (Brammer, 2000, p. 114). Sediment accumulation also expands land forward and sideward, but more importantly progressive compaction of sediments and the weight of overlying sediments result in subsidence lowering the land level. There are no systematic quantitative data on subsidence in Bangladesh, probably because of the difficulties of estimating this process.[3] Some earlier estimates of subsidence ranged between 0.28 inches per year and 0.48 inches per year (7 mm and 12 mm) (Houghton, 2006). These figures appear to be two to three times higher than estimates from carbon dating and other geological dating techniques, which are 0.12—0.14 inches per year (3—3.5 mm) (Islam et al., 1999; Umitsu, 1993). Subsidence of land in Bangladesh has also been attributed to excessive withdrawal of groundwater for irrigation and other purposes. Net SLR is the balance between the absolute rise and some of the competing terrestrial variables. For most of the Holocene epoch (last 10,000 years), the GBM delta has been growing both vertically and seaward, extending its coastline to the current positions from its initial locations around the center of Bangladesh (Morgan and McIntyre, 1959; Rizvi, 1955). During this period, the rivers of Bangladesh had been contributing at least two billion tons of sediment annually (Goodbred and Kuel, 2000), with a significant acceleration in sedimentation rates during the last 5000—7000 years (Uddin, 2009; cited by Khalequzzaman, 2012). However, it should be emphasized here that land-loss/land-gain scenarios in Bangladesh are much more dynamic than an assumed generalization of land loss due to SLR versus land gain due to river sedimentation. For example, most of the new land has emerged in the district of Noakhali, while Bhola Island has lost around 89 mi^2 (230 km^2), attributed largely to changes in the flow pattern of the lower Meghna (Walsham, 2010).

The Way Ahead

The preceding review might seem to imply that land gains along the Bangladesh coast due to river sedimentation are likely to keep pace with SLR. The process is much more complex than such an assumption because of the model-based projections of significant accelerations in the rates of SLR after 2050. Recent rates of SLR along the Bangladesh coast (as we reviewed earlier for 1979–2003) seem to be significantly higher than the regional rates for the Indian Ocean and the Bay of Bengal. Auerbach et al.'s (2015) study on land levels keeping pace with rising sea levels in southwestern Bangladesh includes this short period of accelerated SLR. Since the post-2050 accelerated rates of SLR are likely to be significantly faster than this, a big question is: "Would the rates of land level gains continue to keep pace with post-2050 accelerations in sea level rise?" No one knows the answer. Climate models are not yet able to provide reliable data on regional variability (that superimposes positively or negatively on the global mean rise in sea level) for the next 20, 30, and 50 years. It is therefore very difficult to quantify future SLR in specific regions, and this should be among the priorities for the climate-modeling community (Cazenave and Llovel, 2010). "Much remains uncertain about present and future rates of climate change on the Indian subcontinent and sea-level rise in the Bay of Bengal, and how the predicted changes might actually affect Bangladesh's coastal areas" (Brammer, 2014a, p. 148). In a situation where the future rates of SLR are unknown a prudent sea level management policy should be based on the principle of adaptive management (see Chapter 9). This would allow management authorities to operate within a relatively short time horizon, such as a 5-year or 10-year planning cycle in Bangladesh. An efficient way to deal with rising sea levels would entail a wait-and-see approach and then adopt appropriate adaptation measures depending on how the SLR would unfold with the passage of time. To accomplish this, there is a need for monitoring sea levels more comprehensively by installing additional tide gauges along the Bangladesh coast and through follow-up studies by using satellite altimetry and advanced GPS surveys.

Appendix 1 Tide Gauge Data for the Bangladesh Coast

	Mean Annual Sea Level (mm) (PSMSL Data)		
Year	**Hiron Point**	**Char Changa**	**Cox's Bazaar**
1979		6781	6906
1980		6894	7087
1981		6887	6828
1982		6757	6866
1983	7007	6878	6792
1984	7088	6912	6896
1985	7057	6871	6837
1986	7041	No data	No data
1987	7031	6959	6992
1988	7083	7046	7001
1989	7018	6998	6951
1990	7028	6941	7050
1991	7059	6897	7075
1992	7079	6890	7065
1993	6997	No data	6945
1994	6978	6835	6672
1995	7085	6910	7159
1996	7067	6876	No data
1997	6941	6981	6879
1998	7061	7063	6881
1999	7064	7143	6970
2000	7155	6980	6981
2001	7227		
2002	7034		
2003	7129		

Source: Prepared by the authors based on original sea level data from Permanent Service for Mean Sea Levels (PSMSL): http://www.pol.ac.uk/psmsl.

ENDNOTES

1. In its latest update of *Climate Change and Environmental Risk Atlas*, Maplecroft (2015) has identified 32 "extreme risk countries" in its Climate Change Vulnerability Index (CCVI), which evaluates the sensitivity of populations, the physical exposure of countries, and governmental capacity to adapt to climate change over the next 30 years. Bangladesh ranks as the "1st and most at risk" to climate change, followed by Nigeria, India, Myanmar, Philippines (Maplecroft V., 2015. *Climate Change and Environmental Risk Atlas*. Latest products and reports. http://maplecroft.com/portfolio/new-analysis/2014/10/29/climate-change-and-lack-food-secusity-multiply-risks-conflict-and-civil-unrest-32-count).
2. Hiron Point (Station ID: 1451): 21.78° North and 89.46° East (Southwest coast: on the bank of the Kunga River estuary in the Sundarbans mangrove forest area; in Koyra subdistrict, Khulna district) Charchanga (Station ID: 1496): 22.21° North and 91.05° East (South-central coast: on the east coast of Bhola Island in the Meghna River estuary; in Lalmohan subdistrict, Bhola district). Cox's Bazar

(Station ID: 1476): 21.45° North and 91.83° East (East coast: on the southwest coast of Maheshkhali Island; in Maheshkhali subdistrict, Cox's Bazar district). Source: Permanent Service for Mean Sea Level (PSMSL) (http://www.pol.ac.uk/psmsl).

3. Recent estimates of subsidence of the Ganges delta (west of the GBM estuary) range from 0.16 to 0.28 inches/year (4–7 mm/year) for the Bangladesh coast to about 0.2 inches/year (5 mm/year) near the Indian part of the delta in West Bengal. See Brammer (2014a, pp. 154–155) for additional references and a critical evaluation of subsidence rates along the Bangladesh coast.

REFERENCES

Ahmad, Q.K., 2011. Climate change and freshwater resources of Bangladesh. In: Hossain, M., Selvanathan, E. (Eds.), Climate Change and Growth in Asia. Edward Elgar, Cheltenham, UK, pp. 62–77.

Ahsan, S., Ali, M.S., Hoque, M.R., Osman, M.S., Rahman, M., Babar, M.J., Begum, S.A., Rahman, D.M., Islam, K.R., 2011. Agricultural and environmental changes in Bangladesh in response to global warming. In: Lal, R., Sivakumar, M.V.K., Faiz, S.M.A., Rahman, A.H.M.M., Islam, K.R. (Eds.), Climate Change and Food Security in South Asia. Springer Science + Business Media, New York, pp. 120–134.

Alexander, D., 1993. Natural Disasters. Chapman & Hall, New York.

Ali, A., 1995. A numerical investigation into the back water effect on flood water in the Meghna river in Bangladesh due to the south-west monsoon wind. Estuarine, Coastal and Shelf Science 41 (6), 689–704.

Allison, M.A., 1998. Historical changes in the Ganges-Brahmaputra delta. Journal of Coastal Research 14 (4), 1269–1275.

Auerbach, L.W., Goodbred Jr., S.L., Mondal, D.R., Wilson, C.A., Ahmed, K.R., Roy, K., Steckler, M.S., Small, C., Gilligan, J.M., Ackerly, B.A., January 5, 2015. Flood risk of natural and embanked landscapes on the Ganges-Brahmaputra tidal plain. Nature Climate Change. http://dx.doi.org/10.1038/nclimate2472.

Aulakh, R., February 9, 2013. Bangladesh Faces Mass Migration, Loss of Land from Climate Change. The Toronto Star.

Barnett, J., 2012. On the risks of engineering mobility to reduce vulnerability to climate change: insights from a small island state. In: Hastrup, F., Fog-Olwig, K. (Eds.), Climate Change and Human Mobility: Global Challenges to the Social Sciences. Cambridge University Press, Cambridge, pp. 169–189.

Barry, R.G., Chorley, R.J., 2003. Atmosphere, Weather and Climate, eighth ed. Routledge, London.

Belt, D., 2011. The coming storm. National Geographic 219 (5), 58–83.

Brammer, H., 2014a. Climate Change, Sea-level Rise and Development in Bangladesh. The University Press Limited, Dhaka, Bangladesh.

Brammer, H., 2014b. Bangladesh's dynamic coastal regions and sea-level rise. Climate Risk Management 1, 51–62.

Brammer, H., 2000. Flood hazard vulnerability and flood disasters in Bangladesh. In: Parker, D.J. (Ed.), Floods, vol. I. Routledge, London and New York, pp. 100–115.

Broadus, J.M., 1993. Possible impacts of, and adjustments to, sea-level rise: the case of Bangladesh and Egypt. In: Warrick, R.A., Barrow, E.M., Wigley, T.M.L. (Eds.), Climate and Sea Level Change: Observations, Projections and Implications. Cambridge University Press, Cambridge.

Bryson, R.A., Wendland, W.M., Ives, J.D., Andrews, J.T., 1969. Radocarbon isochrones on the disintegration of the laurentide ice sheet. Arctic and Alpine Research 1 (1), 1–13.

Carney, S., Miklian, J., Hoelscher, K., July/August 2011. Fortress India. Foreign Policy 187, 29–31.

Cazenave, A., Llovel, W., 2010. Contemporary sea level rise. Annual Review of Marine Science 2 (1), 145–173.
CEGIS (Centre for Environment and Geographic Information Services), 2010. Report on Sedimentation. Centre for Environment and Geographic Information Services, Dhaka, Bangladesh.
Chao, B.F., Wu, Y.H., Li, Y.S., 2008. Impact of artificial reservoir water impoundment on global sea level. Science 320, 212–214.
Church, J.A., White, N.J., 2011. Sea level rise from the late 19th to the early 21st century. Surveys in Geophysics 32 (4), 585–602.
Church, J.A., White, N.J., 2006. A 20th century acceleration in global sea-level rise. Geophysical Research Letter 33, L01602. http://dx.doi.org/10.1029/2005GL024826.
Church, J.A., White, N.J., Coleman, R., Lambeck, K., Mitrovica, J.X., 2004. Estimates of the regional distribution of sea-level rise over the 1950 to 2000 period. Journal of Climate 17, 2609–2625.
Coleman, J.M., 1969. Brahmaputra river: channel processes and sedimentation. Sedimentary Geology 3, 129–239.
Cutter, S., 1996. Vulnerability to environmental hazards. Progress in Human Geography 20 (4), 529–539.
FAO (Food and Agriculture Organization), 2010. Feeding Bangladesh's Growing Population Amid Rising Climate Challenges. http://fao.org/news/story/en/item/46153/icode.
Gaan, N., 2005. Environmental scarcity, migration and future sea level rise in Bangladesh: security implications on India. Asian Profile 33 (6), 617–632.
GoB (Government of Bangladesh), 2009. Bangladesh Climate Change Strategy and Action Plan 2009. Government of the Peoples Republic of Bangladesh, Ministry of Environment and Forest, Dhaka, Bangladesh. http://www.sdnbd.org/moef.pdf.
Goodbred, S.L., Kuel, S.A., 2000. Enormous Ganges-Brahmaputra sediment discharge during strengthened early Holocene monsoon. Geology 28 (12), 1083–1086.
Gregory, J.M., 1993. Sea level changes under increasing CO_2 in a transient coupled ocean-atmosphere experiment. Journal of Climate 6, 2247–2262.
Houghton, J., 2006. Global Warming: The Complete Briefing, third ed. Cambridge University Press, Cambridge, UK.
IOM (International Organization for Migration), 2010. Environment, Climate Change and Migration in Bangladesh: Evidence, Dialogue, Strategies. IOM, Dhaka, Bangladesh.
IPCC, 2007. Coastal and low lying areas. In: Climate Change 2007: Working Group II: Impacts, Adaptation and Vulnerability. IPCC Website for Fourth Assessment Report. https://www.ipcc.ch/publications_and_data/ar4/.
Islam, M.S., 2009. The impact of climate change on Bangladesh: responding to a threat to food security for sustainable development. Asian Profile 37 (5), 509–519.
Islam, N., Mahbub, A.Q.M., Nazem, N.I., Angeles, G., Lance, P., 2006. Slums of Urban Bangladesh: Mapping and Census 2005. Centre for Urban Studies, and Chapel Hills, NC: MEASURE Evaluation, University of North Carolina at Chapel Hills, Dhaka, Bangladesh. http://www.cpc.unc.edu/measure/publications/pdf/tr-06-35.pdf.
Islam, M.R., Yamaguchi, Y., Ogawa, K., 2001. Suspended sediment in the Ganges and Brahmaputra rivers in Bangladesh: observations from TM and AVHRR data. Hydrological Processes 15 (3), 493–509.
Islam, M., Begum, S.F., Yamaguchi, Y., Ogawa, K., 1999. The Ganges and Brahmaputra rivers in Bangladesh: basin denudation and sedimentation. Hydrological Processes 13 (7), 2907–2923.
Johnson, B.L.C., 1982. Bangladesh, second ed. Heinemann Educational Books, London.
Khalequzzaman, M., 2012. Geological aspects of climate change: an alternative policy option for the Bengal delta. In: Islam, M.M., Zingel, W. (Eds.), Climate Change Effects and Energy Development. Bangladesh Environmental Network (BEN), Germany, pp. 82–88.

Lal, M., Aggarwal, D., 2000. Climate change and its impact on India. Asia Pacific Journal on Environment and Development 7 (1), 1–11.

Maplecroft V., 2010. Big Economies of the Future—Bangladesh, India, Philippines, Vietnam and Pakistan—Most at Risk from Climate Change. Maplecroft Company, UK. http://maplecroft.com/.

Mondal, M.K., Rajan, S.C., 2009. Climate migrants in Bangladesh: estimates and solutions. In: Rahman, M.H., Noor, M.A., Ahmed, A. (Eds.), Climate Change Impacts and Adaptation Strategies for Bangladesh. BUET, ITN, Dhaka, Bangladesh, pp. 225–235 (cited in Hossain and Selvanathan 2011, p. 146).

Morgan, J.P., McIntyre, W.G., 1959. Quaternary geology of the Bengal basin, east Pakistan and India. Bulletin of the Geological Society of America 70, 319–342.

Nag, S., March 26, 2008. 75 Million Bangladeshis May Inundate India. The Hindustan Times.

Paul, B.K., 2012. Modeling evacuation behavior: the case of 2007 cyclone Sidr, Bangladesh. The Professional Geographer 64 (3), 401–414.

Paul, B.K., 2011. Environmental Hazards and Disasters: Contexts, Perspectives and Management. Willey-Blackwell, Chichester, UK.

Paul, B.K., 2009. Why relatively fewer people died? the case of Bangladesh's cyclone Sidr. Natural Hazards 50, 289–304.

Paul, B.K., Rashid, H., Islam, M.S., Hunt, L.M., 2010. Cyclone evacuation in Bangladesh: tropical cyclones *Gorky* (1991) vs. *Sidr* (2007). Environmental Hazards: Human and Policy Dimensions 9, 89–101.

Penning-Rowsell, E., Sultana, P., Thompson, P., 2011. Migration and Global Environmental Change - CS4: Population Movement in Response to Climate-Related Hazards in Bangladesh: "The 'Last Resort". Government Office for Science, London.

Pethick, J., Orford, J.D., December 2013. Rapid rise in effective sea-level in southwestern Bangladesh: its causes and contemporary rates. Global and Planetary Change 111, 237–245.

Rahman, M.H., Noor, M.A., Ahmed, A., 2011. Climate change impacts and adaptation strategies for Bangladesh. In: Hossain, M., Selvanathan, E. (Eds.), Climate Change and Growth in Asia. Edward Elgar, Cheltenham, UK, pp. 129–146.

Rajan, S.C., 2008. Climate Migrants in South Asia: Estimates and Solutions, a Report by Greenpeace. Greenpeace India, Bangalore, India.

Rashid, H., 2011. Interpreting cyclone disasters in Bangladesh and Myanmar from web-based newspaper discourse: media framing of cyclone vulnerability on the Bay of Bengal coast. The Arab World Geographer 14 (1), 1–32.

Rashid, H., Paul, B., 2014. Climate Change in Bangladesh: Confronting Impending Disasters. Lexington Books, Lanham, MD, Boulder, CO, New York, Toronto, Plymouth, UK.

Rasid, H., 2000. Reducing vulnerability to flood disasters in Bangladesh: compatibility of floodplain residents' preferences for flood alleviation measures with indigenous adjustments to floods. In: Parker, D.J. (Ed.), Floods, vol. II. Routledge, London and New York, pp. 46–65.

Rizvi, A.I.H., 1955. Comparative Physiography of the Lower Ganges and Lower Mississippi Valleys (Ph.D. dissertation). Louisiana State University, Department of Geography, Baton Rouge, LA.

Rob, M.A., 2008. Changing morphology of the coastal region of Ganges delta. In: Amin, S.M.N. (Ed.), Studies on Coastal Environments in Bangladesh. A.H. Development Publishing House, Dhaka, pp. 54–69.

Simon, K.M., James, T.S., Forbes, D.L., Telka, A.L., Dyke, A.S., Henton, J.A., 2014. A relative sea-level history for Arviat, Nunavut, and implications for Laurentide ice sheet thickness west of Hudson Bay. Quaternary Research 82 (1), 186–197.

Singh, O.P., 2001. Cause-effect relationships between sea surface temperature, precipitation and sea level along the Bangladesh coast. Theoretical and Applied Climatology 68 (3), 233–243.

Streetfield, P.K., Karar, Z.A., 2008. Population challenges for Bangladesh in coming decades. Journal of Health, Population and Nutrition 26 (3), 261–272.

Thywissen, K., 2006. Components of Risk: A Comparative Glossary. United Nations University, Bonn.

Uddin, M.N., 2009. Pleistocene-Holocene Sedimentation History in Relation to the Development of the Aquifer System in the Central Part of Bengal Basin (M.Sc. thesis). University of Dhaka, Department of Geology, Dhaka (Cited in Khalequzzaman, 2012).

Umitsu, M., 1993. Late quaternary sedimentary environments and landforms in Ganges delta. Sedimentary Geology 83, 117–186.

Unnikrishnan, A.S., Shankar, D., 2007. Are sea-level-rise trends along the coasts of the north Indian ocean consistent with global estimates? Global Planetary Change 57 (3–4), 301–307.

Walsham, M., 2010. Assessing the Evidence: Environment, Climate Change and Migration in Bangladesh. International Organization for Migration, Dhaka, Bangladesh.

World Bank, 2009. World Development Report 2010: Development and Climate Change. World Bank, Washington, DC.

World Bank, 2000. Bangladesh: Climate Change & Sustainable Development. Report No. 21104. World Bank, Dhaka, Bangladesh.

CHAPTER FOUR

Coastal Landform Changes: Coastal Erosion, Land Accretion and Subsidence

BROADER PERSPECTIVES

Background

The coastal environment around the world is made up of a wide variety of landforms manifested in a spectrum of sizes and shapes ranging from gently sloping beaches to high cliffs and from bars to off-shore islands. Coastal landforms include both depositional (e.g., sand spits, dunes, shore platforms, tombolo, and cuspate forelands) and erosional (e.g., sea cliffs, sea arches, and sea stacks) features. As one of the most dynamic environments, coastal landforms' constant changes are the result of both human activities and physical processes (Masselink and Gehrels, 2014). Over 4 billion people worldwide now live on the coast. With rapid coastal population growth, human impact on coastal regions increased dramatically during the 20th century, and this trend will continue through the 21st century. Human interventions have led to widespread conversion of natural coastal landscapes to agriculture, aquaculture, and silviculture, as well as to industrial and residential uses (Valiela, 2006).

About 23% of the world's population currently lives within 67 miles (100 km) of the coast (Masselink and Gehrels, 2014). In addition, 21 of the 33 megacities can be considered coastal cities (Martinez et al., 2007).[1] The attractiveness of the coastal resources has resulted in disproportionately rapid expansion of economic activity, settlements, urban centers, and tourist resorts. Migration of people to coastal regions is common in both developed and developing nations. Exploitation of coastal resources has led, in many cases, to severely damaged coastal ecosystems and depleted natural resources. Environmentalists suspect that the direct impacts of human activities on the coastal zone have been more significant over the past century than impacts that can be directly attributed to observed climate change (Lotze et al., 2006).

[1] Cities with more than eight million people are called megacities.

Climatic Hazards in Coastal Bangladesh
ISBN 978-0-12-805276-1
http://dx.doi.org/10.1016/B978-0-12-805276-1.00004-1

The other major direct impacts of human interventions on the coastal zones include drainage of coastal wetlands, dune alterations, highway construction, deforestation and reclamation, and discharge of sewage and contaminants into coastal waters. Extractive activities include sand mining and hydrocarbon production, harvests of fisheries and other living resources, introductions of invasive species, and construction of seawalls. Other engineering structures, such as damming, channelization, and diversions of coastal waterways, harden the coast; change circulation patterns; and alter freshwater, sediment, and nutrient delivery.

Among other things, tropical cyclones, storm surges, tsunamis, tidal activities, longshore and rip currents, wind and wave actions, sediment supply, sediment type, tidal inlet dynamics, and morphological feedback are important components of physical processes (Cooper and Pilkey, 2004). Climate and geology are also important factors in the development of and change in coastal landforms. Among climatic factors, rainfall is important because it provides runoff and also is a factor in producing and transporting sediments from inland to the coast. The amount of sediments transported to the coast determines, to a large extent, rate of deposition and erosion. The former occurs when more sediment enters a coastal region than exists. This, in turn, causes the advance of coastline. If the reverse is true when erosion occurs, the coastline retreats (Masselink and Gehrels, 2014).

Similar to rainfall, temperature is an important driver of the physical weathering of sediments and rocks along the coast. This is particularly significant in cold regions where the freezing of water within cracks in rocks causes the rocks to fragment and thereby yield sediments. The increase in sea-surface temperature causes coral beaching. Wind is important primarily because of its relationship to waves. The physical processes are believed to be intensified as climate change progresses over time; climate change will cause permanent inundation of parts of existing coastal areas due to sea level rise (SLR). As a consequence, the high water line will move landward in proportion to the slope of the coastal area. Coastal landforms will also be severely affected by climate change because of an increase of frequency and magnitude of several climatic hazards, such as cyclones and associated storm surges, and coastal flooding. Climate change is also likely to affect rainfall and climate patterns, potentially intensifying and changing seasonal patterns of storms, both resulting in coastal erosion and flooding (Switzer, 2014).

Climate change researchers (e.g., Trenberth et al., 2007; Bindoff et al., 2007) have observed several important climate change-related effects on all coastal zones of the world. Rising carbon dioxide concentrations have produced ocean acidification. Recent trend analyses indicate that tropical cyclones have increased in intensity. Global sea levels rose at 1.7 ± 0.5 mm/year since 1900, while global mean sea surface temperatures have risen about 0.6°C since 1950, with associated atmospheric warming in coastal areas (Zhang et al., 2004; Gibbons and Nicholls, 2006; Bindoff et al., 2007; Nicholls et al., 2007). During the 20th century, sea level rose by 0.2 m; currently, sea level is rising at an accelerated rate and is likely to rise about 1 m in the next 100 years (Masselink and Gehrels, 2014). Table 4.1 shows the effects of climate-driven changes to the coastal region, while the broad factors that cause changes in the morphology of coasts are diagrammatically shown in Fig. 4.1.

Given the actual and potential impacts, the ability to forecast confidently the consequences of the climate change to coastal areas is of paramount importance for two reasons: to design and implement mitigation measures to counter any adverse changes and also to help reduce coastal vulnerability to environmental changes through planning. Although difficult, several models, ranging from very simple to complex ones, are available to predict coastal changes. These models include those based on description of

Table 4.1 Climate Drivers and Their Effects on Coast

Climate Driver	Trend	Effects
Carbon dioxide concentration	Rising since 1750	Ocean acidification
Sea-surface temperature	Rising since 1950	Circulation changes, sea-ice reduction, coral bleaching and mortality, species migration, and algal blooms
Sea level	Rising since 1900	Flooding, erosion, saltwater intrusion, rising groundwater table, and impeded drainage
Storm intensity	Rising	Erosion, saltwater intrusion, and coastal flooding
Storm frequency	Uncertain	Altered storm surges and storm waves
Run-off	Variable	Alterations in flood risk, water quality, fluvial sediment supply, and circulation and nutrient supply

Adapted from Nicholls, R.J., Wong, P.P., Burkett, V.R., Codignotto, J.O., Hay, J.E., McLean, R.F., Ragoonaden, S., Woodroffe, C.D., 2007. Coastal systems and low-lying areas. In: Parry, M.L., Canziani, O.F., Palutikof, J.P., van der Linden, P.J., Hanson, C.E. (Eds.), Climate Change 2007: Impacts, Adaptation and Vulnerability. Contribution of Working Group II to the Fourth Assessment Report of the Intergovernmental Panel on Climate Change. Cambridge University Press, Cambridge, UK, pp. 315–356.

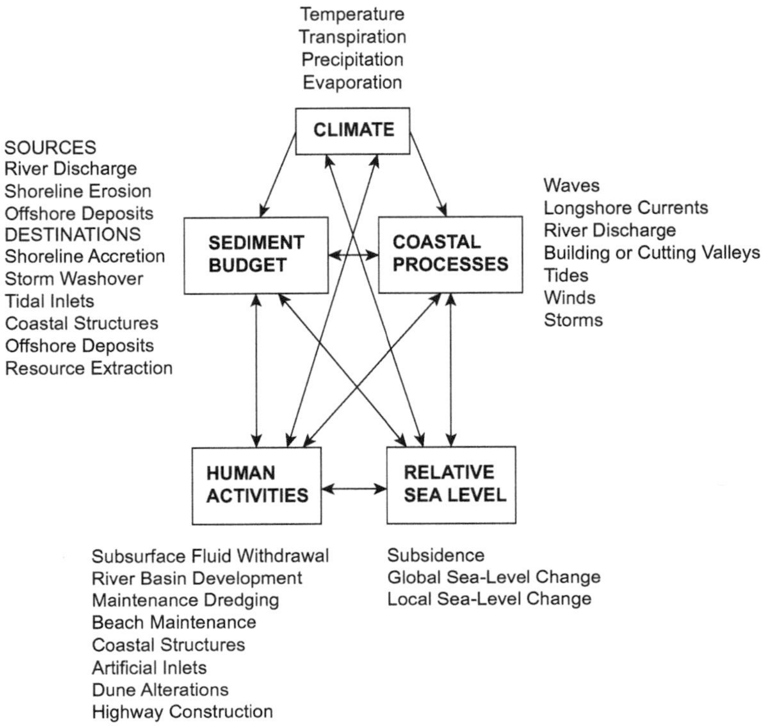

Figure 4.1 The factors that interact to affect coastal environment. *After Williams, S.J., 2001. Coastal erosion and land loss around the United States: strategies to manage and protect coastal resources – examples from Louisiana. In: Coastal Ecosystems and Federal Activities Technical Training Symposium Proceedings Held in Gulf Shores State Park, Alabama, August 20–22, p. 2.*

processes (conceptual models), those that quantify processes (empirical models) and behavior (behavior-oriented models), those that include morphodynamic feedback (process-based morphodynamic models), and those that are a scaled-down version of the real world (physical models). A description of these models is beyond the scope of this chapter, but is included in the introductory chapter of the book edited by Masselink and Gehrels (2014).

Sea Level Rise and Erosion

The two most important consequences of climate change are SLR and increased storminess, both resulting in coastal erosion and flooding. The rise in sea level has both indirect (through dynamic processes) and direct effect on coastal erosion. With a rise in water level, the coastal morphological system will adjust itself to the high water level situations by creating a new coastal profile at a level higher than the presently existing coastal profile. This process, as described by Vellinga (2006), is as follows:

- Coastal water level will move landward,
- SLR will make the coastal profile steeper and erosion will continue until a new dynamic equilibrium is researched at a higher level,
- The natural filling rate of lagoons and tidal basins will increase with an accelerated rise in sea level; the sediments required for the filling will come largely from the upper profile through erosion, and
- Rising sea level will cause a shoaling effect in rivers, leading to reduced sediment yields from rivers; these sediments will not be available to compensate for any erosion in the coastal region.

The effort to predict shoreline behavior related to sea level changes has received less attention. However, a model known as the "Bruun Rule" has been widely used to understand the shoreline response to rising sea level. This is a simple two-dimensional generic geometric model and it has near global application since its original formulation in 1954 (Cooper and Pilkey, 2004). Bruun (1954) maintained that, when considered in the shore-normal dimension, the nearshore zone existed in a profile of equilibrium on the basis of cross-shore profiles in Denmark and California. The profile could be described by the following equation:

$$h = Ay^{2/3} \tag{4.1}$$

where h is water depth, y is the distance offshore, and A is a scaling parameter based on sediment characteristics.

Bruun (1962) concluded that the equilibrium profile would remain unchanged as the shoreline changes in response to a rising sea level. In this model he envisioned that sand was moved from the upper part of the beach profile to deposit on the lower part of the profile. The amount of sediments eroded will be equal to the amount of sediments deposited, assuming that eroded sediments are not lost into the deep sea. A simplified form of Bruun's model is shown in Fig. 4.2. The figure depicts the present coastal profile with respect to the present sea level. Rise in sea level will result in coastal recession because waterline will move landward due to submergence. With coastal recession, erosion will occur until a dynamic equilibrium is reached and a new coastal profile is formed (Islam et al., 1999). In Fig. 4.2, the new coastal profile under the future SLR situation is shown in a dotted line, while the present profile in a continuous line. The eroded and deposited areas are differently shaded in the figure.

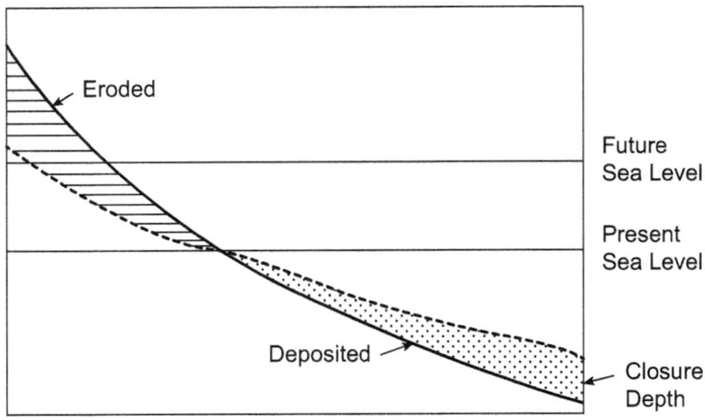

Figure 4.2 The Bruun Rule of shoreline erosion and deposition.

It is worth mentioning that Bruun model apparently works well with beaches on sandy coasts. He further claimed that eroded sand would be held in place by "sediment fence," called the closure depth (Fig. 4.2). The concept is based on the assumption that sand transport on the shoreface occurs solely through the interaction of wave orbitals and sand on the sea floor. At the point of closure, the water is presumed to be sufficiently deep that sediment transport by waves is negligible. Other limiting conditions for the closure depth concept were that no net longshore transport of sediment in or out of the shoreline reaches under consideration and there would be no significant loss to dunes or washover fans in a landward direction (Cooper and Pilkey, 2004).

Several authors (e.g., Swift, 1976; Dean, 1977; Pilkey et al., 1993) modified the concept of the shoreline profile of equilibrium by examining a number of profiles along the US East and Gulf coasts. Clearly it is much more complex than originally proposed by Bruun. Besides wave interaction with the seafloor, many other processes are involved in shaping the shorelines. These processes include wind- and wave-induced upwelling and downwelling, gravity flows such as turbidity currents, tide- and wind-induced currents and interactions between waves, and bottom currents of many origins. However, the major contribution of Bruun was the recognition that shoreline response to SLR was not a simple retreat of a line in the sand but rather was a response of the entire shoreface.

Bruun provided a conceptual basis for several advances that followed. As this understanding of shoreface processes has evolved, the Bruun Rule and its accompanying shoreface profile of equilibrium concept have been utilized in applied coastal studies. These studies form the basis of a considerable number of numerical mathematical models used to predict beach behavior, especially in the US (Thieler et al., 2000), and the Bruun Rule is the basis of theoretical models of coastal response to SLR (Cowell et al., 1995).

Despite its initial breakthrough in identifying the entire shoreface as responsive to SLR, the subsequent advances in scientific understanding have shown the Bruun formulation to be an inadequate descriptor of shoreface response to SLR. The persistence of application of Bruun model is attributed to: (1) appeal of a simple, easy to use analytical model, (2) positive advocacy by some scientists, (3) the simple numerical expression of the model, (4) exclusion of many important variables, (5) restrictive assumptions, and (6) lack of easy alternatives. The Bruun model has no power for predicting shoreline behavior under rising sea level. Because of its lack of predictive power and restrictive assumptions, several researchers (e.g., Cooper and Pilkey, 2004; Zhang et al., 2004) suggest not to use this model.

BANGLADESH CONTEXT

Like all other coasts, the coastal region of Bangladesh is subjected to erosion, deposition, and land subsidence. The Ganges, Brahmaputra, and Meghna (GBM) river system carries about 2.4 billion tons of sediments per year into the Bay of Bengal that interact with dynamic processes in the eastern Bengal shelf, a continental shelf off the coast of Bangladesh leading to an erosion-accretion process in the coastal region (Davies et al., 2003; Kottke et al., 2003; Michels et al., 2003; Subrahmanyam et al., 2008). The bottom topography of the northern part of the Bay of Bengal also contributes to these processes. Tide- and storm-driven processes distribute sediments to major depositional areas on the inner shelf near the Meghna estuary; additionally, some sediments are carried by under currents into the deeper Bay of Bengal and the Indian Ocean (Islam et al., 1999; Michels et al., 2003). The bottom topography of the inner shelf is characterized by a broad U-shaped basin with its south opening to the Indian Ocean. A thick uniform abyssal plain occupies almost the entire Bay of Bengal, gently sloping southward at an angle of 8–10 degree. In many places underwater valleys dissect this plain mass.

The width of the Bengal shelf off the coast of Bangladesh varies considerably. It is less than 66 miles (100 km) off the south coast between Hiron

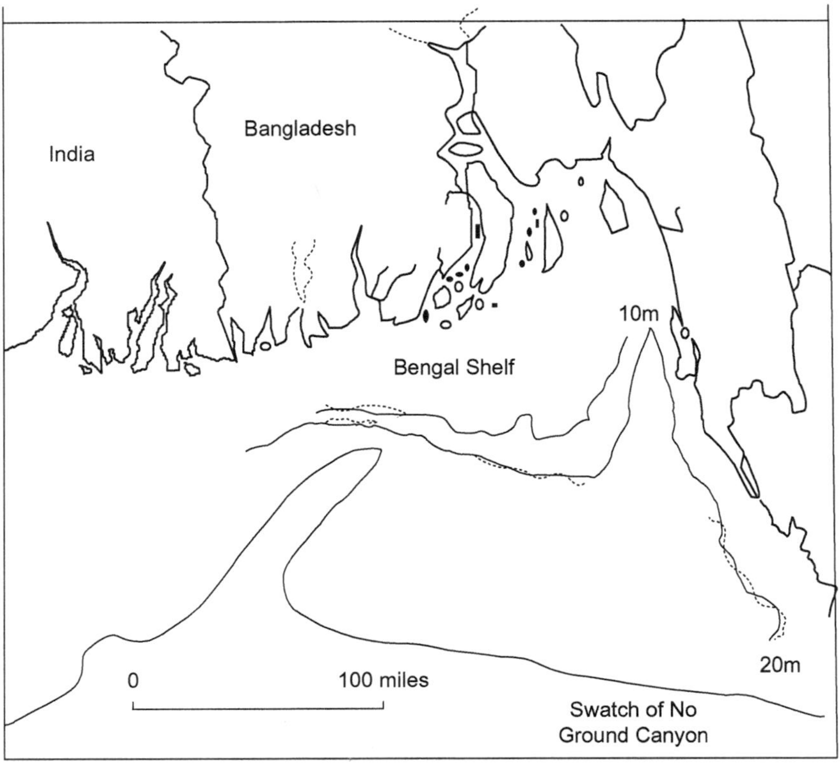

Figure 4.3 Northernmost part of the Bay of Bengal showing the canyon Swatch of No Ground.

Point and the Swatch of No Ground—a submarine canyon, also known as Ganga Through—while more than 166 miles (250 km) off the coast of Cox's Bazar (Fig. 4.3).[2] This canyon controls to a great extent the flow dynamics of the northern Bay of Bengal. It divides the shelf into a western and eastern part, and deeply incises into the shelf about 66 miles (100 km) west of the Meghna river mouth. The inner shelf receives huge amount of sediments, while the outer shelf is sediment starved. Sediments are fine seaward and westward with the thickest accumulation of mud near the submarine canyon. The shallow part (less than 20 m) of the continental

[2] Swatch of No Ground has a comparatively flat floor 5–7 km wide and walls of about 12 degree inclination. It runs NE–SW about 16 miles (24 km) of the coast, partially across the continental shelf. At the edge of the shelf, depths in the trough are about 1200 m. The Swatch of No Ground has a seaward continuation for almost 1333 miles (2000 km) down the Bay of Bengal in the form of fan valleys with levees.

shelf off the eastern coast is covered by sand and the intertidal areas show well-developed sandy beaches. The shallower part of the southern continental shelf off the western and central coast zones is covered by silt and clay, and extensive muddy tidal flats are developed along the shoreline. Some of the shoals and sand ridges present on this part of the continental shelf show an elongation pattern pointed towards the Swatch of No Ground.

The sandbars and ridges near the mouth of the Meghna delta pointing toward the Swatch of No Ground showing sediments are tunneled through this trough into the deeper part of the Bay of Bengal. The Swatch of No Ground feeds the Bengal Deep Sea Fan by turbidity currents.[3] In most of the northern part of the Bay of Bengal, particularly near the Meghna estuary, the water is shallow, causing storm surges and tides. The GBM discharges water and sediment load primarily through this estuary on the eastern shelf and forms the most important sediment source for the inner shelf, the sub-aqueous delta front, and the Swatch of No Ground (Michels et al., 2003). The bottom topography greatly influences the circulation pattern in the Bay of Bengal (Islam et al., 1999).

Apart from the seasonally reversing monsoon winds, the springtime western boundary current (WBC) also affects the circulation of the Bay of Bengal. The WBC's continuous northward-flowing current runs almost in parallel to the west coast from about 12°N to 17°N, which then deflects eastward at the head Bay at around 18°N (Gangopadhyay et al., 2013). Here it breaks into cyclonic and anticyclonic gyres. These gyres may be partly responsible for balancing the northward mass transport by the WBC by a compensating southward transport through the eastern part of the Bay of Bengal (Ali, 1991). There may be some under currents also being generated by the Swatch of No Ground (Curray and Moore, 1974). A persistent warm water zone is also observed along the Chittagong-Myanmar coast. The Bay dynamics is also controlled, particularly at the head Bay, by the river discharge through the GBM river system. As is apparent in satellite imagery, this discharge breaks into two branches: one moves south-westward and seems to be governed by the Swatch as

[3] Bengal Deep Sea Fan is the world's largest submarine fan, also known as Bengal Fan. Together with its eastern lobe, the Nicobar fan, it covers an area of 3106 km^2. It is 2800 to 3000 km long, 830 to 1430 km wide, and more than 16 km thick beneath the northern Bay of Bengal (Davies et al., 2003; Subrahmanyam et al., 2008). Sediments are tunneled to the fan via the Swatch of No Ground. It can be divided into three parts: upper fan, middle fan, and lower fan.

well as the Coriolis force, while the other component flows southward along the Chittagong coast (Ali, 1995).

The coastal landform changes caused by processes discussed above are taking place at different rates within the coastal region of Bangladesh. For example, rates of changes along the Meghna estuary or central coastal zone are much higher than the rest of the coastal region. Landform changes both in the western and eastern coastal zones are generally small or undetectable relative to changes in the central coastal zone (Brammer, 2014a; Allison, 1998). Both erosion and accretion rates are very high in the latter zone.

The coastal area along the Meghna estuary is morphologically one of the most dynamic areas in the world. The enormous sediment loads brought to the estuary by the Ganges and the Brahmaputra rivers and very strong flows in the estuary drive the continuous reshaping of the estuary, migration of channels and banks, and accretion and erosion. The average annual sediment load carried by the rivers to the Bay of Bengal is around 2.4 billion tons. The Ganges and the Brahmaputra are heavily laden with fine sediments. The Ganges carries fine sediments with a heavy clay load, whereas the Brahmaputra particularly transport fine sand and silt in suspension. The Meghna appears to be a relatively low sediment laden river (Hofer and Messerli, 2006).

Land erosion and accretion are common natural phenomena in the coastal region. Both of these phenomena generate socioeconomic anomalies and issues in the coastal region of Bangladesh. While erosion leads to the displacement of people, land capture and illegal settlement are prevalent in areas where the coast is accreting. Contrary to the prevalent notion that Bangladesh experiences a net loss of land every year due to coastal erosion, available data suggest that the country is gaining land, meaning the amount of accretion is higher than the amount of erosion. However, in some areas, such as Bhola Island, the overall rate of erosion is higher than the rate of accretion. The reverse is the case for Noakhali, located in the Meghna estuary.

As indicated, landforms in the coastal region of Bangladesh have been changing due to erosion, accretion, and subsidence. The next two sections cover these three geomorphic processes, and a conclusion section follows.

Erosion and Accretion

Land erosion and accretion are regular natural phenomena in the coastal Bangladesh. While land is eroding one side, new land is accreting on the other side almost simultaneously. Often, continuous sediment accumulation and erosion is observed in the same area. For these reasons, erosion and accretion processes are considered in the same section. Heavy discharge through the GBM river system and associated high discharge currents; wave actions due to strong southwest monsoon winds, high astronomical tides, tropical cyclones and associated storm surges; and sea level variations are the main drivers of erosion and accretion in the coastal areas of Bangladesh. The GBM system drains about 85% of the total volume of water brought into Bangladesh and combined, discharge on the average 35,000 m^3/s within the Lower Meghna (the part downstream of the confluence near Chandpur). There are distinct seasonal fluctuations in flow with extreme discharge in the monsoon period. The peak monthly discharge in August–September is about four times the mean annual while the lean discharge in February–April is about a quarter of the mean annual (MoWR, 2001).

Over the last two centuries, massive changes in the coastal morphology have taken place due to continuous land erosion and accretion along the coastline. In addition to natural causes, human activities along the coast (e.g., land reclamation, port development, salt production, and shrimp farming), offshore islands, and within the river catchments (e.g., construction of dam and dredging) also play an important role in causing erosion and deposition. As indicated, erosion causes loss of both habitable and cultivable land and thus the affected people experience huge economic losses. Other effects of erosion on the coast include loss of biodiversity for increasing turbidity, degradation of water quality, reduction of natural regeneration of mangrove swamps, and siltation in the rivers and their vicinity. However, the extent of erosion and accretion differs within the coastal region. Severe land erosion and accretion have been observed in off-shore islands as well as along the shoreline, particularly in the Meghna estuary. These geomorphological processes adversely affect the ecosystem, navigation, agricultural production, and drainage systems (MoA, 2012). It also affects inland navigational routes in terms of shifting and migration of channels.

Although the available records are scanty, maps drawn by Rennel and satellite images clearly show the changes that have been taking place in the recent past. Human interventions are increasingly playing a leading role in such changes.[4] The morphological dynamics in the coastal Bangladesh have changed due to two structural alterations (construction

[4] James Rennel (1742–1830), a British army major, worked as a surveyor-general of India and prepared several maps of India, including the 1776 map of Bengal and Bihar. His maps were based on land surveys and during his tenure he surveyed an area of about 300,000 mi^2 (780,000 km^2), stretching from the eastern boundaries of Lower Bengal to Agra, and from the Himalayas to the borders of Bundelkhand and Chota Nagpur.

of embankments and cross-dams) and for this reason these two structures are discussed below.

Coastal Embankments

In Bangladesh the construction of coastal embankments started as early as the 17th century on private initiative under the patronage of local land lords (*zamindars*). Systematic development of large-scale embankments for flood control and saline water intrusion in the coastal area started in the early 1960s under the Coastal Embankment Project (CEP) (Ahmed, 2008). The main objective of the construction was to protect the coastal area from tidal inundation. Two other objectives were to (1) increase agricultural production by blocking intrusion of saline water and (2) save lives from floods and storm surges associated with tropical cyclones. However, construction of the embankments resulted in a major morphological change in the coastal region, particularly in the western coastal zone, and disturbed the overall equilibrium (SWMC and BWDB, 2000). There have been reductions in the tidal volume resulting in the reduction in peak tidal discharge velocity. As the siltation in the tidal rivers, small channels, and creeks continues, the bed level and tide level in these water bodies have risen above the surrounding ground levels in many areas. This has caused major drainage congestion and water logging problems (SWMC and BWDB, 2000).

In addition to siltation and water logging, embankments are now gripped in second generation problems. Construction of embankments has caused subsidence inside these structures, thus further causing drainage congestion. Poor maintenance and breach of embankments by powerful shrimp farmers to bring saline water within the polders are making embankments structurally weak (MoA, 2012). Many embankments are in dilapidated conditions in terms of breach slip, erosion, lack of repair, siltation and encroachment of canals and, above all, location in the risk zone. Some of these embankments have been refurbished through a number of projects in the recent past. With the increase of Sea Surface Temperature (SST) in the Bay of Bengal, the waves are damaging the structures and exerting increased energy, which is leading to frequent embankment failures and breaches. For more details on embankment see Chapter 8.

Cross-Dams

Land reclamation by cross-dam construction started in coastal Bangladesh in 1957, with a 9.3 mile (14 km) cross-dam being built in the eastern branch of the Meghna River for the purpose of reclaiming land for agriculture. Cross-dams are closure structures between islands and the mainland, which are meant to accelerate the natural accretion process. A second 10.7-mile (16 km) cross-dam was built in 1964 in the same river branch, after two earlier unsuccessful attempts (World Bank, 2012). Although very slow, the natural process of accretion is giving rise to new *chars* (bars or islands) along major portions of the 475-mile (710 km) long Bangladesh coast. As noted, 2.4 billion tons of sediment passes to the Bay of Bengal each year through the Meghna estuary. Holding of the passing sediments will give rise to about 77 mi^2 (200 km^2) of land each year.

Realizing the huge potential for land reclamation, the Bangladesh Government has been pursuing a number of projects in the coastal areas of Bangladesh since 1977 to address the problems of coastal floods, tidal surges, and erosion and accretion of coastal areas. Land Reclamation Project (LRP), Meghna Estuary Studies (MES), Coastal Embankment Rehabilitation Project (CERP), Delta Development Project (DDP), Char Development and Settlement Projects (CDSP), Integrated Coastal Zone Management Project (ICZMP), and Estuary Development Programme (EDP) are some of the coastal initiatives undertaken by the government (Alam and Uddin, 2013).

The Bangladesh Government has a plan to construct more cross-dams in the Meghna estuary. In 2013, the government initiated a project to build the Urir Char—Noakhali Cross Dam under the climate change resilient fund of the World Bank. This structure extends from the Sandwip Island, situated in the north-east corner of the Bay of Bengal, via the Urir Char, to the Noakhali mainland (World Bank, 2012). The plans to connect Sandwip with the mainland date from 1964. When this project is completed, 24,710 acres (10,000 ha) of land could be reclaimed in seven years (Hasan, 2014).

Since the mid-1970s, the Bangladesh Water Development Board (BWDB) completed a number of feasibility studies under different projects such as the Netherlands-funded Land Reclamation Project (1977—1991), the Meghna Estuary Study (1995—2002), and the Estuary Development Program (2007—2011). These studies made a number of recommendations including commissioning cross-dams for land mass reclamation in the Meghna estuary and connecting Sandwip Island and other chars to the main land (Fig. 4.4). These dams could reclaim about 386 mi^2 (1000 km^2) in 30—35 years (Hasan, 2014). Currently, there is no plan to construct cross-dams beyond Meghna estuary. However, the existing cross-dams

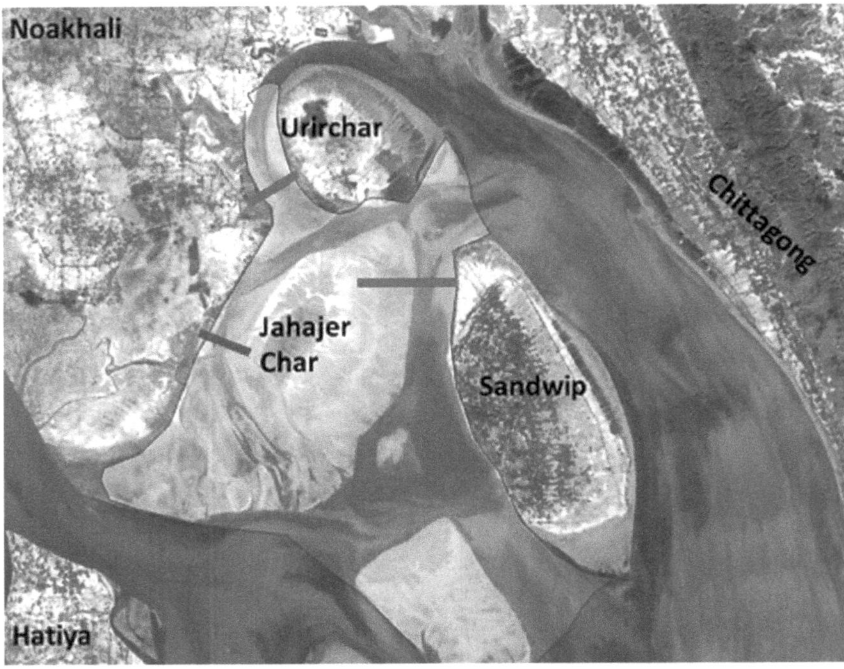

Figure 4.4 Location of cross-dams along Sandwip, Urir Char, and Jahajer Char.

have been affecting local level drainage due to constriction and choking of rivers and rivulets (*khals* or canal) (Ahmed, 2008).

Riverbank Erosion and SLR

As indicated, SLR will also lead to severe coastal erosion in the foreseeable future in Bangladesh (Islam et al., 1999). As a part of climate change impacts, bank erosion of the downstream section of the GBM river system has also increased in recent years due to enhanced hydrological processes (more rainfall, frequent floods, and higher river levels). The two major rivers of the system joined together near Chandpur, a city located close to central coastal zone. The combined river empties into the Bay of Bengal in Bhola district via four principal mounts, Tetulia, Shahbazpur, Hatiya, and Bamni. The combined course from Chandpur to the Bay of Bengal is called lower Meghna and it braids into a number of little rivers including the Pagli, Katalia, Dhonagoda, Matlab, and Udhamodi. All of these rivers flow out from the Meghna and rejoin again at points downstream.

In Bangladesh, riverbank erosion has been a recurrent feature for almost all inland rivers. In the early 1990s, such erosion was reported for 94 of the

462 inland upazilas in about 50 districts.[5] Bank erosion was considered severe and most recurrent in 35 of the 94 upazilas (Elahi, 1991). According to the World Disaster Report published in 2001, 1 million people in Bangladesh were displaced annually and 9000 ha of land were lost by river erosion at the beginning of this century (IFRC and RCS, 2001). With continuation of riverbank erosion to coastal areas, the number of upazilas affected by such erosion has increased. Where the Ganges and Brahmaputra enter Bangladesh, both assume a braided pattern consisting of several channels separated by small islands/sandbars, locally called *chars*. During the last 200 years or more, the river channels have been constantly changing, leading to both erosion and accretion. The former causes loss of land and displaces many residents permanently who live close to the eroded river banks. In contrast to this, new lands generally emerged simultaneously inside the channels not far from the eroded banks. In other words, land that is lost due to bank erosion on one side is gained on the other side (Hofer and Messerli, 2006). However, accessibility to accreted lands by the poor who were displaced by riverbank erosion is limited. Large land owners and local politicians generally gain control of accreted char lands, often through violence.

Inland riverbank erosion occurs throughout the year, with intensification during the monsoon season when heavy rainfall occurs in Bangladesh. During this season, rivers carry huge amounts of water, which often overflows the banks and causes flooding. The bank erosion generally occurs gradually and mostly in a rather unspectacular manner. The river waves regularly hit the river bank, thereby undercutting the loose silt layers. From time to time, a major portion of the river bank collapses and disappears into the river (Hofer and Messerli, 2006).

Although riverbank erosion has been primarily confined to the inland areas, in recent years the process has continued to affect the rivers near the estuary. For this reason, there are few studies available on the extent and impact of such erosion of coastal rivers. It is claimed that during the last three decades the Jamuna, Brahmaputra, and Ganges (called Padma in Bangladesh) rivers combined have eroded 180,000 hectares in land areas. The corresponding figure for the coastal rivers is estimated at about 22,500 hectares (BWDB, 2013). Like their inner counterparts,

[5] An upazila is the lowest administrative unit in Bangladesh. A district is the second largest administrative unit with an average population of about 2.5 million. A district is comprised of 10 or more upazilas.

bank erosion of coastal rivers also destroys cultivable land, dislocates human settlements, and at the same time damages the growing crops and massively disrupts road-linkages and communication infrastructure in the region. It is worth noting that river bank erosion has been a regular phenomenon in off-shore islands, particularly Bhola, Hatiya, Sandwip, and Kutubdia.

A huge amount of water and sediments discharges into the Bay of Bengal through the Meghna estuary. The SLR will push the coastline as well as the river mouth landward and is likely to modify the discharge current and hence the erosion. The SLR will cause a steeper coastal profile; as a consequence the gradient flow is likely to increase, which will further lead to coastal erosion. Additionally, the inundation due to SLR may increase the nonlinear effect and hence increase the tidal range, thereby increasing the erosion effect. Activity of the tropical cyclones and associated storm surges will intensify with climate change. This will aggravate the erosion problem in the coastal Bangladesh.

Several researchers examined as well as compared shoreline changes both by variability of sediment supply and changes in SLR. Using Bruun formula, Ali and Ahmad (1992) studied the loss of land due to SLR in the western coastline of Bangladesh. They used the following formula to calculate a shoreline recession through erosion:

$$x(e+h) = ab \tag{4.2}$$

or

$$x = ab/(e+h) \tag{4.3}$$

If the values of the parameters on the right hand side of second equation are known then the erosion due to SLR can be calculated. All the parameters used in the above two equations are explained below:

a = rise in water level due to SLR

b = shelf width before SLR

x = shoreline recession

e = increase in elevation of shore after SLR

h = elevation of the shore before SLR

$x(e+h)$ = the amount of eroded sediments after SLR

Using $e = 1.0$ m, $h = 20$ m, $b = 60{,}000$ m (typical to the western coastal zone of Bangladesh), and $a = 1.0$ cm/year, Ali and Ahmad (1992) found that the formula gives a shoreline recession through erosion of about 2.9 m/year. This value will be larger in Meghna estuary where b is larger. The value of x for the western coastal zone is about 3000 times the SLR.

This figure seems to be very large compared to the figures from Florida coast and the coasts of Belgium and Denmark. For this reason Ali and Ahmad (1992) concluded that Bruun's formula is not applicable to Bangladesh.

However, the conclusion by Ali and Ahmad (1992) was based on the rough estimate for the west coast of Bangladesh where the continental shelf is very wide and gently sloping. In fact, Bruun's formula is valid and more applicable for a coast with a steep bottom profile. It is also more applicable for a sandy shore. Such a situation exists on the east coast of Bangladesh, which contains several beaches.[6] Considering this point Islam et al. (1999) applied Bruun's formula for the eastern sandy coast of Bangladesh. They calculated "b" and "h" from the Admiralty Charts and elevation "e" from the profiles. The recession distance "x" was calculated for three values (30 cm, 75 cm, and 1.0 m) of SLR. They selected six points along the east coastal zone. They found that, on the average, a recession of 0.87 cm occurs per 1 cm rise in sea level, meaning recession distance through erosion due to SLR is 87 times the SLR. They further reported that the recession rate is more or less uniform throughout the coastal area under study.

Islam et al. (1999) claimed that agricultural land will experience the largest loss due to SLR in their study area. They calculated that about 5.8 km^2 and 11.2 km^2 will be lost due to SLR of 30 cm and 75 cm, respectively. The corresponding loss of agricultural land will be about 5.5 km^2 (94%) and 10.8 km^2 (96%). Bruun's model, as applied by Islam et al. (1999) for the sandy eastern coast of Bangladesh, seems to work better than when applied by Ali and Ahmad (1992) and Ali (1995) to the western coast.

Erosion-Accretion Balance

Although scattered, several studies have been conducted on the erosion-accretion balance for the Bangladesh coast. Most of these studies reported

[6] The Bangladesh coast contains several beaches of varying sizes. All these beaches are located in the eastern and central zones. The largest of these beaches is Cox's Bazar beach in the eastern coastal zone. It is the world's longest unbroken sandy sea beach (75 miles or 120 km) sloping gently down to the blue water of the Bay of Bengal. It runs from Cox's Bazar in the north to Badar Mokam in the south. The beach, particularly its southern part, is less vulnerable to tropical cyclones and associated storm surges. The Myanmar Trench, a submarine canyon, plays a significant role in the dynamics of this region (Islam et al., 1999). Several other (Patenga and Fouzdarhat beach, Inani beach, and Parki beach) relatively small beaches are located along the eastern coastal zone. The noteworthy rivers that cut across the eastern coast are Karnaphuli, Sangu, Matamuhuri, and Naaf. Kuakata is the only sandy beach located in the western part of the central coastal zone. The beach stretches for 15 miles (23 km) and is about two miles (3 km) wide.

conflicting claims. One of the earliest studies was conducted by Miah (1975) who conducted a qualitative discussion on the erosion-accretion problem in the Meghna estuary covering the period 1779–1975. Similarly, Jabbar (1979) examined this problem by comparing the 1931 Survey of India map and a 1977 Landsat map. He reported that during the study period (1931–1977) a net accretion of 439 km^2 took place in the mainland. He claims that this accretion was the result of the building of a cross-dam for reclamation of land in the Meghna estuary. A net erosion of 291 km^2 was observed in Bhola island and 192 km^2 in Sandwip island. During the study period, a number of small *chars* such as Char Udaukal and Char Clark were completely eroded and several new chars including Char Dhal, Char Shabani, and Nijum Dwip were formed.

Pramanik et al. (1981) examined seven different maps of the coastal areas between 1779 and 1979. They reported that while the islands Hatiya, Sandwip, Shahbazpur, Manpura, and others decreased in size, the mainland gained area due to construction of cross-dams. They further reported that around 1950 Hatiya island got larger, but then divided into two islands, and the northern one joined the mainland. In his second study, Parmanik (1983) found that the coastal region showed a net accretion of land by about 213 km^2 between 1972 and 1979. Erosion was observed in the north-eastern part of Bhola, northern part of Hatiya, and north-western part of Sandwip. Contrary to this finding, an SPARRSO (1987) study reported a loss of about 844 km^2 between 1960 and 1984.

Another study by Pramanik (1988) reported that major erosion occurred in the south-eastern and southern part of Sandwip, the northern part of Hatiya, and the north-eastern part of Bhola between 1972 and 1987. Siddique (1988) also provided erosion rates for two offshore islands. The rate was about 150 m/year between 1940 and 1982 for Hatiya. The Sandwip island had a rate of erosion of about 200 m/year between 1913 and 1963, and increased to 350 m/year during 1963–1984. Another SPARRSO (1993) study reported that 858 km^2 was eroded along the coast between 1976 and 1990, and 808 km^2 was accreted, resulting in a net erosion of 50 km^2. Findings of these early studies on accretion and erosion clearly show neither the accretion nor erosion rates as alarming. Both compensated each other to a reasonable extent. Allison (1998) calculated annual net gains of 5.7 mi^2 (14.8 km^2) between 1792 and 1840 and of 1.7 mi^2 (4.4 km^2) between 1840 and 1984. This historical evidence of net annual land gains in the Meghna estuary suggests that land gain might exceed land loss resulting from the slow rates of SLR projected for the 21st century.

Like the earlier studies, more recent studies on accretion and erosion also provide conflicting claims of erosion-accretion balance. Allison and Kepple (2001) estimated erosion-accretion balance in Bangladesh and concluded that the balance has been tilted towards net erosion. Erosion and accretion rates in the Meghna estuary were calculated by comparing the positions of bank lines digitized from several Landsat satellite images taken during the period 1973 to 2000 (MES, 2001). Erosion/accretion of intertidal areas, i.e., mudflats, is not included in this analysis. The net change for the period shows an overall gain for the estuary of about 196 mi^2 (508 km^2). The average annual gain for the entire study period was 7.3 mi^2 (11 km^2).

A study by Rob (2008) claims that the southeastern land mass of the active delta has migrated 50 miles (80 km) seaward during a period of 200 years (1780–1980) (also see Hossain, 2012). Although there was a net gain of land, there were also considerable land losses in the Meghna estuary. Alam and Uddin (2013) claim that Bangladesh reclaimed 386 mi^2 (1000 km^2) of new land in the Meghna estuary by building two dams in 1957 and 1964. Another study reported that Bangladesh gained 637 mi^2 (1650 km^2) of land in the Meghna estuary between 1973 and 2008 and it had lost 405 mi^2 (1050 km^2) to erosion in the estuary, with a net gain of 232 mi^2 (600 km^2) in the period (Hasan, 2014).

A comparison of Landsat images taken in 1984 and 2007 showed a net land gain of 174 mi^2 (451 km^2) in the Meghna estuary in the period, accounting for an average annual growth rate of 7.6 mi^2 (19.6 km^2) (Brammer, 2014a). Based on remote sensing data, Alam and Uddin (2013) concluded that more accretion than erosion of land occurred in the coastal areas and offshore islands of Bangladesh during the 1977–2010 period (Table 4.2). The results of the morphological changes during the 34-year period in the coastal areas of Bangladesh show a net gain of 54 mi^2

Table 4.2 Land Erosion and Accretion in the Coastal and Offshore Islands of Bangladesh, 1977–2010

Erosion and Accretion (in km^2)	1977–1989	289–2000	2000–2010	Total
Erosion	92	269	130	491
Accretion	359	74	197	630

Adapted from Alam, M.S., Uddin, K., 2013. A study of morphological changes in the coastal areas and offshore islands of Bangladesh using remote sensing. American Journal of Geographic Information System 2 (1), 15–18, p. 17.

($139\ km^2$) of land. Other studies (e.g., Caroagal et al., 2011; Rashid et al., 2011; Morner, 2010) support the above finding. The study of Rashid et al. (2011) reported that the GBM deltaic coast has progressed seaward in the south and southeastward in the early stage of deltaic development, but during 1750–2000 period progression took place towards the south and southwestward.

Erosion-accretion balance favors erosion for almost all off-shore islands of Bangladesh. All these islands are threatened by continuous erosion. Sandwip Island has lost $70\ mi^2$ ($180\ km^2$) in the last 100 years. During the last 40 years, Bhola Island has been decreased to $1312\ mi^2$ ($3400\ km^2$) from $2470\ mi^2$ ($6400\ km^2$) in 1960. The eastern coast of the island has been experiencing severe erosion. If the erosion continues at the current rate, the island will completely disappear over the next four decades. Hatiya has become a much smaller island than it was a few decades ago (MoA, 2012). The island has been reduced from $1000\ km^2$ to only $21\ km^2$ in the past 250 years (Miah et al., 2010). Erosion is observed mainly on the northern part of the coastal islands and accretion on the southern part (Sarwar and Islam, 2013). Rapid erosion and rising sea levels due to climate change and climate variability are increasingly threatening the existence of most of the off-shore islands.

Kutubdia Island, once $97\ mi^2$ ($250\ km^2$), lost around 65% during the last 100 years and more than 60% of its population had to migrate to other areas (MoA, 2012).[7] Using remote sensing (Landsat Multispectral Scanner images of 1972 and 1978, Landsat Thematic Mapper (TM) images of 1989 and 2009, and Landsat 8 images of 2013, and GIS (spatial video and GPS) data), Rahman (2015) recently calculated that Kutubdia island lost $3.5\ mi^2$ ($9\ km^2$) of area due to erosion during the 1972–2013 time period and gained $0.14\ mi^2$ ($0.35\ km^2$) of land due to accretion during the above time period (Fig. 4.5). At the current rate of erosion, Kutubdia Island will literally cease to exist within 30 years, along with dozens of other islands off the coast of Bangladesh (Vidal, 2013).

Implications of Coastal Erosion and Accretion

As noted coastal erosion is common in Bangladesh year round, but it is more severe during the southwest monsoon period (June–October) or any episodic events such as tropical cyclones, storm surges, or sea swells. Results

[7] Kutubdia is an island of eastern coastal zone with a population of 125, 279 in early 2011 (BBS, 2011).

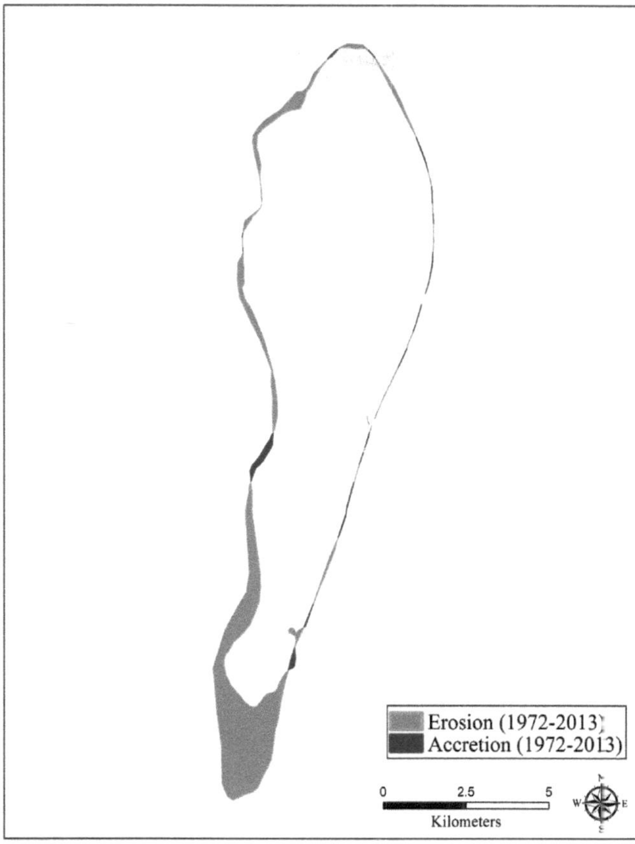

Figure 4.5 Land erosion and accretion in Kutubdia Island. *Reproduced from Rahman, M.K., 2015. Environmental and Social Vulnerabilities and Livelihoods of Fishing Communities of Kutubdia Island, Bangladesh (Ph.D. dissertation). Department of Geography, Kent State University, OH, p. 88.*

of erosion are often devastating and people who live in coastal Bangladesh experience frequent displacement due to loss of habitat. Every year thousands of coastal residents are displaced from their ancestral homes and lose their belongings due to erosion. People are often displaced up to 14 times in their lives. The poorest are the worst victims of erosion because they are unable to relocate themselves following such an event. Unless they are properly rehabilitated, they cannot escape hardship, and physical and emotional suffering. Other effects of erosion include loss of biodiversity for increasing turbidity, degradation of coastal water quality, and siltation in the rivers.

Erosion takes away fertile crop land, the major source of living for the poor and marginal farmers. Embankment failure often inundates crop fields and reduces production substantially. This causes scarcity of food as well as employment opportunities, particularly for poor families. In desperate attempts for survival, one or more members of uprooted families migrate to nearby or distant urban centers. Most of these migrants live in slums and squatter settlements of large cities and they not only face physical insecurity and poor housing, but also live with ill health and constant threat of eviction. Such migration is very common from the offshore islands. For example, Vidal (2013) claims that because of coastal erosion and embankment failure, at least 80,000 people have left Kutubdia Island since 1986. Nearly half of these migrants have been living in the Cox's Bazar neighborhood, a coastal city located 50 miles (75 km) south of the island. This neighborhood is popularly known as Kutubdia Para because most of its residents migrated from the island between 1986 and 1997 to escape land erosion and other extreme natural events (Rahman et al., 2015).

Similar to erosion, tropical cyclones also cause migration from affected areas. Although the magnitude of migration differs from one event to another, some residents of cyclone-affected areas, particularly young males, invariably migrate to nearby or distance cities. It is considered a strategic step to cope with the adverse effects of such a devastating event. Kartiki (2011), who studied migration movement of people in the aftermath of Cyclone *Aila* in Bangladesh, reported that available livelihood opportunities in the destination often encourage household members to leave their place of origin. A survey of 288 survivor households of Aila reveals that the male members of 35% of the sampled households had moved cities immediately after the emergency relief works ended, i.e., four weeks after the cyclone (Mallick and Vogt, 2012).

Whether near or far, temporary or permanent, migration of male family members, particularly head of the household, due to coastal erosion and/or cyclones have serious social implications. Without having the head of household around, the woman suffers from social oppression and physical and sexual harassment. More severe is the fact that sometimes the married migrant man abandons the family, settles down in the new city, and starts a new life there. In such cases, the woman faces greater hardship than ever and she has to bear responsibility for the entire family. Such a woman is often forced to earn money by working as a maid-servant to the local landlords or even begging. Sometimes employers molest women employees, forcing them to be used for sexual desire (Mallick and Vogt, 2012). Additionally, abandoned women find it difficult to arrange marriages for their adolescent daughters (Ahmed, 2008).

Despite the considerable risk of living in the coastal region, the newly accreted lands attract many landless households from noncoastal districts as well as internally displaced poor people of coastal districts (Mallick and Vogt, 2012). However, there are many problems with the new lands, including their instability due to less compactness and bonding. These lands can be protected from further erosion. The new lands become productive very slowly and may take decades even after empoldering. These lands, according to the Bangladesh Government policy, are declared as *khas* (state) lands and handed over to the Ministry of Forestry and Environment for 20 years for the purpose of afforestation. During this period, the Ministry should plant mangrove forests as part of the coastal protection system. As accretion progresses, the land behind the mangrove belt is converted into agricultural land for settlement.

After the defined time period, the land ownership goes to the Ministry of Land and the Ministry has the overall responsibility for the allocation of accreted lands to eligible families. However, at the early stage of land development, local influential persons or absentee landlords lay claim to the land planted by the Ministry of Forest and Environment long before the most vital living conditions are met. These influential people rent out occupied lands to poor and homeless families. Often competing factions led by powerful people, locally called *jotdars*, determine life in these new areas.

The newly accreted lands are generally cultivated for at least three to four years before habitation and good crops persuade the landless households to settle in these lands. Capturing large amount of *khas* lands usually bring *jotdars* into the public eye through press and media. Therefore, instead of grabbing land openly, jotdars often sell state land to poor landless families at a relatively cheap price. In the absence of formal institutions being involved in the accreted lands in the coastal region, jotdars have developed their own system of selling government land. Jotdar mediators invite landless households from eroded areas to take shelter and to settle on newly accreted lands. It is worth mentioning that the new and eroded land in the coastal zone is widely geographically dispersed and there is no necessary connection between a household or community experiencing loss of its land, and access to fresh land in other parts of the coastal zone. Thus, initial migration takes time, but once it begins, it continues through chain migration. The first group of migrants invites their relatives and friends to settle close to them.

In this way they create a society that provides support and security for the newly settled communities.

Land allocation is a complicated job that faces a long story of foreseeable and unforeseeable obstacles, ranging from manipulation by powerful local people to political influence, bureaucratic inefficiency, and widespread practice of corruption by government officials. Several public projects are involved in the allocation of accreted lands. For example, the Char Development and Settlement Project (CDSP), financed by the Government of the Netherlands, formally settled thousands of families in the coastal belt and offshore islands of three coastal districts, namely Noakhali, Laxmipur, and Chittagong. The main goal of the project is reduced poverty and hunger for poor people living on newly accreted coastal chars. This is being accomplished through the development of improved and more secure rural livelihoods in agriculture, provision of legal title to land, and through provision of climate resilient infrastructure. There are a number of nongovernmental organizations (NGOs) who work with the landless households in coastal Bangladesh to establish their rights and access to khas lands.

LAND SUBSIDENCE

As a dynamic physical process, land subsidence has been taking place in coastal Bangladesh along with rest of the Ganges-Brahmaputra delta. This delta consists of thick Holocene depositional bodies (Goodbred and Kuehl, 2000a; Stanley and Warne, 1994). Such massive material loading is accompanied by sediment compaction and crustal sinking, i.e., subsidence experienced as a relative sea level (RSL) rise (Syvitski et al., 2009). Accurate knowledge of the rate of subsidence is as indispensable as predictions of the rate of future SLR, because both effects can lead to land lowering. However, estimating past subsidence rates is quite a challenging task because paleo-sea-level indicators, which are reliable for precise dating, are rare.

Boreholes drilled in the coastal region suggest that the region consists of deposits of several tens of meters of thick Holocene fine-grained sediments (Goodbred and Kuehl, 2000b). These deposits have been subjected to significant compaction-related subsidence (Steckler et al., 2010). Previous reconstructions of land accretion rates in the lower outer delta calculated long-term Holocene subsidence rates of $\leq$5 mm/year (Allison and Kepple, 2001; Goodbred and Kuehl, 2000a; Stanley and Hait, 2000). However, none were able to resolve historical subsidence rates accurately (Hanebuth et al., 2013).

The subsidence rate seems to be relatively rapid in the southern part of the western coastal zone and the western part of the central coastal zone. Like the Ganges-Brahmaputra delta, the costal zones show evidence of three different types of subsidence: tectonic, anthropogenic, and that resulting from the compaction of the peat layer. Anthropogenic land subsidence in the coastal Bangladesh includes coastal modifications, introduction of aquaculture, embankment and polder construction, and groundwater extraction (Hanebuth et al., 2013). Under natural conditions the subsidence is compensated by sediment brought into the system by rivers, tides, and coastal currents. However, manmade projects, especially the coastal embankment and polder projects, are altering the sedimentation pattern of the coastal zones.

The diurnal high tides play the most important role in sedimentation in coastal Bangladesh. Under natural conditions, such subsidence was counteracted by sedimentation from the rivers at high tides. However, the areas being protected by embankments and polders from salt water intrusion and flood water are receiving very little to no sediments from outside. Uncompensated subsidence is causing widespread waterlogging and drainage problems in the coastal region. With the reduced rate of sediment deposition, it is more likely that the subsidence rate has increased since the early 1960. However, there is no reliable information either on the historical or current subsidence rate (Brammer, 2014b).

Using high-resolution satellite altimetric data, Syvitski et al. (2009) calculated that the Ganges delta is sinking at a rate of 18 mm per year. They attributed this sinking to construction of embankments and abstraction of underground water for irrigation purposes as well as abstraction of gas. The rate calculated by Syvitski et al. (2009) may not be fully applicable for the coastal region of Bangladesh because the coastal region is a part of the Ganges delta and subsidence rates are not uniform within a large area like the Ganges delta (Brammer, 2014a). Also, the withdrawal rate of underground water is much lower in the coastal region compared to the rest of the delta.

Measurements of plinth levels of three old structures (a 15th century mosque in Bagerhat, a 400-year-old Hindu temple in the Sundarbans, and a 200-year-old temple in Khepupura) located in the western coastal zone show that long-term subsidence rates in this zone have not exceeded 1–2.5 mm per year (Sarker et al., 2013), much lower than the estimates of Syvitski et al. (2009). If subsidence had occurred at the 7.1 inches (18 mm) per year rate given by Syvitski et al. (2009), these buildings would

Table 4.3 Land Subsidence Rates on the Ganges Tidal Floodplain Calculated from Radio-Carbon-Dated Buried Organic Material

Site No.	Material	Depth (m)	Radio Carbon Age	Calendar Age (years BP)	Depth/Calendar Age (mm/year)
1	Buried tree	1.5	455 ± 45	614 ± 14	2.4
2	Marine shell	1.4	1200 ± 30	730 ± 50	1.9
3	Wood	0.7	90 ± 40	227 ± 48	3.1
4	Crab claw	2.2	127 ± 7	309 ± 8	7.1
5	Peat	1.5	570 ± 50	584 ± 64	2.6
6	Peat	1.1	910 ± 50	820 ± 111	1.3
7	Peat	3.2	2300 ± 40	2378 ± 29	1.3
8	Peat	5.8	2690 ± 40	3192 ± 17	1.8

Adapted from Brammer, H., 2014a. Bangladesh's dynamic coastal regions and sea-level rise. Climate Risk Management 1, 51–62, p. 57 and Allison, M.A., Khan, S.R., Goodbred, S.L., Kuehl, S.A., 2003. Stratigraphic evolution of the late Holocene Ganges-Brahmaputra lower delta plain. Sedimentary Geology 155, 317–342.

now be 7.9–24.9 ft (2.4–7.6 m) below sea level (Brammer, 2014a). Calculation of depths to radio-carbon dated organic materials in the Sundarbans mangrove forest where embankments have not interfered with sedimentation show subsidence rates of 0.05–0.28 inches (1.3–7.1 mm)/year (Table 4.3), from which probably 0.05 inch (1.3 mm)/year can be subtracted for global SLR.

Radiocarbon dating of 300-year-old submerged salt-producing kilns on the coast in the east of the Sundarbans suggests subsidence rates in that area of 5–6 mm/year (Hanebuth et al., 2013).[8] These kilns were built just above the previous winterly spring high-tide level, but are currently located ~155 (plus or minus 15) cm below the corresponding modern level. The kilns were ultimately fired ~300 years ago (1705 plus or minus 35 AD), as shown by optically stimulated luminescence dating (Hanebuth et al., 2013).

The evidence provided above suggests that subsidence is occurring in the coastal region, but at a lower rate than the rest of the country.[9] An accurate calculation of subsidence rates is difficult because it depends not only on deposition rate, but also on compaction rate of deposited sediments. As indicated, land subsidence in the western coastal zone is not caused by gas abstraction and only to a limited extent by groundwater abstraction in the

[8] Subsidence of this specific coastal area by 5–6 mm/year in combination with future SLR will lead to enhanced saltwater intrusion, worsen the effects of storm surges, and reduce the arable areas as well as the drainage during monsoon flooding.

[9] One estimate suggests that Dhaka is sinking 13.91 mm per year (Revkin, 2014).

few urban areas such as Khulna. Much of the Ganges delta is underlain by Holocene, Pleistocene, and Tertiary sandy sediments that are not subject to compaction with water abstraction; and Holocene peat layers underlying large parts of the tidal floodplain have remained saturated since their formation and so have not shrunk by drying out. It is most likely that most of this subsidence is due to tectonic subsidence of the Bengal Basin within which the region lies, possibly complicated by folds and faults within the basin (Stanley and Hait, 2000; Steckler et al., 2008).

Contrary to subsidence, parts of coastal Bangladesh also experience uplift. A study by Khan et al. (2005) observed uplift on the eastern coast of Bangladesh. They calculated that the coast uplifts at a rate of 1.9 inches (3.6 mm)/year in the coastal island of Maiskhali.

CONCLUSIONS

Coastal landform changes result from interactions of land erosion, accretion, and subsidence. The Bangladesh government has adopted measures to reduce erosion and accelerate accretion in the coastal region. This will help in stabilizing the region, particularly the coastline and offshore islands. Further efforts to enhance the process of land accretion in the coastal area and offshore islands of Bangladesh should be strengthened to save millions from drowning. However, the government seems to have no active plan to disperse sediments in areas enclosed by embankments and polders. The present protection of the coastal region (except the Sundarbans) by these structures has almost completely stopped new land aggradation inside the polders. In areas of high subsidence rates, it will be increasingly difficult to drain the polders, as well as protect them against storm surges. Thus, a controlled sedimentation system is needed to keep the western coastal zone above sea level and in habitable form.

At the same time, much more research is needed for comprehensive understanding of the Ganges-Brahmaputra delta building process, along with dynamics of land gain, loss, and subsidence in the coastal zones. Based on the information presented in this chapter, the following recommendations are made to arrest erosion and help accretion in the coastal belt of the country:

- Action is needed to prevent and discourage settlement in highly erosion-prone areas and to make a database of people displaced by erosion. These people should be prioritized in the allocation of government land.

- Proper maintenance, strengthening, and increasing the height of the existing embankments and polders along the coastal belt and design hydraulic engineering projects to arrest erosion and accelerate accretion are needed.
- Establishing monitoring and developing tools of prediction of morphological changes in the coastal region is imperative.

REFERENCES

Ahmed, A.U., 2008. Assessment of Vulnerability to Climate Change and Adaptation Options for the Coastal People of Bangladesh. Practical Action, Dhaka.

Alam, M.S., Uddin, K., 2013. A study of morphological changes in the coastal areas and offshore islands of Bangladesh using remote sensing. American Journal of Geographic Information System 2 (1), 15–18.

Ali, A., 1995. Numerical investigation into the retardation of flood water outflow through Meghna river in Bangladesh due to SW monsoon wind. Estuarine, Coastal and Shelf Science 41, 689–704.

Ali, A., 1991. Satellite observation and numerical simulation of western boundary current in the Bay of Bengal. In: Presented at the 12th Asian Conference on Remote Sensing Held in Singapore, October 30–November 5.

Ali, A., Ahmad, A.A.Z., 1992. Impact of sea level rise on other disasters in Bangladesh. In: Presented at the IOC/UNEP Workshop on Impact of Sea Level Rise Due to Global Warming for the South Asian Region Held in Dhaka, Bangladesh, December 16–19.

Allison, M.A., 1998. Historical changes in the Ganges-Brahmaputra delta front. Journal of Coastal Research 14, 1269–1275.

Allison, M., Kepple, E., 2001. Modern sediment supply to the lower delta plain of the Ganges-Brahmaputra river in Bangladesh. Geo-Marine Letters 21, 66–74.

Allison, M.A., Khan, S.R., Goodbred, S.L., Kuehl, S.A., 2003. Stratigraphic evolution of the late Holocene Ganges-Brahmaputra lower delta plain. Sedimentary Geology 155, 317–342.

BBS (Bangladesh Bureau of Statistics), 2011. Bangladesh Population and Housing Census 2011. Dhaka.

BWDB (Bangladesh Water Development Board), 2013. Main River Flood and Bank Erosion: Risk Management Program. Final Report, vol. 1. Main Report. Dhaka.

Bindoff, N.L., Willebrand, J., Artale, V., Cazenave, A., Gregory, J., Gulev, S., Hanawa, K., Le Quere, C., Levitus, S., Nojiri, Y., Shum, C.K., Talley, L.D., Unnikrishnan, A., 2007. Observations: oceanic climate change and sea level. In: Climate Change 2007: The Physical Science Basis. Contribution of Working Group I to the Fourth Assessment Report of the IPCC. Cambridge University Press, New York.

Brammer, H., 2014a. Bangladesh's dynamic coastal regions and sea-level rise. Climate Risk Management 1, 51–62.

Brammer, H., 2014b. Climate Change, Sea-Level Rise and Development in Bangladesh. The University Press Limited, Dhaka.

Bruun, P., 1954. Coast Erosion and the Development of of Beach Profiles. In: U.S. Army Beach Erosion Board Technical Memorandum, vol. 44. U.S. Army Engineering Waterways Experiment Station, Vicksburg, Mississippi.

Bruun, P., 1962. Sea-level rise as a cause of shore erosion. Journal of Waterways and Harbors Division 88, 117–130.

Caroagal, I.F.M., Khan, Z.H., Rahman, M.M., 2011. Land formation and erosion in the estuary. In: de wilde, K. (Ed.), Moving Coastlines: Emergence and Use of Land in the Ganges-Brahmaputra-Meghna Estuary. University Press Limited, Dhaka.

Cooper, J.A.G., Pilkey, O.H., 2004. Sea-level rise and shoreline retreat: time to abandon the Brunn Rule. Global and Planetary Change 43, 157–171.
Cowell, P.I., Roy, P.S., Jones, R.A., 1995. Simulation of large-scale coastal change using a morphological behaviour model. Marine Geology 126, 45–61.
Curray, J.R., Moore, D.G., 1974. Sedimentary and tectonic process in the Bengal Deep-Sea Fan and geosyncline. In: Bank, C.A., Drake, C.L. (Eds.), The Geology of Continental Margins. Spinger-Verlag, New York, pp. 617–627.
Davies, C., Best, J., Collier, R., 2003. Sedimentology of the Bengal shelf, Bangladesh: comparison of late Miocene sediments, Sitakund anticline, with the modern, tidally dominated shelf. Sedimentary Geology 155, 271–300.
Dean, R.G., 1977. Equilibrium Beach Profiles: U.S. Atlantic and Gulf Coasts, Department of Civil Engineering, Ocean Engineering Report No. 12. University of Delaware, Newark.
Elahi, M.K., 1991. Impacts of riverbank erosion and floods in Bangladesh. In: Elahi, K.M., Ahmed, K.S., Mafizuddin, M. (Eds.), Riverbank Erosion, Flood and Population Displacement in Bangladesh. Jahangirnagar University, Savar, Dhaka, pp. 1–10.
Gibbons, S.J.A., Nicholls, R.J., 2006. Island abandonment and sea-level rise: an historical analog from the Chesapeake Bay, USA. Global Environmental Change 16 (1), 40–47.
Gangopadhyay, A., Raj, G.N.B., Chaudhuri, A.H., Babu, M.T., Sengupta, D., 2013. On the nature of meandering of the springtime western boundary current in the Bay of Bengal. Geophysical Research Letters 40, 2188–2193.
Goodbred, S.L., Kuehl, S.A., 2000a. Large sediment supply, active tectonism, and eustasy on margin sequence development: late Quaternary stratigraphy and evolution of the Ganges-Brahmaputra delta. Sedimentary Geology 133, 227–248.
Goodbred, S.L., Kuehl, S.A., 2000b. Enormous Ganges-Brahmaputra sediment discharge during strengthened early Holocene monsoon. Geology 28, 1083–1086.
Hanebuth, T.J.J., Kudrass, H.R., Linstadter, J., Islam, B., Zander, A.M., 2013. Rapid coastal subsidence in the central Ganges-Brahmaputra delta (Bangladesh) since the 17th century deduced from submerged salt-producing kilns. Geology 41, 987–990.
Hasan, M., April 13, 2014. Land area not assessed since independence: Govt. ignores land reclamation potential in Meghna Estuary, sea. New Age.
Hofer, T., Messerli, B., 2006. Floods in Bangladesh: History, Dynamics and Rethinking the Role of the Himalayas. United Nations University Press, Tokyo.
Hossain, M.M., 2012. Storm Surges and Coastal Erosion in Bangladesh – State of the System, Climate Change Impacts and 'Low Regret' Adaptation Measures (A Master thesis Franzius-Institut, Hannover, Germany).
IFRC (International Federation of Red Cross) and RCS (Red Crescent Societies), 2001. World Disaster Report: Focus on Recovery. IFRC and RCS, Geneva.
Islam, S.M.R., Huq, S., Ali, A., 1999. Beach erosion in the eastern coastline of Bangladesh. In: Huq, S., Karim, Z., Asaduzzaman, M., Mahtab, F. (Eds.), Vulnerability and Adaptation to Climate Change for Bangladesh. Kluwer Academic Publishers, Dordrecht, pp. 71–92.
Jabbar, M.A., 1979. Local Accretion in the Coastal Area of Bangladesh. Report of Bangladesh Landsat Programme. Dhaka.
Kartiki, K., 2011. Climate change and migration: a case study from rural Bangladesh. Gender and Development 19 (1), 23–38.
Khan, M.S.H., Prakash, B., Kumar, S., 2005. Soil-landform development of a part of the fold belt along the eastern coast of Bangladesh. Geomorphology 71 (3–4), 310–327.
Kottke, B., Schwenk, T., Breitzke, M., Wiedicke, M., Kudrass, H.R., Spiess, V., 2003. Acoustic facies and depositional processes in the upper submarine canyon Swatch of No ground (Bay of Bengal). Deep Sea Research II 50, 979–1001.

Lotze, H.K., Lenihan, H.S., Bourgue, B.J., Bradbury, R.H., Cooke, R.G., Kay, M.C., Kidwell, S.M., Kirby, M.X., Peterson, C.H., Jackson, J.B., 2006. Depletion, degradation, and recovery potential of estuaries and coastal seas. Science 312 (5781), 1806–1809.

Mallick, B., Vogt, J., 2012. Cyclone, coastal society and migration: empirical evidence from Bangladesh. International Development Planning Review 34 (3), 217–240.

Martinez, M.L., Intralawan, A., Vazquez, G., Perez-Maqueo, O., Sutton, P., Landgrave, R., 2007. The coasts of our world: ecological, economic and social importance. Ecological Economics 63, 254–272. Washington, DC: AGU and Wiley.

Masselink, G., Gehrels, R., 2014. Introduction to coastal environments and global change. In: Masselink, G., Gehrels, R. (Eds.), Coastal Environments & Global Change, pp. 1–27.

MES (Meghna Estuary Study), 2001. Hydrological Dynamics of the Meghna Estuary. Bangladesh Water Development Board, Dhaka.

Miah, G., Bari, N., Rahman, A., 2010. Resource degradation and livelihood in the coastal region of Bangladesh. Frontier of Earth Science in China 4 (4), 427–437.

Miah, M.M., 1975. Changing morphology of the Ganges-Meghna delta. In: Proceedings of Bangladesh National Seminar on Remore Sensing, Dhaka.

Michels, K.H., Suckow, A., Breitzke, M., Kudrass, H.R., Kottke, B., 2003. Sediment transport in the self canyon "Swatch of No ground" (Bay of Bengal). Deep Sea Research II 50 (5), 1003–1022.

MoA (Ministry of Agriculture), 2012. Master Plan for Agricultural Development in Southern Region of Bangladesh. Government of Bangladesh, Dhaka.

MoWR (Ministry of Water Resources), 2001. Hydro-morphological Dynamics of the Meghna Estuary. Bangladesh Water Development Board, Bangladesh, Dhaka.

Morner, N.A., 2010. Sea level changes in Bangladesh: new observational facts. Energy and Environment 21 (3), 213–249.

Nicholls, R.J., Wong, P.P., Burkett, V.R., Codignotto, J.O., Hay, J.E., McLean, R.F., Ragoonaden, S., Woodroffe, C.D., 2007. Coastal systems and low-lying areas. In: Parry, M.L., Canziani, O.F., Palutikof, J.P., van der Linden, P.J., Hanson, C.E. (Eds.), Climate Change 2007: Impacts, Adaptation and Vulnerability. Contribution of Working Group II to the Fourth Assessment Report of the Intergovernmental Panel on Climate Change. Cambridge University Press, Cambridge, UK, pp. 315–356.

Pilkey, O.H., Young, R.S., Riggs, S.R., Smith, A.W.S., Wu, H., Pilkey, W.D., 1993. The concept of shoreface profile of equilibrium: a critical review. Journal of Coastal Research 9 (1), 255–278.

Pramanik, M.A.H., 1983. Remote Sensing Applications to Coastal Morphological Investigations in Bangladesh (Ph.D. thesis). Jahangirnagar University, Savar, Dhaka.

Pramanik, M.A.H., 1988. Methodologies and techniques of studying coastal systems. In: Presented at the National Development and Management Held in Dhaka, October 3–4.

Pramanik, M.A.H., Ali, A., Rahman, A., 1981. An assessment of the land accretion and erosion in the Meghna Estuary. In: Presented at the Third National Geographical Conference Held in Dhaka.

Rahman, M.K., 2015. Environmental and Social Vulnerabilities and Livelihoods of Fishing Communities of Kutubdia Island, Bangladesh (Ph.D. dissertation). Department of Geography, Kent State University, OH.

Rahman, M.K., Paul, B.K., Curtis, A., Schmidlin, T.W., 2015. Linking coastal disasters and migration: a case study of Kutubdia island, Bangladesh. The Professional Geographer 67 (2), 218–228.

Rashid, S., Alam, M.S., Shamsuddin, S.D., 2011. Development of island in the Meghna Estuary over the past 250 years. In: Ahmed, R., Shamsuddin, S.D. (Eds.), Climate Change: Issues and Perspectives for Bangladesh. Sahitya Prkash, Dhaka.

Revkin, A.C., March 28, 2014. Rising seas + dams + aquifer pumping = delta blues. The New York Times.
Rob, M.A., 2008. Changing morphology of the coastal region of Ganges delta. In: Amin, S.M.N. (Ed.), Studies on Coastal Environments in Bangladesh. A.H. Development Publishing House, Dhaka, pp. 54–69.
Sarker, M.H., Choudhury, G.I., Akter, J., Hore, S.K., 2013. Bengal Delta Not Subsiding at a Very High Rate. http://www.cegisbd.com/pdf/bengaldelta.pdf.
Sarwar, M.G.M., Islam, A., 2013. Multi hazard vulnerabilities of the coastal land of Bangladesh. In: Shaw, R., Mallick, F., Islam, A. (Eds.), Disaster Risk Reduction: Methods, Approaches and Practices. Springer, Tokyo, pp. 121–141.
Siddiqui, M.H., 1988. Land accretion and erosion in the coastal area. In: Presented at the National Workshop on Bangladesh Coastal Area Resource Development and Management Held in Dhaka, October 3–4.
SPARRSO (Space Research and Remote Sensing Organization), 1993. Monitoring of Changes in Coastal Zone Area Based on Landsat MSS and Infrared Aerial Photographs. Dhaka.
SPARRSO (Space Research and Remote Sensing Organization), 1987. Report on Pilot Project on Remote Sensing Application to Coastal Zone Dynamics in Bangladesh. Dhaka.
Stanley, D.J., Hait, A.K., 2000. Holocene depositional patterns, neotectonics and Sundarban mangroves in the western Ganges-Brahmaputra delta. Journal of Coastal Research 16 (1), 26–39.
Stanley, D.J., Warne, A.G., 1994. Worldwide initiation of Holocene marine deltas by deceleration of sea-level rise. Science 265, 228–231.
Steckler, M.S., Akter, S.H., Seeber, L., 2008. Collision of the Ganges-Brahmaputra delta with the Burma arc: Implications for earthquake hazard. Earth Planet Science Letter 273, 367–378.
Steckler, M.S., Nooner, S.L., Akhter, S.H., Chowdhury, S.K., Bettadpur, S., Seeber, L., Kogan, M.G., 2010. Modeling earth deformation from monsoonal flooding in Bangladesh using hydrographic, GPS, and Gravity Recovery and Climate Experiment (GRACE) data. Journal of Geophysical Research 115, B08407. http://dx.doi.org/10.1029/2009JB007018.
Subrahmanyam, V., Krishna, K.S., Ramana, M.V., Murthy, K.S.R., 2008. Marine geophysical investigations across the submarine canyon (Swatch-of-No-Ground), northern Bay of Bengal. Current Science 94 (4), 507–513.
Syvitski, J.P.M., Kettner, A.J., Overeem, I., Hutton, E.W.H., Hannon, M.T., Brakenridge, G.R., Day, J., Vörösmarty, C., Saito, Y., Giosan, L., Nicholls, R.J., 2009. Sinking deltas due to human activities. Nature Geoscience 2, 681–686.
Swift, D.J.P., 1976. Continental self sedimentation. In: Stanley, D.J., Swift, D.J.P. (Eds.), Marine Sediment Transport and Environmental Management. Wiley, New York, pp. 311–350.
Switzer, A.D., 2014. Coastal hazards: storms and tsunamis. In: Masselink, G., Gehrels, R. (Eds.), Coastal Environments & Global Change. AGU and Wiley, Washington, DC, pp. 104–127.
SWMC (Surface Water Modeling Center) and BWDB (Bangladesh Water Development Board), 2000. Late Monsoon Flood in the Southwest Region of Bangladesh. Ministry of Water Resources, Dhaka.
Thieler, E.R., Pilkey, O.H., Young, R.S., Bush, D.M., Chai, F., 2000. The use of U.S. coastal engineering: a critical review. Journal of Coastal Research 16 (1), 48–70.
Trenberth, K.E., Jones, P.D., Ambenje, P., Bojariu, R., Easterling, D., Tank, A.K., Parker, D., Rahimzadeh, F., Renwick, J.A., Rusticucci, M., Soden, B., Zhai, P., 2007. Observations: surface and atmospheric climate change. In: Climate Change

2007: The Physical Science Basis. Contribution of Working Group I to the Fourth Assessment Report of the IPCC. Cambridge University Press, New York.

Vellinga, P., 2006. Beach and Dune Erosion During Storm Surges (Doctoral thesis). Delft University of Technology, Delft, The Netherlands.

Valiela, I., 2006. Global Coastal Change. Blackwell Publishing, Malden, MA.

Vidal, J., January 7, 2013. Sea change: the Bay of Bengal's vanishing islands. Rapid erosion and rising sea levels are increasingly threatened the existence of islands off the coast of Bangladesh and India. The Gurdian 29.

Williams, S.J., 2001. Coastal erosion and land loss around the United States: strategies to manage and protect coastal resources – examples from Louisiana. In: Coastal Ecosystems and Federal Activities Technical Training Symposium Proceedings Held in Gulf Shores State Park, Alabama, August 20–22.

World Bank, 2012. Annual Report 2011: Bangladesh Climate Change Resilience Fund (BCCRF). Dhaka.

Zhang, K., Douglas, B.C., Leatherman, S.P., 2004. Global warming and coastal erosion. Climate Change 64, 41–58.

CHAPTER FIVE

Salinity Intrusion and Impacts

INTRODUCTORY CONCEPTS

Primary Versus Secondary Salinization

Seawater intrusion into coastal water bodies and accumulation of salts in coastal soils, aquifers, and wetlands are natural processes as waves, storm surges, tides, and currents interact with different components of the coastal environment. Accumulation of salts in the coastal environment by such natural processes is called primary salinization because they occur naturally in response to glacial and interglacial cycles that occur over a time scale of approximately 100,000 years or longer in some cases (Herbert et al., 2015). In contrast, increase in salt concentrations caused by anthropogenic forces, such as climate change, is known as secondary salinization (Herbert et al., 2015). The latter can occur on a time scale as short as decades or even more rapidly. Salinity intrusion in coastal Bangladesh due to global warming–induced sea level rise (SLR) is an example of secondary salinization. Increased salinity in coastal Bangladesh may also be considered as a secondary impact of climate change, since it is consequent upon SLR as its primary impact. With rising sea levels along the Bangladesh coast the following components of the coastal zones are at risks of salinity intrusion: (1) coastal rivers draining fresh-water from upstream areas, (2) tidal inlets and creeks penetrating saltwater inland, (3) coastal wetlands (both freshwater and saltwater marshes), (4) coastal soils, and (5) coastal aquifers.

Methods of Salinity Measurements

The main chemical composition of seawater is as follows: Sodium (Na^+), Potassium (K^+), Calcium (Ca^{++}), Magnesium (Mg^{++}), Chloride (Cl^-), Carbonate (CO_3^-) or Bi-Carbonate (HCO_3^-), and Sulphate (SO_4^-) (Baten et al., 2015). Among these, sodium chloride (salt) is the most abundant chemical compound in seawater, consisting of 35 parts per thousands. Because of this dominance of salt in seawater, salinity or brackishness is its signature characteristic. Salinity in water can be measured in two ways: (1) total dissolved solids (TDS) and (2) electrical conductivity of saltwater. The TDS can be measured in the laboratory by evaporating saltwater and

Climatic Hazards in Coastal Bangladesh
ISBN 978-0-12-805276-1
http://dx.doi.org/10.1016/B978-0-12-805276-1.00005-3

weighing in the left-over solids, normally expressed in mg/L. Soil salinity is most commonly measured by the electrical conductivity of a saline solution (for example, a sample of soil water), i.e., the ability of the solution to transmit an electrical current through it. The modern metric unit (SI unit) for conductivity is deci-siemens per meter (dS/m). Electrical conductivity (S) is the reciprocal of resistance (or ohm), which can also be expressed as S = ohm (Ω)/1 (www.dictionary.com/browse/siemens).

Soil Salinity Classes

Soil salinity levels vary widely from one environment to another. Therefore, their classification may also be site-specific. Based on generic data on electrical conductivity of a salt solution through different soil depths, Alberta (Canada) Agriculture and Forestry classifies saline soils into four or five categories (Agri-Facts, 2001). Compared to nonsaline soils (<2 dS/m), strongly saline soils have conductivity values of 8–16 dS/m. In very strongly saline soils, conductivity values may exceed 16 dS/m. As electrical resistance increases from the surface to a depth of 2–4 feet (60–120 cm), conductivity values increase correspondingly (mostly double) (see also Table 7.1 in Chapter 7).

GEOGRAPHY OF SALINITY INTRUSION IN COASTAL BANGLADESH

Recent Increases in Soil Salinity

Recent increases in soil salinity in coastal Bangladesh due to saltwater intrusion have already reduced crop productivity, which constitutes a major threat to the livelihoods of the people of coastal Bangladesh. Field data from soil surveys conducted by the Soil Resources Development Institute (SRDI) indicate that soil-affected areas in coastal Bangladesh have increased from about 2 million acres in 1973 to 2.6 million acres in 2009—an increase of about 1% per year (Table 5.1). A striking characteristic of this historical trend is that the rates of expansion of areas with higher salinity levels (8–15 to >15 dS/m) were significantly higher than increases (and decreases) in areas with low and moderate salinity. Thus, low salinity areas (2–4 dS/m) increased at a rate of 0.5% per year and moderate salinity areas (4–8 dS/m) decreased at a rate of −1.3% per year. In contrast, both high salinity areas and very high salinity areas increased at alarming rates of 12.63% per year and 5.76% per year, respectively. The Bangladesh Rice Research Institute (BRRI) has developed several salt-tolerant varieties of rice that can grow in a wide range of soil salinity (4–16 dS/m). Soil salinity with electrical

Table 5.1 Salt-Affected Areas in Coastal Bangladesh: 1973, 2000, and 2010

Salinity Classes: S1-S4	**Salt-Affected Areas: Acres (Hectares)[a]**			
[b]Soil Conductivity: dS/m	**1973**	**2000**	**2009**	**Rates of Increase or Decrease per Year**
S1 (low salinity): 2–4 dS/m	710,106 (287,370)	716,011 (289,760)	811,567 (328,430)	0.5%/year
S2 (moderate salinity): 4.1–8 dS/m	1,053,730 (426,430)	716,011 (289,760)	811,567 (328,430)	−1.3%/year
S3 (high salinity): 8.1–15 dS/m	197,066 (79,750)	831,706 (336,580)	869,044 (351,69)	12.63%/year
S4 (very high salinity): >15	98,595 (39,900)	215,327 (87,140)	251,849 (101,920)	5.76%/year
Total: S1-S4	2,059,497 (833,450)	2,522,324 (1,020,750)	2,610,071 (1,056,260)	0.99%/year

[a]Original data in hectares in parenthesis. Data in acres are converted from hectares.
[b]Soil salinity classes are based on Islam and Gregorio (2013). See also Table 7.2 (Chapter 7 in this volume).
Source: Prepared by authors based on data in MoA (Ministry of Agriculture), 2012. Master Plan for Agricultural Development in Southern Region of Bangladesh. Government of Bangladesh: Ministry of Agriculture, Dhaka, p. 43.

conductivity of 16 dS/m seems to provide a threshold for most of the new hybrid varieties (see Chapter 7).

Geographical Variations in Soil Salinity

Aggregate data in Table 5.1 mask major geographical variations in soil salinity that may be summed up as follows: (1) major differences in the extent of salt-affected areas among three coastal zones and (2) discontinuous areas of salinity within each of the three zones. The details of geographical distribution of soil salinity in coastal Bangladesh, which have been presented in a map in a government planning document (MoA, 2012, Fig. 3.5), display a set of spatial patterns of salinity areas. Two such patterns are dominant throughout each of the zones:

- Several bands of high salinity areas, especially along the Bay of Bengal coastline
- Extensive irregular patches (enclosed areas) of high, moderate, and low salinity

Highlighting these patterns, we have summarized geographical variations in soil salinity in coastal Bangladesh in Table 5.2. Physical settings seem to be a major determinant of soil salinity within each of the coastal zones. In the southwestern zone, about one-half of each of the three districts (Satkhira, Khulna, and Bagerhat) is covered with the Sundarbans forest. Soil salinity is not a concern for most of the salt-resistant species of this mangrove forest. Main salt-affected areas of this zone are located outside the forest and immediately north of it in the form of three continuous patches with high to very high soil salinity. Salinity levels decrease inland displaying at least four major patches of moderate salinity adjacent to high salinity areas. Moving further north, moderate salinity areas are succeeded by at least five large patches of low salinity (Table 5.2).

In the vast coastal plain of the central zone there are clear differences in soil salinity between the coastal belt (the so-called exposed coast) and inland areas. There are at least two coastal belts of high soil salinity along the Bay of Bengal coastline in Barguna and Patuakhali districts (Table 5.2). All coastal belts do not necessarily have high salinity levels due to a variety of factors. Probably, freshwater discharges from adjacent rivers play an important role. Thus, at least five coastal belts in Barguna, Patuakhali, Bhola, Lakshmipur, and Nokhali (one in each of the districts) have moderately saline soils. Areas of low soil salinity are limited to inland parts of these districts. A case study on soil salinity in Kalapara (a coastal subdistrict in Patuakhali district)

Table 5.2 Geographical Variations in Soil Salinity in Coastal Bangladesh

Coastal Zones	High to Very High Salinity (8–15 dS/m or Higher)[a]	Moderate Salinity (4–8 dS/m)	Low Salinity (<4 dS/m) Including Nonsaline Areas
Southwestern zone	Inland areas (districts: Satkhira, Khulna, Bagerhat) three main areas north of the Sundarbans[b]	Inland areas (districts: Satkhira, Khulna, Bagerhat) at least four areas adjacent to high salinity areas	Inland areas (districts: Satkhira, Khulna, Bagerhat) at least five areas adjacent to (mostly north of) moderate salinity areas
Central zone	Exposed coasts (districts: Patuakhali, Bhola) at least two coastal belts	Exposed coasts (districts: Barguna, Patuakhali, Noakhali) at least four coastal belts	Inland areas (districts: Pirojpur, Barguna, Patuakhali, Bhola, Lakshmipur, Noakhali)
Eastern zone	Discontinuous areas (districts: Chittagong, Cox's Bazar):	An isolated area (district: Cox's Bazar)	Exposed coast (district: Chittagong)
	Three discontinuous areas: Sandwip island, an elongated patch adjacent to Cox's Bazar, a narrow belt along the Naf River	A large oval shaped area surrounding Cox's Bazar	A narrow coastal belt along the Bay of Bengal coastline

[a]Salinity levels are based on Islam and Gregorio (2013).
[b]All salinity areas are approximate only since these have been interpreted by authors based on color shades in the legend of the MoA map (MoA, 2012, Fig. 3.5).
Source: Prepared by authors based on Fig. 3.5 in MoA (Ministry of Agriculture), 2012. Master Plan for Agricultural Development in Southern Region of Bangladesh. Government of Bangladesh: Ministry of Agriculture, Dhaka.

Table 5.3 A Case Study of Coastal Versus Inland Variations in Soil Salinity: Asasuni Versus Kalapara

Salinity Class: S1-S4	Salt-Affected Areas: Acres (Hectares)	
Soil Conductivity: dS/m	**Asasuni (22.55 N and 89.16 E) (Inland Subdistrict in Satkhira District in Southwestern Zone)**	**Kalapara (21.98 N and 90.24 E) (Coastal Subdistrict in Patuakhali District in Central Coast)**
S1 (low salinity): 2−4 dS/m	2249 (910)	77,591 (31,400)
S2 (moderate salinity): 4.1−8	18,286 (7400)	29,900 (12,100)
S3 (high salinity): 8−15	16,012 (6480)	18,286 (7400)
S4 (very high salinity): >15	7660 (3100)	5683 (2300)
Total: S1-S4 (percent of subdistrict area)	44,207 (17,890) 45%	131,460 (53,200) (100%)

Original data in hectares in parenthesis. Data in acres are converted from hectares.
Source: Prepared by authors based on data in Nahar, N., Uddin, M.K., Alam, A.K.M.M., 2011. Impact of salinity on soil properties of coastal areas in Bangladesh. AGRIVITA: Journal of Agricultural Science 33 (2), 161−173.

and Asasuni (an inland subdistrict of Satkhira district) demonstrates typical coastal versus inland differences in the extent of soil salinity. As data in Table 5.3 (column 3) indicate, the entire subdistrict of Kalapara, i.e., 100% of its land suffers from varied levels of soil salinity. In contrast, about 45% of Asasuni suffers from salinity problems.

In the eastern zone, soil salinity is limited to discontinuous and isolated areas (patches) and a narrow coastal belt. There are three discontinuous areas of high soil salinity (Table 5.2). A large oval-shaped area of high and moderate soil salinity surrounds Cox's Bazar Entire island of Sandwip suffers from high soil salinity. A narrow belt of high salinity occurs along the Naf River paralleling the narrow Teknaf peninsula.

Geographical Variations in Groundwater Salinity

Groundwater salinity poses a major threat to the drinking water quality in coastal Bangladesh. It is also a problem for irrigation water drawn from shallow tube-wells. The ideal drinking water salinity level is 1.6 dS/m, which is equivalent to 1070 ppm (NRW/Queensland, 2016). Salinity level up to 3 dS/m may be acceptable for drinking water, but any further increases in salinity may pose a health risk. Irrigation water salinity limits for most of the mildly salt-tolerant crops range between 1.3 and 5.2 dS/m (NRW/ Queensland, 2016). Salinity levels in most of the aquifers in coastal

Bangladesh not only exceed the upper limit for drinking water (i.e., 3 dS/m) but far exceed the upper limits for most of the salt-tolerant crops (MoA, 2012). Therefore, in Table 4.4 we have classified groundwater salinity into four classes based on a general guideline for agricultural saltwater classification (Agrifacts, 2001). The ranges of data for each of the four classes (odd intervals) differ slightly from the above guideline because we have adjusted each of the "salinity ranges" (intervals) based on the legend of the MoA map (MoA, 2012, Fig. 3.5).

One of the striking characteristics of groundwater salinity (summarized in Table 5.4) is that salinity levels in coastal aquifers seem to be too high compared to salinity thresholds for both drinking water (2–3 dS/m) and irrigation water (up to 5 dS/m). Data in Table 5.4 are more similar to soil

Table 5.4 Geographical Distribution of Groundwater Salinity in Coastal Bangladesh

Geographical Regions	Salinity Levels (EC: dS/m)[a] % of District Area[b]			
Coastal Districts	**Low (Up to 4.69 dS/m)**	**Moderate (4.7–8.8 dS/m)**	**High (8.9–16 dS/m)**	**Very High (>16 dS/m)**
Southwestern zone (coastal districts): Satkhira, Khulna	15%[b]	35	50	
Bagerhat	10	50	20	20
Central zone (coastal districts): Barguna				100
Patuakhali			70	30
Bhola			100	
Lakshmipur		10	90	
Noalkali			100	
Feni			100	
Central zone (inland districts): Pirojpur			90	10
Jhalokati			100	
Barisal	90	10		
Eastern zone: Chittagong		60	40	
Cox's Bazar		40	60	

[a]Salinity levels are based on (1) "salinity ranges" (intervals) in the legend of the MoA map (MoA, 2012, Fig. 4.4) and (2) a general classification of agricultural saltwater classification (Agrifacts, 2001).
[b]All percentage data for district areas under different salinity levels are approximate only since these have been interpreted by authors based on color shades in the legend of the MoA map (MoA, 2012, Fig. 3.5).
Source: Prepared by authors based on Fig. 3.5 in MoA (Ministry of Agriculture), 2012. Master Plan for Agricultural Development in Southern Region of Bangladesh. Government of Bangladesh: Ministry of Agriculture, Dhaka.

water salinity. Heavy rainfall plays a paradoxical role in groundwater salinity dynamics. Because of heavy rainfall most of the coastal aquifers have good recharge characteristics. Consequently, the water table tends to be high and most of the aquifers are shallow. At places, the water table rises to the surface and the aquifers remain saturated up to the surface level throughout the rainy season. Counteracting the freshwater effect of rainfall, tidal effects saturate top soils with brackish water in tidal plains and in other areas with strong tidal effects. Rainfall helps in leaching saline soil water from the top soil into the groundwater. During the dry season (November through February) salts from soil water start concentrating in soils peaking in April or May when evaporation exceeds rainfall. Leaching of salts from the soil profile starts in May or June with the beginning of rainfall. Not only shallow aquifers receive salts from overlying soil profiles by leaching, but saline groundwater contributes to soil salinity above by capillary action. Because of such exchanges of saline water between the soil surface and the groundwater most of the shallow aquifers remain saline throughout the year in several districts that experience strong tidal effects, such as Bagerhat, Barguna, Pirojpur, Jhalokati, Patuakhali, and Noakhali. Shallow aquifers extend up to depths of 60–110 feet (18–33 m). Besides salinity, in some areas shallow aquifers are also contaminated with high levels of arsenic.

It appears from Table 5.4 (summarized from the original MoA map) that large areas of each of the 11 coastal districts suffer from varied levels of groundwater salinity. Entire land areas (100%) of the following four coastal districts suffer from very high to high salinity problems: Barguna, Bhola, Noakhali, and Feni. Even some of the inland districts (of coastal Bangladesh), such as Jhalokati and Pirojpur, suffer from high groundwater salinity problems attributable to their shallow aquifers and strong tidal influences. Most of the other inland districts of coastal Bangladesh (not reported in this table) experience only low groundwater salinity. Overall, groundwater salinity is much higher in the southwestern and central coastal zones than in the eastern zone. Groundwater salinity levels decrease landward from the coastal belt both in the southwestern and central zones. Thus, groundwater salinity levels are either moderate or low in most of the inland aquifers situated closer to the northern boundaries of each of the districts.

The southwestern zone is notorious for its excessive soil salinity. However, its groundwater salinity seems to have greater geographical variations than in the central coastal zone. Whereas 70%–100% of the land areas in each of the five coastal districts experience high groundwater salinity, aquifers in the southwestern zone display greater ranges of salinity, varying from

Table 5.5 A Case Study of Soil and Groundwater Salinity in Two Sample Villages in Khulna District: Mirzapur and Baradanga

	Mirzapur		Barodanga	
Season (Months)	**Top Soil: 0–6 inches (0–15 cm) Salinity: dS/m**	**Groundwater Salinity: dS/m**	**Top Soil: 0–6 inches (0–15 cm) Salinity: dS/m**	**Groundwater Salinity: dS/m**
Premonsoon (MAM)	6.9–8.74	1.75–1.81	10.07–12.09	1.65–1.70
Monsoon (JJASO)	1.52–4.59	1.8–2.04	2.48–5.52	1.71–1.98
Winter (NDJF)	1.98–6.78	1.8–1.88	2.54–9.38	1.6–1.69

Source: Prepared by authors based on original data in Mondal, M.K., Bhuiyan, S., Franko, D.T., 2001. Soil salinity reduction and production of salt dynamics in the coastal riceland of Bangladesh. Agricultural Water Management 47 (1), 9–23.

low to moderate and high. A case study by Mondal et al. (2001) provides evidence of low groundwater salinity in the northern part of Khulna district (Table 5.5). Two sample villages, Mirzapur and Barodanga, are located west of Khulna City (latitudes 22.47–22.49°N and 89.24–89.34°E) in Dumuria thana (subdistrict) in the district of Khulna. Water samples were collected from shallow observation wells (40–82 feet or 12–25 m). The water tables (groundwater levels) rise close to the surface during the monsoon season and fall to their lowest levels in October and November. Compared to the bleak pictures presented in the MoA (2012) map and summarized in Table 5.4, both soil salinity and groundwater salinity in these sample villages were significantly lower than in the rest of Khulna district. The groundwater in two observation wells had low salinity levels, which were almost ideal for drinking and potable water. Another significant finding of this study was that there were very little seasonal variations in groundwater salinity, which increased only slightly during the monsoon season, presumably because of accelerated leaching by heavy rainfall. In contrast, soil salinity varied significantly between the premonsoon and monsoon seasons. Soil salinity builds up during the winter months peaking in March, April, and May and declines significantly during the monsoon season when rainfall removes salts from soils and washes them away.

SALINITY DYNAMICS IN COASTAL BANGLADESH

Multivariate Problems of Salinity Intrusion

So far we have focused narrowly on salinity dynamics between soils and groundwater. Salinity intrusion in coastal Bangladesh is driven by much more complex interactions among several interrelated natural and anthropogenic processes. Coastal waters play a fundamental role in salinity dynamics, especially in the coastal belt (exposed coasts) adjacent to the Bay of Bengal coastline. Circulations of coastal waters are dependent upon several interrelated variables, such as the volumes of freshwater flow from upstream areas, inland penetration of saline water from the Bay of Bengal, meteorological conditions (e.g., low pressure systems, winds, cyclones, and storms surges), and anthropogenic factors. Among them, volumes of freshwater flow perhaps play the most significant role in determining salinity levels in coastal waters. Although salinity intrusion is a multivariate problem, climate change—induced SLR is the most pressing cause of salinity in coastal Bangladesh. Reductions in freshwater inflows from the trans-boundary Ganges River in India, siltation of the southern tributaries of the Ganges, and siltation of other rivers of the southwestern coastal zone following the construction of the polder system in the 1960 and 1970s have also contributed to salinity increases (Dasgupta et al., 2014). We have classified these interrelated causes into three broad groups: natural, socioeconomic, and political processes. Fig. 5.1, based on Seal and Baten (2012, p. 15), illustrates leading causes of salinity intrusion in coastal Bangladesh.

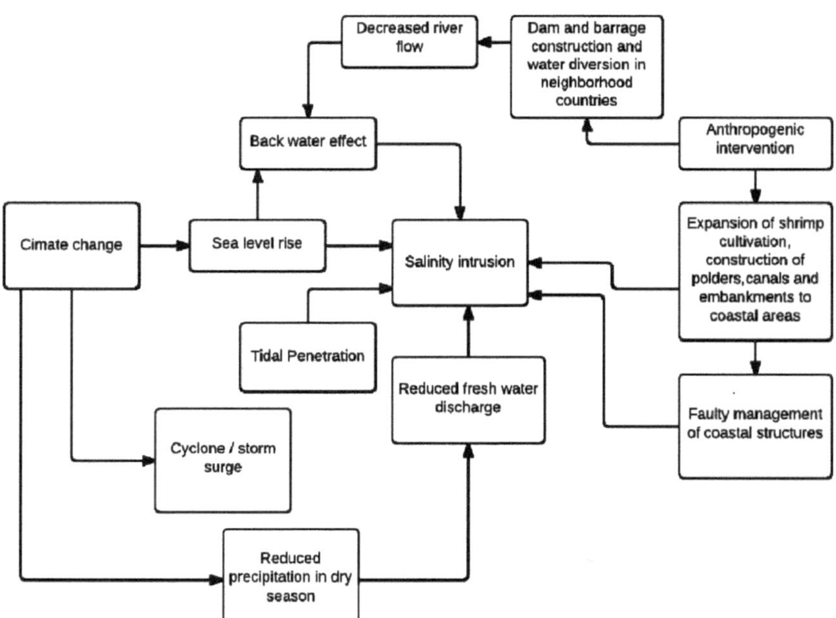

Figure 5.1 Causes of salinity intrusion in coastal Bangladesh. *Source: After Seal, L., Baten, M.A., 2012. Salinity Intrusion in Interior Coast: A New Challenge to Agriculture in South Central Part of Bangladesh. Unnayan Onneshan, Dhaka, p. 15.*

Natural Causes of Salinity

The natural causes include geographical location, sedimentation, astronomical tides, and climatic hazards, such as storms and cyclones, storm surges, and rising sea levels. The coastal zone of Bangladesh forms the southeastern-most landmass of the extended "Himalayan Drainage Ecosystem." Enormous amounts of monsoon rainfall and snowmelt in the Himalayan region are drained through Bangladesh discharging into the Bay of Bengal. Water salinity in the coastal zone is highly dependent on this discharge. The annual average discharge of approximately 1.5 million cfs (cubic feet per second) (42,475 m^3/s) is characterized by significant seasonal variations. The peak flow accounting for about 80% of the total discharge occurs in the summer months, whereas the remaining low flows (lean flows) occur in winter months (Mahmuduzzaman et al., 2014). In response, salinity in coastal waters (especially in lower reaches of major rivers) decreases in the summer and increases in the winter in the coastal region of Bangladesh.

The two Himalayan Rivers, the Ganges and the Brahmaputra (locally called Jamuna in Bangladesh), are among some of the leading sediment-bearing rivers in the world. The Meghna, third largest river of Bangladesh, also carries substantial amounts of sediment but the total volume is much smaller than that of either the Ganges or the Brahmaputra. The part of the sediment load carried by these three rivers is deposited on the river beds, often creating new islands/sand bars called chars. Excessive deposition of sediments and emergence of shoals and chars reduce freshwater areas in the estuary, resulting in increased salinity in tidal rivers. Sedimentation in the tidal rivers is also the main reason for the water logging problem. These sediments have blocked the rivers and caused upstream drainage congestion and flooding with saline water.

Tropical cyclones originating in the Bay of Bengal frequently make landfall along the Bangladesh coast and the coasts of neighboring India and Myanmar. As a result, coastal areas of the country experience storm-surge flooding generated by tropical cyclones. Sometimes this saline water flooding lasts for a while. For example, the coastal part of Satkhira district remained flooded for more than a year following cyclone Aila, which made landfall in May 2009. Rabbani et al. (2013) noted a sharp increase in salinity in Sathkhira district after cyclone Aila.

Storm surge–induced floods occur through April to June and September to November. Approximately 4633 mi^2 (12,000 km^2) of coastal land is prone to cyclonic storm surge floods. Intrusion of storm surges in the southwestern coastal zone can be up to 37 mi (55 km), while in the eastern coastal zone it can be up to 10 mi (15 km). Estimated surge depths of a 100-year return period are 26 feet (7.8 m), 17 feet (5.2 m), and 15 feet (4.5 m) at the entrance of the lower Meghna estuary in the central coastal zone, at the southwestern coastal zone, and at Chittagong coast in the eastern coastal zone, respectively (Ahmad et al., 2001). Climate change research postulates increased frequencies of higher-magnitude cyclonic storms and surges in coastal Bangladesh.

According to another study sponsored by the World Bank (2000), climate change is likely to result in higher temperatures and fluctuations in precipitation. The temperature increases will be higher in winter than in summer. These differential increases are likely to cause higher evaporation rates from river surfaces in winter in the upcoming years. Moreover, fluctuating precipitation would likely reduce precipitation in the dry season. The cumulative impacts of such climatic changes are likely to result in lower river flows, which in turn are likely to increases river water salinity.

The areas adjacent to estuaries and tidal rivers in the southwestern and central coastal zones experience tidal flooding twice a day. During the spring tide, which occurs fortnightly, a large area is flooded by tidal water. Tide is experienced up to 150 mi (225 km) inland in the wet season and 217 mi (325 km) inland in the dry season. The average tidal range at Hiron Point, located in the southern-most point of the southwestern coastal zone, is approximately 10 feet (3 m). Tidal fluctuation is felt even around 200 mi (300 km) inland in this coastal zone. A considerable area in the western coastal zone is below the high water level of spring tide (Ahmad et al., 2001).

Salinization in coastal Bangladesh is also caused by backwater effect, which is a special type of saline water intrusion at the mouth of the river when fresh water is not sufficiently strong enough to counteract the tide water moving towards the river from the Bay of Bengal. This effect is caused mainly by three reasons: (1) southwest monsoon wind, (2) astronomical tides, and (3) storm surges. The southwest monsoon wind primarily affects the southwestern and central coastal zone pushing seawater into the Meghna estuary. Similarly, storm surges are also responsible for the backwater effect in the mouth of the Meghna estuary (Mahmuduzzaman et al., 2014). In fact, the backwater effect is commonly pronounced in the Meghna estuary through which about 90% of the river water in Bangladesh discharges into the Bay of Bengal. This occurs particularly during flood season. Consequently, inland floodwater continues to accumulate, bringing in more areas under inundation and prolonging the flood duration. This aggravates the flood situation in the country. The Meghna estuary is also subjected to salinity intrusion due to decreasing fresh water in the dry season, resulting in saline water intrusion into the river. Increased salinity in river water also increases groundwater salinity, making it unfit for irrigation and drinking/potable water. It occurs when ground water levels in coastal aquifers are drawn down faster than their recharge rates. Finally, seawater intrusion and saltwater distribution in coastal aquifers are likely to be fundamentally changed by rising sea levels (Rasmussen et al., 2013; IPCC, 2007).

Effects of Shrimp Ponds

Socioeconomic systems that are responsible for saltwater intrusion in coastal Bangladesh include two dominant activities: shrimp cultivation and inadequate infrastructures (e.g., roads, embankments, polders, and sluice gates) and their poor management. Shrimp farming, introduced in the southwestern zone in the 1980s, has evolved into a lucrative commercial industry, which now includes such high demand species as the brackish water shrimp (*bagda chingri*) and freshwater shrimp (*galda chingri*). Shrimp farming expanded rapidly because of the spectacular rise in international demand for brackish water shrimp and for its ability to generate very high economic returns.[1] Consequently, shrimp cultivation has taken a massive horizontal expansion and engulfed almost the entire coastal region of Bangladesh. Shrimp farming increased from 49,420 acres (20,000 ha) in 1980 to 605,395 acres (245,000 ha) in 2010 and is still increasing (Belton et al., 2011).

The increase in brackish water shrimp cultivation is mostly concentrated in Satkhira, Khulna, Bagerhat, and Cox Bazar districts because of favorable geophysical conditions. Shrimp farms in these districts are mainly located within polders.[2] Despite its commercial success, shrimp aquaculture has caused havoc on saltwater intrusion in the surrounding agricultural lands (Hossain et al., 2013). In Khulna and Satkhira districts, farms largely alternate between shrimp and rice production, but in Cox's Bazar, they alternate between shrimp and salt production. The average farm size of 93% of all shrimp farms inside polders is less than 25 acres (10 ha). Shrimp cultivation requires saline water as an input to the shrimp ponds/farms, locally known as *ghers*. A gher is an earthen mini-polders within which shrimp is cultivated in imprisonment. This saline water needed for shrimp cultivation is drawn from adjacent rivers and canals and are thus dependent on the tides for water supply. However, high tides often overflow ghers and cause considerable damage to shrimp production. As a result of rapid expansion of shrimp cultivation, the availability of croplands reduced and the extent of salinity in both surface and groundwater increased (Mahmuduzzaman et al., 2014).

Besides decreases in croplands, shrimp farming also damaged embankments constructed in the coastal region to protect croplands from intrusion of saline water. Shrimp cultivators build unauthorized sluice gates to allow saline water to enter shrimp farms in monitored volumes, which is necessary for shrimps to grow. In contrast, they deliberately prevent floods from moving downstream towards their shrimp ghers by closing smaller bridges and culverts, and elevating local roads. Intrusion of brackish water weakens the possibility of cultivating other crops, forcing neighboring farmers also to take up shrimp farming (if they can afford it), or find alternative sources of employment (Kartiki, 2011). It is important to note that shrimp cultivation is a capital-intensive enterprise. It requires expensive fertilizers and pesticides because shrimps are subject to regular outbreaks of viral diseases. Thus a handful of very rich large farmers reap the full benefits of growing shrimp in coastal areas of Bangladesh.

Shrimp hatcheries and shrimp fields are also vulnerable to SLR. Though SLR will help shrimp farming in Bangladesh by introducing salinity in the coastal area, it is also harmful to the sector. Permanent inundation due to SRL will overflow shrimp ponds, allowing the shrimp into open water. An unexpected flood that ravaged the western part of western coastal zone in 2000 caused loss of shrimp worth US $230 million (Sarwar, 2005).

Wealthy and influential shrimp farmers also bring saline water through low lift pumps (LLPs). This practice affects the salinity balance inside the polder and causes damage to crops in surrounding fields. They also operate the sluice gate. They control the water in the immediate catchments and catch fish. They open and close the gate according to their will to allow fish to come in (MoA, 2012). Ironically, polders are constructed to protect land from saline water intrusion, but shrimp growers bring saline water inside the polder through controlling sluice gates.

Effects of Polders

In the early 1960s, the Government of East Pakistan (now Bangladesh) introduced polders in coastal areas with the technical help from the Government of the Netherlands.[3] To date, the GoB has constructed more than 3345 mi (5017 km) of embankments under the Coastal Embankment Project (CEP) to boost agricultural production by preventing saltwater intrusion into rice fields from by tidal flooding[4]. Retardation of storm surges was also a secondary purpose of polders although their low elevations (normally 3–6 feet or 1–2 m above the rice fields) have not been effective in reducing the impacts of high-magnitude surges. Polders have sluice gates to drain excess rainwater during the monsoon season. Before CEP, some areas were protected by low embankments constructed by the local people. However, these embankments would allow spills during high tides and were not effective in hindering tidal flooding and flushing processes.

Construction of the coastal polders resulted in a major morphological change in the coastal areas and disturbed the overall equilibrium (SWMC and BWDB, 2000). The Bangladesh Water Development Board (BWDB) formerly known as Water and Power Development Authority is in charge of maintaining and conducting the rehabilitation project of the polders. The BWDB has so far constructed 139 polders since the 1960s (Fig. 5.2).[5] However, the state of maintenance of coastal structures is generally poor. As a result, many of these structures are dilapidated. The BWDB has categorized 51 polders as "most vulnerable" and another 55 polders as "medium vulnerable" (MoA, 2012). Additionally, most of the sluice gates have been damaged and/or have ceased functioning, allowing saline water inland. The situation is further deteriorated because powerful shrimp farmers often breach the embankment to promote shrimp culture that resulted in crop damage. Breaching of embankments makes such structures weak and as a

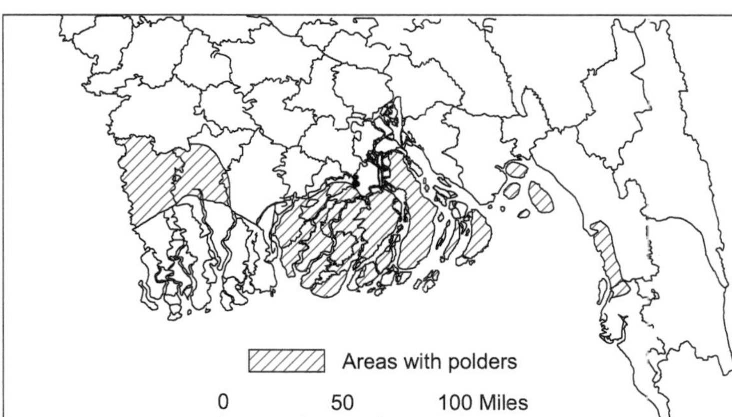

Figure 5.2 Areas with relatively large number of polders in coastal Bangladesh. *Source: After MoA (Ministry of Agriculture), 2012. Master Plan for Agricultural Development in Southern Region of Bangladesh. Government of Bangladesh: Ministry of Agriculture, Dhaka, p. 51.*

result embankments are easily damaged due to tidal pressure, particularly during full moon (Mahmudazzaman et al., 2014).

Effects of Unauthorized Local Projects

A number of political factors are also responsible for salinity increases in coastal Bangladesh. Numerous human activities, such as untimely water use, unplanned expansion of shrimp culture in areas where the potential is low, and poorly maintained infrastructure are causing salinity intrusion. Additionally, local people are also involved in unauthorized activities that may directly or indirectly intensify salinity intrusion. For example, some coastal residents occupy canals. They construct earthen cross-dams (*bundh*) on canals for aquaculture. As a result, the canals lose much of their drainage function (MoA, 2012). All of these activities reflect a weak water governance system at the local level. The BWDB, the GoB agency that is supposed to be responsible for water management initiatives is severely underfunded and therefore fails to protect the environment from such unauthorized projects. The local government has no jurisdiction to protect coastal polders, embankments, roads, and any kind of infrastructure as well as unplanned shrimp culture (Mahmuduzzaman et al., 2014).

Effects of Engineering Interventions in Upstream River Basins

There are 57 rivers that originate outside the boundary of Bangladesh. Of these rivers 54 originate in India and three in Myanmar (Ahmad et al., 2001). Among the 54 rivers originating in India, more than 25 of them face one or more upstream diversions primarily in dry months. Consequently, upstream freshwater flows from these rivers have decreased drastically in the dry winter season, causing intrusion of saline water from the Bay of Bengal. This intrusion increases salinity in soil and water of the coastal region since salinity intrusion in coastal areas is dependent upon a balance between upstream freshwater flow and salinity of coastal waters.

Among all river engineering projects in India, regulation of the Ganges flow by the Farakka Barrage in West Bengal, which was completed in 1975, has seriously affected salinity problems in downstream distributaries of the Ganges in Bangladesh, especially in the dry season in southwestern coastal zone. The Farakka Barrage was constructed on the Ganges River near the India–Bangladesh border for the main purpose of diverting a certain portion of the Ganges flow into the Hoogli River in West Bengal during the dry season. This withdrawal is one of the basic reasons for reduced flows in rivers of the southwestern zone. The downstream impacts of the Farakka Barrage on the hydrologic regimes of the Ganges have been catastrophic. The minimum flow of the Ganges in Bangladesh has dropped to as low as 5300 cfs (150 m^3/s) at the Hardings Bridge in 1995 from its normal pre-Farakka low of 68,000 cfs (1920 m^3/s). Following prolonged bilateral negotiations between Bangladesh and India the Ganges Water Treaty was signed in 1996. The treaty could not restore pre-Farakka normal minimum flow of the Ganges but at least the average minimum flow at Hardinge Bridge was increased to 22,000 cfs (619 m^3/s) (Dasgupta et al., 2014; Mirza, 1998) (Fig. 5.3). To make the matters worse, contrary to flow reductions during the dry season, discharges during the monsoon season (July–October) have increased, which have exacerbated flood problems in upper parts of the southwestern zone (Mirza, 1998). In short, upstream flow regulation by the Farakka Barrage has worsened both downstream flood problems in Bangladesh during the monsoon season and salinity problems in the dry season.

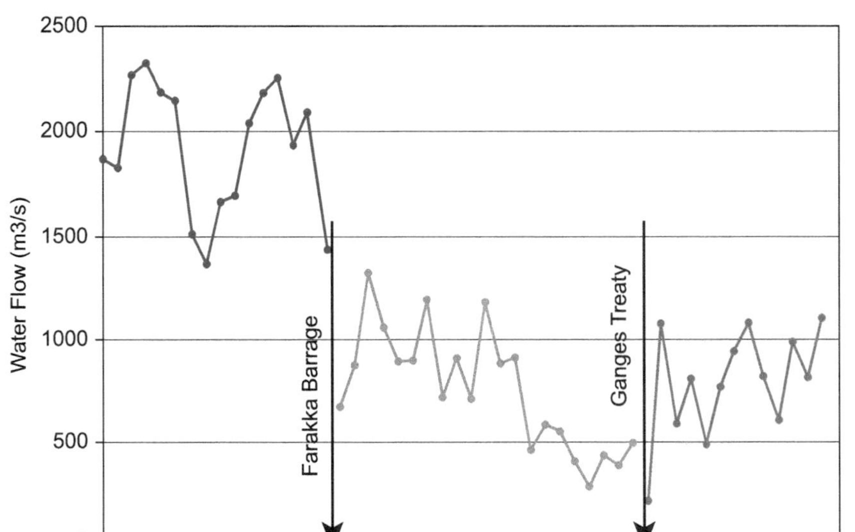

Figure 5.3 Annual daily minimum flow (m^3/s), Hardings Bridge (The Ganges River). *Source: Based on information provided by Dasgupta, S., Kamal, F.A., Khan, Z.H., Choudhury, S., Nishat, A., 2014. River Salinity and Climate Change: Evidence from Coastal Bangladesh. World Bank, Washington, DC, p. 38.*

SALINITY IMPACTS

Impacts on Agriculture

At the beginning of this chapter we have summarized salinity impacts on soils and groundwater, which constitute by far the most significant impacts in coastal Bangladesh. In this section we focus on other direct and indirect impacts of increased salinity. Threat to crop productivity is a major concern among potential impacts of salinity on agriculture. Crop production has already decreased at the current concentrations of salts in coastal soils and aquifers. In particular, a significant loss of yield has been observed in the dry season when rainfall is inadequate and incapable of lowering salinity in surface water and leaching salts from soils. The most common response of crops and plants to increased salinity has been reductions in growth. In low to moderate concentrations, salinity affects crop production by lowering the soil water potential and increasing the concentration of salt in the root zone. Low water potential indicates that the plant cannot extract adequate amount of water from the soil and maintain adequate cell rigidity (turgor) at very low soil water condition. This effect is known as osmotic effect, which is linked directly to crop productivity and the quantity of water passed through the plant by water transpiration (Bauder et al., 2007). When electrical conductivity (EC) of dissolved solids in water increases, less water is likely to be available for plants.[6] Yield potential of plants, therefore, deteriorates with increases in EC values in irrigation water.

In the changing scenario of SLR, it has been predicted that the increasing concentration of salinity in water and soil will create more pressure on the farmer by reducing yield and thus threatening their livelihood, income generation, and food security. Increase in salinity will also decrease the availability of agricultural land in coastal Bangladesh. Most crops are sensitive to salinity caused by high concentration of salts in the soil. Soil salinity causes soil degradation and reduces soil fertility, which is an important factor for crop production. In general, the coastal regions of Bangladesh are quite low in soil fertility. Most of the soils are moderately to strongly alkaline, and the pH values of the surface soils are lower than those of the subsurface soils. At places, soils with higher pH values and deficiencies in micronutrients are expected. Coastal soils are also poor in organic matter content.

Salinity also decreases the terminative energy and germination rate of some plants (Rashid et al., 2004; Ashraf et al., 2002). The amount of salt accumulation in the root zone of soil affects crop and plant growth, particularly at the critical stages of growth (Yadav et al., 2009). Ali (2006) investigated the loss of rice production in a village in Satkhira district and found that rice production in 2003 was 1151 metric tons lower than in 1985, corresponding to a loss of 69%. Of the total decreased production, 77% was due to conversion of rice fields into shrimp ponds and the remaining 23% was because of loss of yields. Practicing shrimp cultivation reduces rice production in neighboring fields due to soil degradation. Additionally, conversion of rice fields for shrimp cultivation is also decreasing rice production area and thus also reducing its total production. The decrease of both rice production area and yield are evident in almost all coastal districts, particularly those located in the southwestern zone.

According to a World Bank (2000) study, a net reduction of 0.5 million metric tons of rice production is likely to take place due to a one foot (0.3 m) SLR in coastal Bangladesh. In the fiscal year (FY) 1997–1998, rice production area decreased by 1% compared to the FY 1993–1994, whereas the total rice production (yield) decreased by 26% during the same period (Islam, 2004). Sarwar (2005) reported that rice production in the coastal area (16%) was lower compared to the production area (24%). The coastal region is notable for its production of pulses, oil seeds, and vegetables. The areas and productions of these crops and other crops grown in the region are likely to fall gradually similarly to rice, with increases in salinity due to SLR. Soil salinity is also believed to be responsible for lowland use rates as well as cropping intensity in the coastal region.

Another World Bank study predicts potential impacts of climate change–induced increases in soil salinity on the output of high yielding varieties (HYV) rice (Dasgupta et al., 2014). The findings of this study indicate that increases in soil salinity may lead to a decline in the yield of HYV rice by 15.6% and reduce farmers' income significantly in coastal areas. The study further found that many areas in the coastal region had already suffered large losses in yields and substantial price reductions from rising salinity, and this would likely continue with rising sea levels and corresponding increases in soil salinity.

Since the introduction of shrimp farming and the subsequent reductions in crop areas, particularly in rice crop areas, there is a severe shortage of grazing lands in coastal areas. Decreasing rice production means decreasing fodder production. Ali (2006) blamed fodder shortage for a declining livestock population from 630 in 1985 to 168 in 2003 in a village in Satkhira district. Decreased agriculture production in the coastal region would cause food shortage, leading to health hazards or even famine.

FAO (2011) sets the threshold value for different crops in response to salinity and their yield potentiality. Threshold value is the value of the soil salinity level at which plants start experiencing the impact of salinity in terms of loss of yield. Different crops have different threshold values. The higher the threshold value, the more a crop is tolerant to salt; the lower the threshold value, the less tolerant a crop is to soil salinity. In a household survey in an inland village (within coastal Bangladesh), Seal and Baten (2012) found that 72% of 60 respondents perceived that salinity was most prevalent in their farms during the dry season. They maintained that they experienced a decline in crop production in the last 5–6 years. According to the study participants, salts intruded in soils when they irrigated their lands. The respondents indicated that they used water from adjacent smaller rivers, which they had obtained from tidewater by LLPs during the irrigation period. The study further reported that most of the respondents in the study area already experienced yield reductions due to increased soil salinity. A small number of respondents reported that they had changed crop varieties to cope with increasing salinity. According to them the most pressing problem in the study area was the loss of yields. However, these losses were not uniform for all food crops (Seal and Baten, 2012).

Impacts on Freshwater

Another potential impact of increased salinity is the reduction of freshwater availability, which will drastically decrease the supply of drinking water in both rural and urban areas of the coastal region as well as the availability of water for irrigation for the dry season agriculture. Industries located in the coastal region will also be affected by a decreased supply of fresh surface water. Salinity restricts the usability of river water for industrial purposes and causes leakage in equipment and production loss. At the same time, production costs increase because the producers must import fresh water to avoid corrosion and leakage. For instance, the Pakshi paper mill closed in 1993 due to salinity intrusion in the north of Khulna (Mirza, 1998).

Another study by the World Bank estimates that several coastal districts will be most severely affected by increases in river salinity in a changing climate (Dasgupta et al., 2014). The study finds that in the best-case future scenario of least change in salinity from the March 2012 baseline, fresh water (0–1 ppt) river zones will be lost entirely in four districts (Barguna, Jhalokati, Khulna, and Patuakhali). In the worst-case scenario (most change from the baseline, March 2012), over 90% of fresh river water in Pirojpur,

Bagerhat, and Barisal will be at risk. Under this scenario, river water will no longer be suitable for dry season agriculture in Barguna, Bhola, Jhalokati, Khulna, and Patuakhali districts. Even in the best-case scenario, fresh water scarcity will be severe, as 98% of rivers in Khulna and 97% of rivers in Bhola will be adversely affected (Dasgupta et al., 2014).

Impacts on Coastal Fisheries

Based on water salinity, fisheries in coastal Bangladesh are of three types: freshwater fisheries in rivers (with low salinity: <0.5 ppt), brackish water fisheries along the coastline and estuaries (with moderate salinity: 0.5–29 ppt), and marine fisheries (with high salinity: >29 ppt). Between 140,000 and 160,000 households are directly or indirectly dependent on fisheries-based livelihoods (Minar et al., 2013). Changes in river and estuary salinity and scarcity of freshwater will affect productivity, abundance, and distribution of many capture fisheries in the coastal region. It is also likely to affect adversely wild habitats of some of the popular fresh water fish species of Bangladesh, such as *catla* (*Catla catla*), *rui* (*Labeo rohita*), and *mrigal* (*Cirrhina mrigala*) and jumbo freshwater prawn, especially during the dry season. The most popular fish in Bangladesh, *hilsa* (*Tenualosa ilisha*) (a variety of shad) has been affected severely by increased saltwater intrusion in coastal rivers. *Hilsa* is both a marine and a freshwater fish as it migrates inland through estuaries for spawning. It is harvested mainly as a freshwater river fish at the peak of its flavor. Marine varieties of *hilsa* are not very popular in Bangladesh. One of the consequences of increased saltwater intrusion in coastal rivers has been that the production of inland freshwater *hilsa* has declined by about 20% during the last two decades, whereas marine water production increased three-fold (Hussain and Hoq, 2008).[7] Major *hilsa* catches have gradually shifted from inland to marine waters. With the intensification of climate change, similar situation might occur for other marine fish species.

Reductions of fishing days in the Bay of Bengal due to extreme weather events such as cyclones and floods have also been anticipated (Ahmed and Neelormi, 2008). Frequencies and magnitudes of these events are likely to increase due to impacts of climate change. Reductions in fishing operations, in turn, would decrease production of coastal fisheries, resulting in increased livelihood vulnerabilities for coastal fishing communities.[8] Climate change is also likely to exacerbate nonclimatic pressures on coastal fisheries such as overfishing and pollution. Fish processing and traditional fish drying costs may also increase. Fish drying is very sensitive to variations in temperature and rainfall. Impacts on catch and processing would ultimately influence employment, income, and nutrition of fishery-dependent households and communities (Islam et al., 2014).

Reductions in the catches of wild fish are also expected for many smaller indigenous freshwater fish species. With increased salinization, optimum river area for *catla*, *rui*, and *mrigal* is likely to decrease in most coastal districts, along with a decrease in habitats for smaller indigenous fresh water fish species, which are the most common fish consumed in rural areas and the most important source of dietary protein (Belton et al., 2011; Thilsted, 2010, 2012).[9] Rural poor can afford to buy only smaller varieties of fish since these are cheaper than larger fish (Roos et al., 2007). Anticipated decline in biodiversity of freshwater, low-value, wild fish species with increased river salinity may have significant implications for nutrition of the rural poor as about 60% −80% of animal protein intake in the country comes from consumption of both smaller and larger fish (Alam and Thomson, 2001; World Bank, 2000). Thus, decreased fisheries in coastal water bodies would cause protein scarcity among almost all coastal population. It is also likely that intrusion of salinity into freshwater will adversely impact the livelihoods and incomes of fishing communities that currently rely on freshwater capture fisheries. However, expansion of brackish water aquaculture and the shrimp sector may offer alternative livelihood opportunities for rural people in the future.

Poor health status will increase poverty in the coastal area. At the same time, poverty will boost health hazards because of insufficient access to medicine, health care, and nutrition. If the coastal fisheries decrease, Bangladesh's foreign exchange earnings will drop significantly because the frozen food industry is the second largest foreign exchange earning sector after garment industries in Bangladesh, and the industry is dependent on coastal fisheries. Insufficient foreign exchange earnings will also increase unemployment as well as poverty (Sarwar, 2005).

Loss of Biodiversity

Saline water intrusion and retention of saline water in coastal areas are likely to damage biodiversity (disappearance or extinction of indigenous flora and fauna) and cause loss of genetic diversity due to monoculture. Loss of biodiversity in the coastal region has also been occurring due to expansion of shrimp culture in crop fields as well as in parts of the Sundarbans and other coastal mangrove forests. Nearly 50% of Bangladesh's forests are in the coastal region. In the shrimp farm areas, selective species of shrimps are cultured intensively, and all other natural aquatic species have declined or become extinct (MoA, 2012).

The Sundarbans is very rich in biodiversity. SLR has already increased soil and surface water salinity in the Sundarbans. Further increases in salinity with rising sea levels are likely to alter habitat patterns of this forest. Sundari (*Heritiera fomes*), the single most dominant and valuable tree of the Sundarbans, is thought to suffer from Top dying disease because of increased salinity (Kausher et al., 1993). If the present rate of salinization continues, the Sundari tree will disappear in next 40 years. Aquatic organisms will migrate inward, because of increased salinity (Dasgupta et al., 2014).

The Sundarbans is the last habitat of Royal Bengal Tiger and it is the home to a number of unique and globally or nationally endangered species of plants like rare Sundari, Gewa, Passur; animals species like endangered Royal Bengal Tiger, vulnerable Pallas, Fishing Eagle and Masked Fin foot, and critically endangered River Terrapin; rare species of shark; and very rich avifauna (Sarwar, 2005). At this transitional zone between freshwater and saline water, many fish species such as *Peneaus monodon*, *Macrobrachium rosenbergii*, *Lates calcarifer*, *Metapeneaus Monoceros*, and *Pangaisus pangaisus* depend on the aquatic habitat of the Sundarbans for spawning and juvenile feeding (Sarwar, 2005). The Sundarbans is also a habitat for some other important species like wild boar, spotted deer, barking deer, rhesus macaque, jungle cat, leopard cat, otter, squirrels, and the Indian porcupine (Sarwar, 2005). The forest also supports habitats for marine turtles, crocodiles, frogs, and fresh water dolphins. Any reductions in the Sundarbans habitats for these species due to increased salinity would constitute a major threat to its biodiversities.

Increased salinity in the Sundarbans would reduce plant height considerably and thus productivity of the forest will be severely constrained. Total commercial wood volume per unit area of forest land would decline with increasing soil and water salinity. It is estimated that increased salinity would cause loss of over 50% of commercial wood from the Sundarbans (Minar et al., 2013).

Other Impacts

Salinity increases in the coastal region are likely to have major health impacts. Drinking saline water causes waterborne diseases that primarily affect children and women. Common likely health problems include hypertension, increased blood pressure, scabies, and other skin diseases (Seal and Baten, 2012). Scarcity of potable freshwater also leads to gastrointestinal infections and kidney diseases, such as kidney stones and rheumatism. It has negative impacts on domestic cattle in terms of deterioration of the milk productivity and reproductive health (Seal and Baten, 2012).

Coastal infrastructures, particularly roads, are vulnerable to penetration of saline water. This penetration might cause blistering and cracking of the road surface. It is estimated that maintenance expenditure for paved roads in coastal Bangladesh would increase by about 252% because of increased ground water salinity (Dasgupta et al., 2014).

STRATEGIES FOR REDUCING SOIL SALINITY

In this section we focus on different methods of reducing soil salinity. Leaching salt from soil is a widely used practice for soil salinity management, and this can be done either naturally or artificially.[10] Leaching is a process by which soluble constituents of soil are removed by liquid, such as water, passing through it (Grattan, 2002). Naturally, rainfall helps leach salt from soil surface but in the dry season rainfall is not sufficient to drain salt from the root zone. For this reason, irrigation of fresh water may be applied to crop fields. This means more water than regular water requirement of the crop is needed. However, there is a high risk of increasing salinity content in the root zone if irrigation is done with saline water (Seal and Baten, 2012).

Deep-water irrigation is required in the case of high concentration of salinity in root zone. Amacher et al. (2000) recommend application of 6 inches of water in certain vegetable root zones if the roots are 12 inches deep. This application is capable of leaching salinity by 50%, 12 inches of water is capable of leaching salinity by 80%, and 24 inches by approximately 90%. Irrigation type and adequate drainage also help in reducing soil salinity. Simple surface water run-off is unable to leach soil salinity effectively. Instead, water should be drained through soil. For this reason, deep tillage is essential to ensure internal drainage as it is very important to break up the restraining layer that delays water flow in the crop field. In this regard, sprinkler irrigation provides a better control of water application rates. However, this type of irrigation is almost nonexistent in Bangladesh (Seal and Baten, 2012).

For level fields, controlled flood irrigation will be very efficient in removing salt from soil. Quality of irrigation water influences crop production as well as soil salinity management. Irrigation water contains various dissolved minerals, which are largely determined by the source. Surface water is the principal source of irrigation water in coastal Bangladesh. Coastal farmers often release water from local canals into their fields to flush out the salt that would otherwise build up in the soil. A system of drainage ditches allows brackish water from irrigated field to drain away.

It may be stressed here that slight variations in the microrelief of crop field leads to salt accumulation in the raised spots. Thus land should be properly leveled to prevent accumulation of water in the low-lying patches with shallow ground water tables and to facilitate uniform drainage of excess water. This helps in applying irrigation water uniformly in the field and in facilitating uniform germination of seeds and better growth of crops (Haque, 2006; Hossain et al., 2012). As rainwater does not contain salt, it could be used as irrigation water. However, this would require storage of rainwater in ponds in summer months for subsequent irrigation use in the winter.

The leaching requirement of different crops varies according to their thresholds or tolerance of soil salinity. Leaching requirements also depend on crop yield. Higher yields require greater leaching. Leaching thresholds for different crops may be used as a guide for farmers to choose ideal crops in their respective geographical areas for optimal production in specific physical settings. There is a close relationship between salt tolerance level and the amount of irrigation water that should be applied for reducing salinity level. The crops, which have a higher tolerance level, are likely to have lower leaching requirements and therefore will require less irrigation water. On the contrary, a low tolerance value requires high leaching, which means that more water is needed to increase the yield potentiality. This, in turn, results in high production cost. Moreover, salinity in irrigation water also influences soil salinity and its pH. Higher pH in soil creates deficiencies of nutrients like phosphorus. To reclaim the nutrients, appropriate fertilizer needs to be applied against the deficiency of certain nutrients. For example, to address the deficiency disease named "lime—induced chlorosis," an iron-deficiency disease, iron sulphate should be applied in the soil (Gale et al., 2001). High pH due to high salinity usually causes carbonate and bicarbonate salt to be insoluble and therefore creates calcium and magnesium ion deficiency with the dominance of sodium ion. To overcome this problem, gypsum or calcium sulfate could be applied to release calcium ion and replace sodium ion (Gale et al., 2001).

Timing of irrigation is another important factor in managing salinity. A hot, dry day is likely to evaporate more water from soil and therefore, cause more salt to be concentrated in the crop field. Therefore, for irrigation, a cool, humid day or night is preferable. The frequency of irrigation should be increased in hot and dry weather to address salinity stress.

Another way to reduce soil salinity in the coastal region is to cultivate crops in the dry season. Farmers of this region often keep crop fields fallow during this season. This leads to high salinity in soil due to evaporation of soil moisture. Since ground water is largely saline and present at shallow depths it is necessary to lower the water table and maintain it below the critical depth to prevent salt effect on crops. This can be done by installing subsurface drainage systems. This technology is effective to reduce soil salinity but somewhat expensive (Hossain et al., 2012). The salinity can also be managed by constructing embankments of suitable height with the provision of sluice gates to remove excess water and also to prevent ingress of saline water during high tides.

Based on a survey among 360 households in several sample villages, the Bangladesh Center for Advanced Studies reported that 39% of the respondents had introduced salt-tolerant rice varieties in the study area to cope with soil salinity problems and another 27% flushed out salts from their rice fields by irrigating them with freshwater from nearby streams and ditches (Rabbani et al., 2013). Nearly one-tenth of the respondents reported that they had tried some other measures to reduce salts from their fields, such as by raising seed beds with additional soils and applying solid sugar or sugar solution and gypsum in the rice fields. Frequent tillage of the crop fields also helps in reducing salinity during cultivation.

CONCLUSION

The literature review and the supporting data presented in this chapter suggest that salinity levels in soils, surface water, and groundwater in coastal Bangladesh have been increasing over past several decades. Climate change will likely accelerate the salinization process through SLR and increased evaporation from higher temperatures. Rising sea level is likely to penetrate seawater further inland than its current positions. Cyclones and storm surges are also likely to increase the height of tidewater into coastal rivers. With increasing tide levels river water salinity is also likely to increase. Furthermore, decreasing rainfall trends in the winter would probably reduce natural leaching of soil salinity. With higher rates of evaporation in the dry season, salinity problems are likely to be more acute in the dry season than in the wet season.

The evidence provided in this chapter further suggests that present salinity concentrations have already threatened crop productions in the coastal region. Significant losses of crop yields have been observed, particularly in the dry season. With rising sea levels, it has been anticipated that increasing salinity levels in soils and water would result in significant pressures for farmers by decreasing crop yields and thus threatening their livelihoods, income generation, and food security.

Although permanent inundation of parts of coastal Bangladesh may be unavoidable in the long run even with some of the major engineering interventions, in the short-term and the medium-term soil salinity in coastal areas may be managed reasonably well by changing methods of drainage from agricultural fields, improved irrigation methods, and applications of appropriate and adequate fertilizers (Bauder et al., 2007). Introduction of drip irrigation systems may go a long way for alleviating soil salinity problems. There is also a need for educating farmers about varied salt tolerance levels of different crops so that they could choose appropriate crops for cultivation. After all, development and adaptation of salt-tolerant rice and other crops and fish species (reviewed in Chapter 7) seem to be the necessary steps for combating salinity problems in coastal Bangladesh prior to its complete inundation by SLR.

ENDNOTES

1. Bangladesh produces about 2.5% of the global shrimp trade, and shrimp exports are the second largest foreign exchange earner of the country after garment industry.
2. Freshwater shrimp is now being cultivated in almost all coastal districts where sweet water is available.
3. A polder is a low-lying tract of land enclosed by embankments or dikes. It forms an artificial hydrological entity, meaning it has no connection with outside water other than through manually operated devices.
4. Today about 3 million acres (1.2 million ha) of land (15% of the country's total arable land) under agriculture is within the coastal embankment system (Dasgupta et al., 2014).
5. At least 49 polders are sea facing (Dasgupta et al., 2014).
6. Electrical conductivity (EC) is a property that is used to describe how well materials allow electrons to flow. The more salts in water the more conductive the solution is, resulting a higher electrical conductivity. Conductivity is the reciprocal of resistivity, meaning the higher the conductivity, the lower the resistivity. EC is expressed either as micro-Siemens per cm (μS/cm) or as deci-Siemens per meter (dS/m) (Seal and Baten, 2012; Agri-Facts, 2001).
7. *Hilsa* accounts for about 15% of the total fish production in Bangladesh (Hissain and Hoq, 2008).
8. The marine fisheries sector in Bangladesh supports livelihoods of over half a million fishers and their household members (DoF, 2012).
9. The nutritional contribution of small fish species is generally high. And, as many small fish species are consumed whole, they provide a significant percentage of recommended intakes of calcium, vitamin A, iron, and some minerals (Thilsted, 2010, 2012).
10. Leaching is a process by which soluble constituents of soil are removed by liquid, such as water, passing through it (Grattan, 2002).

REFERENCES

Agri-Facts, 2001. Salt Tolerance of Plants. Agri-facts: Practical Information for Alberta's Agriculture Industry. Government of Alberta. Alberta Agriculture and Forestry, Canada. www.agric.gov.ab.ca/.

Ahmad, E., Chowdhury, J.U., Hassan, K.M., Haque, M.A., Khan, T.A., Rahman, S.M.M., Salehin, M., 2001. Floods in Bangladesh and their processes. In: Nizamuddin, K. (Ed.), Disaster in Bangladesh: Selected Readings. DRTMC, Dhaka, pp. 9–28.

Ahmed, A.U., Neelormi, S., 2008. Livelihood of Coastal Fishermen in Peril. In Search of Early Evidence of Climate Change Induced Adverse Effects in Bangladesh. Centre for Global Change, Dhaka.

Alam, M.F., Thomson, K.J., 2001. Current constraints and future possibilities for Bangladesh fisheries. Food Policy 26, 297–313.

Ali, A.M.S., 2006. Rice to shrimp: land use/land cover changes and soil degradation in south-western Bangladesh. Land Use Policy 23 (4), 421–435.

Ashraf, M.Y., Sarwar, G., Ashraf, M., Afaf, R., Sattar, A., 2002. Salinity induced changes in α-amylase activity during germination and early cotton seedling growth. Biologia Plantarum 45 (4), 589–591.

Amacher, M., Katherine, O., Perry, H., 2000. Soil Vital Signs: A New Soil Quality Index (SQI) for Assessing Forest Soil Health. Research Pap. RMRS-RP65-WWW. Department of Agriculture, Fort Collins, CO, USA, Rocky Mountain Research Station.

Baten, M.A., Seal, L., Lisa, K.S., 2015. Salinity intrusion in interior coast of Bangladesh: challenges to agriculture in south-central coastal zone. American Journal of Climate Change 4, 248–262.

Bauder, T.A., Waskom, R.M., Davis, J.G., 2007. Irrigation Water Quality Criteria, Crop Series Irrigation, No. 0.506. http://cospl.coalliance.org/fez/eserve/co:6315/ucsu2062205062007internet.pdf.

Belton, B., Karim, M., Thilsted, S., Kondker, M., Collis, W., Phillips, M., 2011. Review of Aquaculture and Fish Consumption in Bangladesh. The World Fish Center.

Dasgupta, S., Kamal, F.A., Khan, Z.H., Choudhury, S., Nishat, A., 2014. River Salinity and Climate Change: Evidence from Coastal Bangladesh. World Bank, Washington, DC.

DoF (Department of Fisheries), 2012. National Fisheries Week 2012. Government of Bangladesh, Dhaka.

FAO (Food and Agriculture Organization of the United Nations), 2011. Water Quality for Agriculture. http://www.fao.org/DOCREP/003/T0234E/T0234E02.htm.

Gale, J., Koenig, R., Barnhill, J., 2001. Managing Soil pH in Utah. Utah State University, Logan.

Grattan, S.R., 2002. Irrigation Water Salinity and Crop Production. University of California, Oakland, CA.

Haque, S.A., 2006. Salinity problems and crop production in coastal regions of Bangladesh. Pakistan Journal of Botany 38 (5), 1359–1365.

Herbert, E.R., Boon, P., Burgin, A.J., Neubauer, S.C., Franklin, R.B., Ardón, M., Hoppensperger, K.N., Lamers, L.P.M., Gell, P., 2015. A global perspective on wetland salinization: ecological consequences of a growing threat to freshwater wetlands. Ecosphere 6 (10), 1–43. http://onlinelibrary.wiley.com/doi/10.1890/ES14-00534.1/epdf.

Hossain, M.S., Uddin, M.J., Fakhruddin, A.N.M., 2013. Impacts of shrimp farming on the coastal environment of Bangladesh and approach for management. Reviews in Environmental Science and Biotechnology 12 (3), 313–332.

Hossain, M.L., Hossain, M.K., Salam, M.A., Rubaiyat, A., 2012. Seasonal variation of soil salinity in coastal areas of Bangladesh. International Journal of Environmental Sciences 1 (4), 172–178.

Hussain, M.G., Hoq, M.E., September–December 2008. Impacts of climate change on coastal and marine fisheries Resources in Bangladesh. Bay of Bengal News 34–39.
IPCC (Intergovernmental Panel on Climate Change), 2007. IPCC Working Group Report. Cambridge University Press, Cambridge.
Islam, M.R., 2004. Where Land Meets the Sea: A Profile of the Coastal Zone of Bangladesh. The University Press Limited, Dhaka.
Islam, M.R., Gregorio, G.B., 2013. Progress of salinity tolerant rice variety development in Bangladesh. SABRAO Journal of Breeding and Genetics 45 (1), 21–30.
Islam, M.M., Sallu, S., Hubacek, K., Paavol, J., 2014. Vulnerability of fishery-based livelihoods to the impacts of climate variability and change: insights from coastal Bangladesh. Regional Environmental Change 14, 281–294.
Kartiki, K., 2011. Climate change and migration: a case study from rural Bangladesh. Gender and Development 19 (1), 23–38.
Kausher, A., Kay, R.C., Asaduzzaman, M., Paul, S., 1993. Climate Change and Sea-Level Rise: The Case of the Coast. Briefing Document No. 6. Bangladesh Unnayan Parishad, Dhaka.
Mahmuduzzaman, M., Ahmed, Z.U., Nuruzzaman, A.K.M., Ahmed, F.R.S., 2014. Causes of salinity intrusion in coastal Bangladesh. International Journal of Plant Research 4 (4A), 8–13.
Minar, M.H., Hossain, M.B., Shamsuddin, M.D., 2013. Climate change and coastal zone of Bangladesh: vulnerability, resilience and adaptability. Middle East Journal of Scientific Research 13 (1), 114–120.
MoA (Ministry of Agriculture), 2012. Master Plan for Agricultural Development in Southern Region of Bangladesh. Government of Bangladesh: Ministry of Agriculture, Dhaka.
Mondal, M.K., Bhuiyan, S., Franko, D.T., 2001. Soil salinity reduction and production of salt dynamics in the coastal riceland of Bangladesh. Agricultural Water Management 47 (1), 9–23.
Mirza, M.M.Q., 1998. Diversion of the Ganges water at Farakka and its effects on salinity in Bangladesh. Environmental Management 22 (5), 711–722.
Nahar, N., Uddin, M.K., Alam, A.K.M.M., 2011. Impact of salinity on soil properties of coastal areas in Bangladesh. AGRIVITA: Journal of Agricultural Science 33 (2), 161–173.
NRW/Queensland, 2016. Measuring Salinity. Queensland Government, Natural Resources and Water, Science Notes, Land Series L137 (Web-Based Fact Sheet). www.qld.gov.au.
Rabbani, G., Rahman, A., Mainuddhin, K., Shoef, I.J., 2013. Loss and Damage from Salinity Intrusion in Satkhira District, Coastal Bangladesh. Bangladesh Centre for Advanced Studies (BCAS), Dhaka.
Rashid, M.M., Hoque, A.K.F., Iftekhar, M.S., 2004. Salt tolerances of some multipurpose tree species as determined by seed germination. Journal of Biological Sciences 4 (3), 288–292.
Rasmussen, P., Sonnenborg, T.O., Goncear, G., Hinsby, K., 2013. Assessing impacts of climate change, sea level rise, and drainage canals on saltwater intrusion to coastal aquifers. Hydrology and Earth System Sciences 17, 421–443.
Roos, N., Wahab, M.A., Hossain, M.A.R., Thilsted, S.H., 2007. Linking human nutrition and fisheries: incorporating micronutrient dense, small indigenous fish species in carp polyculture production in Bangladesh. Food and Nutrition Bulletin 28 (2), S280–S293.
Sarwar, M.G.M., 2005. Impacts of Sea Level Rise on the Coastal Zone of Bangladesh. Lund University, Lund, Sweden.
Seal, L., Baten, M.A., 2012. Salinity Intrusion in Interior Coast: A New Challenge to Agriculture in South Central Part of Bangladesh. Unnayan Onneshan, Dhaka.
SWMC (Surface Water Modelling Center) and BWDB (Bangladesh Water Development Board), 2000. Late Monsoon Flood in the Southwest Region of Bangladesh 2000. SWMC and BWDB, Dhaka.

Thilsted, S.H., 2010. Improved Management, Increased Culture and Consumption of Small Fish Species Can Improve Diets of the Rural Poor. The World Fish Center, Consultative Group on International Agricultural Research (CGIAR), Dhaka. http://pubs.iclarm.net/resource_centre/WF_3165.pdf.

Thilsted, S.H., 2012. The potential of nutrient-rich small fish species in aquaculture to improve human nutrition and health. In: Subasinghe, R.R., Arthur, J.R., Bartley, D.M., De Silva, S.S., Halwart, M., Hishamunda, N., Mohan, C.V., Sorgelos, P. (Eds.), Farming the Waters for People and Food (Proceedings of the Global Conference on Aquaculture 2010, Phuket, Thailand, 22–25 September 2010). FAO, Bangkok, pp. 57–73. Rome and NACA.

World Bank, 2000. Bangladesh: Climate Change and Sustainable Development. World Bank, Dhaka.

Yadav, J.S.P., Sen, H.S., Bandyopadhyay, B.K., 2009. Coastal soils-management for higher agricultural productivity and livelihood security with special reference to India. Journal of Soil Salinity and Water Quality 1 (1–2), 1–13.

CHAPTER SIX

Land Use Change and Coastal Management

Land use change is a process by which human activities transform the natural landscape, referring to how land has been used, usually emphasizing the functional role of land for economic activities. Land use changes are often nonlinear and might trigger feedbacks to the system, stress living conditions, and threaten people with vulnerability (Kasperson et al., 1995). Therefore, not only do land use change trajectories need assessment, but also possible future conditions must be projected for certain assumptions, both basic to ensuring sustainable conditions (GLP, 2005; Veldkamp and Lambin, 2001). Land use changes are generally grouped into two broad categories: conversion and modification. Conversion refers to the intended shift from one cover or use type to another, while modification involves maintaining the cover or use type in the face of changes in its attributes (Baulies and Szejwach, 1998).

In land cover and land use change (LCLUC) literature, land use/cover conditions are broadly grouped into four categories: undisturbed, frontier, agricultural/managed, and industrial/urban (Mustard et al., 2004). Undisturbed landscapes are dominated by natural cover types, where change occurs primarily through natural disturbance with little anthropogenic interference. Frontier landscapes see transformations in natural cover, usually through extensive anthropogenic land uses (e.g., conversion to agriculture, forest regrowth through resource extraction). In agricultural/managed landscapes, management either matches or supersedes nature in function, such as rangelands or cultivated lands sustained by intensive inputs. Land covers may be relatively stable, and changes are slow. Urbanized/industrialized landscapes are dominated by residential, commercial, and industrial land cover, and highly managed vegetation for services and recreation (e.g., parks, sports fields, and managed natural areas), but few land resources are used (Mustard et al., 2004).

Most of the world's lands can be categorized using this broad framework, or some version of it, and significant portions have transitioned one or more times from the undisturbed state. A region's transformation between these broad categories can be tracked when LCLUC histories are sufficiently

Climatic Hazards in Coastal Bangladesh
ISBN 978-0-12-805276-1
http://dx.doi.org/10.1016/B978-0-12-805276-1.00006-5

long and well documented. Many natural landscapes transition through all four states, though the timelines are elastic, and no given region is fated to experience all conditions. Moreover, the timeline does not flow in one direction; processes (e.g., abandonment land cover) may revert from managed to undisturbed given enough time. The most profound impacts on land cover will occur during transitions between conditions (Mustard et al., 2004).

Land use change is driven by synergetic factor combinations of resource scarcity, leading to an increase in pressure on use of resources, changing opportunities created by markets, outside policy intervention, loss of adaptive capacity, and changes in social organization and attitudes (Lambin et al., 2003). The specific drivers of land use change vary by geographic scale and surely do in terms of their relative roles (Mustard et al., 2004). At the global scale, deforestation, soil degradation, overgrazing, rapid population growth, and urbanization and industrialization, among other things, are associated with land use changes. Land use changes also determine, in part, the vulnerability of places and people to climatic, economic, or socio-political perturbations (Kasperson et al., 1995). When aggregated globally, land use/cover changes significantly affect central aspects of earth system functioning. Drivers of land use change differ not only over time, but also from country to country as well as within regions of a given country.

Not all effects of land use change are negative. Many forms of land-use/cover changes involve increases in agricultural production, resource use efficiency, and wealth and well-being. Over the last few decades, particularly since the formation of the Intergovernmental Panel on Climate Change in 1988, many researchers have undertaken studies on land cover change. Their attempt has facilitated our understanding of the causes of land use change and predictive models of land use/cover change. The primary aim of this chapter is to provide an analysis of coastal land use changes in Bangladesh by presenting temporal variations of land use changes in the coastal Bangladesh along with contemporary uses of coastal land. The chapter also includes a discussion of drivers of coastal land use changes as well as a discussion of coastal management efforts undertaken in Bangladesh since its independence in 1971.

LAND USE CHANGE

Besides human settlement, the coastal region of Bangladesh is used for farming, forests, water bodies, aquaculture, infrastructures, and industries, including salt manufacturing. The region includes critical terrestrial and aquatic habitats, like natural and planted mangroves, hills, wetlands, and tidal flats. As noted, the region is vulnerable to different natural disasters. Most of the people of coastal region are rural dwellers in depressed socioeconomic condition. Moreover, the population has increased rapidly in the region. All of this has shaped the current land uses, which are diverse, conflicting, and competitive (Brammer, 2002). Demand for new land uses (tourism, export-processing zones, and others) is increasing, along with encroachment and conversion of land from one use to another (Islam, 2006). Before presenting current land use pattern, temporal variations in land use in the coastal region will be discussed briefly. After covering current land use, drivers of land use change will be covered.

Temporal Variation in Land Use

Land uses in coastal Bangladesh have undergone many changes in the past several centuries. The coastal region has been the center of development and economic activity for hundreds of years. Modern expansion of these activities, the main forces behind recent land use changes in the coastal region, began in the early 1960s and intensified in the 1970s. Chittagong and surrounding areas in the eastern coastal zone is famous as a center of economic activity because of the Chittagong port. The port's location and its natural harbor made it an important center of trade and business even as far back as the ninth century AD when Arab merchants found it a lucrative center for commerce and business. By the beginning of the 15th century, the port was an important trading center. In the 16th century, Portuguese sailors took great interest in the area around Chittagong (ASB, 2004).

Chittagong port was established at its present location in 1887. As a major seaport of Bangladesh, it provides a major gateway for the country's trade with the outside world. From 2012 to 2013, Chittagong Port handled more than 43.37 million metric tons of cargo, which is about 92% of the total maritime trade of Bangladesh. The growth in the GDP of Bangladesh economy is around 6–7%, but container traffic growth through Chittagong Port is about 14%, double the GDP rate. To meet the challenges of globalization and the liberalization of world trade and economy, Chittagong Port has undertaken many ambitious projects to enhance its capacity, improve efficiency, and quality of services, and develop adequate facilities to become a world class regional port (CPA News, 2016). The improvement and modernization of the port led to the rapid growth of the Chittagong city, which is now the second largest city of Bangladesh.

The southwestern coastal zone, particularly the Khulna region, was an agrarian frontier, dividing forest from agriculture for several centuries after AD 1200. Land reclamation and human settlement were encouraged in this region by Islamic religious leaders during the Bengal Sultanate period (1204–1575). During the Mughal period (1565–1717), state intervention and development in this region increased (ASB, 2004). Since the 17th century, local landlords have constructed small dykes and embankments around individual plots of land to limit saline water overflow and prevent crop damage. This traditional construction of embankments using local efforts almost ceased in 1947 when the British India broke into two independent countries and Bangladesh became part of Pakistan. Land use before independence was dominated by rice cultivation, especially locally adapted low-yielding rice varieties. In very limited areas of southwest, traditional shrimp cultivation was practiced (Mukerjee, 1938).

As noted in the previous chapters, the coastal area saw efforts at large-scale development in the 1950s and 1960s; construction of polders and cross-dams, as well as mangrove plantations and social forests. The primary purpose of empolderment was to increase agricultural production in the coastal region. Using cross-dams to form new land was also attempted in the central coastal zone from 1957 to 1965 (Islam, 2006 and 2004). During this time, the World Bank supported the establishment of mangrove plantations in the newly accreted land. This is an attempt to protect the hinterland from cyclones and storm surges. Dominant land use during the 1960s was cultivation of local varieties of rice. In the 1970s, new high yielding varieties (HIVs) of rice were introduced in the coastal region. Other uses of land remained the same: salt farming, mangrove forest, and traditional shrimp farming (Islam, 2005).

After Bangladesh became independent in 1971, rapid population growth occurred in the coastal region, and the consequence was intensified agricultural land use. Establishing mangrove plantations in the newly accreted land continued. To increase food production, HIV crop varieties saw cultivation, and large-scale polderization continued to protect the agricultural land from the intrusion of saline water. However, many polders in the southwest coastal zone were turned into large shrimp culture *ghers* (ponds), so saline water was deliberately allowed to enter the polders. During 1990s, however, this zone started to see massive drainage congestion and water logging because of heavy siltation outside the polders, unplanned construction of roads and other infrastructure, changes in land use pattern, and expansion of shrimp cultivation (Islam, 2005).

This evidence all suggest that land use in coastal Bangladesh is changing over time. This trend likely will continue. The key changes are: (1) reduction of cultivable land, (2) intensification of agricultural land use, (3) conversion of agricultural land to settlement and infrastructure, (4) conversion of forest land to aquaculture and settlement, and (5) conversion of cropland to aquaculture.

Present Land Use

Available data suggest that in coastal Bangladesh nearly 29% is used for settlements, water bodies, reserved forest, and roads, about 1% for salt manufacturing, with the remaining 70% or 13.7 million acres (5.6 million ha) under cultivation. Coastal land uses have changed greatly over the last several centuries, and subject to further change due to the direct and indirect effects of global warming—induced sea level rise. The current major uses of land in coastal Bangladesh are described below.

Agriculture

Like rest of Bangladesh, growing crops dominate agricultural land use in the coastal area. Nearly 5 million acres (2 million ha) of land is devoted to cultivating crops. Rice accounts for nearly 80% of the cropped area in the coastal region and the region provides approximately 16% of all rice production in the country. Among the three rice varieties, (transplanted) *aman* rice accounts for highest percentage in the entire coastal area, followed by *aus* and *boro* rice.[1] Other crops grown in the coastal area include wheat, different types of pulses, and vegetables, all grown in the winter period or *rabi* season.[2] Although jute is the main cash crop in Bangladesh, it is almost nonexistent in the coastal area although it is grown in a few coastal districts, including Satkhira.

Flooding, physiography, soil salinity, drainage congestion, and irrigation facilities all affect land used for agriculture in the coastal region. Crop cultivation is limited to the wet season because soil salinity is high during the dry season. The coastal region produces a relatively high proportion of pulses, oilseeds, and betel nuts and leaves. Despite their agro-ecological sensitivity, a number of fruits and vegetables grow well in the region. About 89% of the guava produced in the country comes from this region. The region is also famous for some minor fruits with high nutrition and the potential for processing, among them, sapota, velvet apple (gab), hog plum, and date palm (MoA, 2012).

The coastal region lags behind the country in the intensity of agricultural (crop) land use. Nearly 50% of the net cropped area is cultivated once a year and 40% twice a year. The remaining land remains either fallow and/or is not cultivated at all. Cropping intensity runs 159% while the rest of Bangladesh sits at 176%.[3] The principal reasons for this lower intensity are soil and water salinity, subsidence, and water logging (MoA, 2012).

The pattern of agricultural land use is not uniform everywhere in the coastal area. Exposed and nonexposed coastal districts show one significant difference. The net cultivated area in exposed coastal districts is much less than nonexposed coastal districts. Nearly 40% of total land area is considered net cultivated area in the exposed districts as opposed to 68% in the nonexposed districts. Thus, the cropping intensity between these two types of coastal districts differs significantly. It is also reflected in the area under single, double, and triple cropping. The percentage of single cropped area is much higher in the exposed coastal districts than nonexposed districts. The same is also true for double and triple cropped area.

Like the difference between exposed and nonexposed coastal districts, cultivated and cropped areas also differ among three coastal zones. The net cropped area is lowest in the western coastal zone and highest in the central zone. Because of the Sundarbans, land for cultivation is very limited in the western zone. Within the coastal region, three coastal districts (Gopalganj, Pirojpur, and Barisal) have a history of practicing an agricultural system called soil-less agriculture or floating agriculture/gardens. This type of agriculture is practiced where agricultural land is submerged for extended periods during the monsoon season. It involves a floating layer of water hyacinth, straw, or rice stubble to which is added upper layers of small and quick-rotting waterworts that are good green manure. The floating raft is strengthened with bamboo, and bamboo poles fix it in position to avoid damage caused by wave action or drifting (Irfanullah et al., 2011). This floating raft can then be moved to any submerged location for agricultural purposes. Summer vegetables are generally grown using floating agriculture.

Forest

Nearly 50% of the national forests are in the coastal region; they comprise both natural forests, including the Sundarbans, and planted forest, adding up to 6243 mi^2 (10,046 km^2) (Ahmed, 2011). Tropical evergreen, mixed evergreen and deciduous forests, and mangrove plantations are the major forest types. The region is home to the world's largest expanse of natural (the Sundarbans) and planted mangroves. Apart from the Surdarbans, natural mangroves can be found in both the Chakaria and Teknaf areas of the southern coastal zone. The former has mangrove plantations that cover about 18,533 acres (7500 ha); it is under great threat because of excessive human interference and the expansion of shrimp cultivation (Rahman and Hossain, 2015; Sarker et al., 2010). No major natural forest area is found in the central coastal zone. Homestead forests are a prominent feature of the rural coastal area. This type of forest is part of agroforestry or social forestry. Agroforestry is in its early stages, though homestead plantations, strip plantations along roads, and mangrove forestry in accreted land have progressed significantly (MoA, 2012).

Rivers and Perennial Water Bodies

Rivers, canals, and permanent water bodies are large aquatic systems of the coastal Bangladesh. As noted, a vast network of tidal rivers and creeks, *beels* (natural depressions in floodplains that are not embanked), *baors* (dead river sections), flood lands and ponds, and low-lying marshy lands provide opportunities for both capture and culture fisheries. Among the major beels are Beel Dakatia (22,239 acres or 9000 ha) in Jessore and Khulna districts and Bhutiar Beel (13,170 acres or 5330 ha) in Bagerhat district. The main land uses are pond aquaculture and shrimp cultivation. Ponds are man-made, used for various purposes, and come in many different sizes, shapes, and depths. The total area of ponds in the coastal area comes to 286,636 acres (116,000 ha), representing about 40% of all ponds in the country in the mid-2000s (Ahmed, 2011). The exclusive economic zone (EEZ) in the Bay of Bengal also supports marine fisheries.

Shrimp Farming

Shrimp farming in the coastal Bangladesh began in the early 1970s. From less than 49,420 acres (20,000 ha) of brackish water ponds in 1980, the area under cultivation expanded to 605,395 acres (245,000 ha) in 2010 (Belton et al., 2011). This represents 1125% increase in area under shrimp cultivation over the last 30 years. The major brackish water shrimp (*bagda*)—producing districts are Bagerhat, Satkhira, Pirojpur, Khulan, Cox's Bazar, and Chittagong. Recently, especially in the Bagerhat and Pirojpur districts, farmers have begun shrimp farming in their paddy fields. Freshwater prawn (*galda*) farming has also grown rapidly over the last 30 years. Freshwater shrimp is now being cultivated in almost all *upazilas* (subunits of the district) of coastal region where sweet water is available; the practice is intensively

Figure 6.1 Shrimp *ghers* in Satkhira district. *Photographs by Salim Momtaz, reproduced with permission.*

concentrated in Bagerhat, Khulna, and Jessore districts. The largest concentration of bagda and galda farms is in the greater Khulna region—Khulna, Satkhira, and Bagerhat districts (Belton et al., 2011).

Traditionally brackish water shrimp farming in coastal region began by trapping tidal waters in coastal enclosures known as "*gher*" (Fig. 6.1). Many coastal polders constructed in the southwestern coastal zone to protect agricultural land from inundation by salt water were turned into large shrimp *ghers* (also see Chapter 7). Although gher farming is often considered (two distinct types: brackish water shrimp farming and freshwater prawn farming) in practice, farmers often stock shrimp, prawn, and white fish in the same system. These mixed farming systems undergo seasonal changes in salinity regimes. A high saline period occurs from January to July. During this time, brackish water shrimp and euryhaline fish species are cultured and harvested. A period of lower salinity occurs from August to December, during which freshwater fish and prawn may be grown with brackish water fish species and/or shrimp. Slightly salt-tolerant transplanted aman rice may also be cultivated in elevated parts of the fields (Table 6.1).

The gher is normally suitable for one crop of transplanted aman rice from August to December, when the water and soil salinity are low (Fig. 6.2). Freshwater for irrigation is not available in coastal region during the dry months (January—May) because both surface water and ground water are highly saline. Agricultural crop production from January to July is difficult, and the soil is acidic in many places. When exposed to the sun, the soil

Table 6.1 Types of *Gher* Farming Systems Based on Species Combination and Integration With Agriculture

System Type	Species Combination
Monoculture	Only shrimp cultured
Shrimp and rice (gher) culture	Shrimp – February to August Rice – August to January Rice and shrimp sometimes cultured together (February to June)
Shrimp and salt culture	Salt – January to March Shrimp – March to November
Shrimp and prawn gher culture with rice and fish	Shrimp – February to August Prawn – July to January Carp and rice culture integrated with shrimp and prawn

Source: Belton, B., Karim, M., Thilsted, S., Jahan, K.E.M., Collis, W., Phillips, M., 2011. Review of Aquaculture and Fish Consumption in Bangladesh. The World Fish Center, Penang, Malaysia, p. 20.

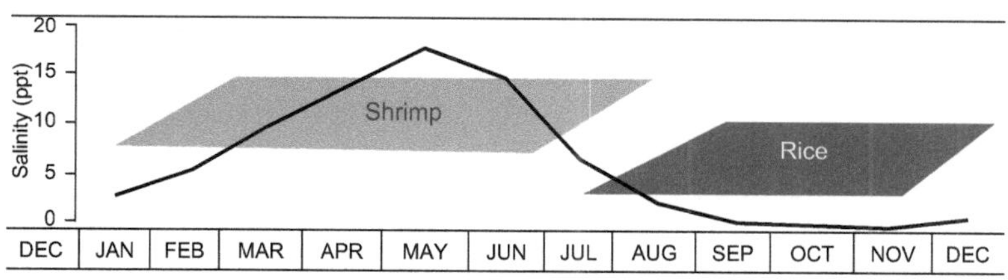

Figure 6.2 Shrimp and rice cultivation in the same field at different times of the year in western coastal zone. *Source: Based on ICM (Integrated Coastal Management), 2010. Best Practices and Lessons Learned from the South Asian Countries of the Bay of Bengal Large Marine Ecosystem (BOBLME); India, Bangladesh, Maldives and Sri Lanka. Workshop Report, Colombo, p. 14.*

acidity increases still further, reducing soil productivity (Belton et al., 2011; ICM, 2010). Instead of keeping the land fallow during high salinity periods (January through July), many farmers find it profitable to use the low-lying and adequately submerged lands for shrimp farming and rice during low salinity period (August through December) (Fig. 6.2).

Based on level of intensity, shrimp farming can be grouped into three categories:

1. *Extensive culture*: shrimp depend entirely on naturally produced organisms in the ponds for their growth;
2. *Improved extensive or semiintensive culture*: depends on both natural food and fertilizer applications to the pond water, as well as supplementary feeds to enhance shrimp growth rate; and

3. *Intensive culture*: depends entirely on artificial feeding and use of intensive management practices, i.e., aeration, draining water between cycles, and adjustments to water quality.

Land previously used for agriculture and mangroves in the coastal region was transformed for shrimp farming, often forcibly, by wealthy and powerful shrimp cultivators. This has resulted in wide-scale land use conflicts and social unrest (Islam, 2006).

Salt Production

At present, the main salt production areas are concentrated in the districts of Chittagong and Cox's Bazar, both in the eastern coastal zone. Chittagong produces about 95% of all salt in the country. About 15% of rural households in this district are salt farmers (Al Mamun et al., 2014). Other salt producing districts are Noakhali and Barisal in the central zone, including the adjoining off-shore islands, and Khulna and Satkhira districts in the southwestern coastal zone. Salt is produced on at least 100,000 acres (40,469 ha) of land and employs more than 50,000 people.

Other Land Uses

The second (Chittagong), third (Khulna), and ninth (Barisal) largest cities of Bangladesh, along with at least 74 municipalities and urban centers are in the coastal region. Thus, a large area of the region is devoted to urban use. Both area and population of all these cities have increased consistently over time. As noted in Chapter 4, the coastal region has a number of beaches, that attract both domestic and foreign tourists. Popular destinations for these tourists are Cox's Bazar, Kuakata beach, St. Martin Island, and the Chakaria Sundarbans. Cox's Bazar is the only well-developed beach town and is the most visited beach by tourists.

St. Martin is the only coral reef island in Bangladesh. It is a small island in the north eastern part of the Bay of Bengal. It is about 5 miles (8 km) west of the northwest coast of Myanmar, at the mouth of the Naf River and covers only 5 mi^2 (8 km^2) with about 5500 permanent residents.

St. Martin Island has a vast variety of ingenious wildlife: coral, mollusk, fish, amphibians, turtles, snails, birds, and mammals. Coconut is the most important cash crop. The island is under tremendous pressure from migrating populations, tourism, coral exploitation, shell and turtle destruction, and overcutting of vegetation. If the island is to remain home for its inhabitants and a precious national treasure, vigorous land-use planning and control are essential.

The Chakaria and Teknaf mangrove forests have been all but destroyed by salt and shrimp farmers, who expanded salt fields and shrimp cultivation by clearing the forests. The only way to preserve the Chakaria and Teknaf mangrove forests is strict land-use controls.

The coastal land uses in Bangladesh have significant spatial variation, particularly between exposed and the nonexposed coastal zones. The districts in the exposed zone have more natural resources (like rivers, forests, and shrimp areas) as well as more inhabitants in both urban and rural areas than nonexposed districts. In the interior, nonexposed districts, land is intensively used for agriculture and aquaculture. Population density in the less-exposed interior is higher than in the exposed coastal districts.

Drivers of Land Use Change

Most coastal lands are suitable for more than one use. Hence, even the smallest piece of land on newly emerged islands in the estuaries suffers the pressure of land use change. This pressure has not only forced changes in land cover but also intensified existing land uses even in endangered areas. Construction of roads and embankments on wetlands, and floodplains, and the interruption of former feeder channels have led to a loss of resilience in the coastal ecosystem (Messerli et al., 2000). Thus, the major drivers of land use changes in the coastal Bangladesh must be understood.

Population Pressure

According to the World Bank, the annual population growth rate in Bangladesh was 1.2% in 2014, but in the coastal region, the rate is higher than in the rest of the country. With a rapidly increasing population, the availability of per capita land is not only declining, but land is converted from productive purposes like crop cultivation to nonproductive purposes like habitation and settlement. This has also led to more intense use of agricultural land, with frequent cultivation, heavier use of labor, and planting high yield salt-tolerant crops. Moreover, inheritance laws contribute to fragmentation of already small agricultural plots, thus reducing effective crop area as well as the use of improved technology. This, in turn, makes intense agricultural production less likely.

Degradation of Land

Coastal land is also continuously degraded or lost to erosion, saline intrusion, water logging, prolonged inundation, and other anomalies, including frequent natural disasters. Land degradation reduces agricultural productivity. Construction of embankments and other water infrastructure, particularly in the central coastal zone, has also contributed to land degradation. In the central coastal zone, land and coastal river bank erosion displace thousands of people every year. Along with erosion, this zone sees some land accretion. However, the new land cannot be immediately used; such land takes two to three decades before it can be productively used. Natural forces of land degradation are often modified by human actions. For example, the southwestern coastal zone has suffered severe ecological catastrophe because of the construction of Farakka barrage in the early 1970s by India near the India-Bangladesh border, which has restricted the flow of water from the Ganges River.

Emergence of Commercial Shrimp Farming

Increased demand and the high price for shrimp at the international markets have led to increased commercial shrimp farming in coastal Bangladesh. The southwestern coastal zone has a history of traditional shrimp farming, but construction of embankments and polders provided further opportunities to expand and intensify shrimp production in this zone. Driven by commercial interests, owners of shrimp farms have been encroaching on agricultural and forest lands, often forcibly (Haque, 2004). As a result, the land use patterns around existing shrimp farms have gradually changed to shrimp farming (Islam, 2006). This has created land use conflicts between powerful land owners and landless and/or land-poor farmers (Ahmed, 2011).

As agricultural lands were converted to shrimp farms, sharecroppers and landless wage laborers lost their livelihoods, and they began to resist the introduction of shrimp cultivation in their areas. This often resulted in violence (Ahmed, 2011). Apart from social unrest, encroachment has also caused wide-scale environmental pollution. Several studies have reported less land for cattle grazing and less drinking water in the coastal region because of the expansion of shrimp farming (Ahmed, 2011; Islam, 2006).

Changes in Socioeconomic Condition

As indicated, the economy and society in coastal Bangladesh are becoming more diverse as the overall economy has developed and become more global. Societal changes have the most visible impact on the land and common property uses. Land is highly concentrated in few hands and landless households have become larger. Local resources like fuel wood, common grazing land, and fishing grounds, once available to the poorest, are dwindling (Rozario, 2002). Limited access to the agricultural and common property land, along with displacement due to erosion push more and more coastal residents below the poverty line. Fortunately, opportunities for nonfarm employment have increased and other opportunities help the poor evade the poverty trap over time. One such opportunity is the evolving tenurial relationships where sharecropping arrangements are giving way to fixed-rent tenancy and medium-term leasing arrangements (Saha, 2002).

Development Interventions

Interventions in both public and private development also contribute to land use change in coastal Bangladesh. Over time, the number of upazila has increased in the coastal region, resulting in increased public services, new infrastructure, and mobility of people and goods. To promote the economy and protect ecology, the government has declared special zones like the Export Processing Zone, Ecological Critical Areas, and World Heritage Sites. Land and water management projects have also contributed to reducing conflicting land uses, maximizing potential land use, and facilitating integrations of land uses in the coastal region. Public interventions thus far implemented in coastal Bangladesh have also had some negative effects on its environment and the livelihoods of the people. For example, a few years after embankments were constructed, drainage congestion, drying of water creeks, soil and water salinization, and flash flooding have occurred (Choudhury et al., 2004).

To accommodate diverse land uses, changed patterns of land use, and land suitability for different uses, in the 1980s, land zoning was implemented as a management approach (Islam, 2006). The Coastal Zone Policy (CZPo) of 2005 proposed zoning coastal land on the basis of suitability: the most suitable zone, a moderately suitable zone, and an unsuitable zone. This policy stresses sustainable resource use and proposes to enforce conservation regulations using alternative employment opportunities (MoA, 2012).

Several studies (e.g., Quader et al., 2004; Hossain and Lin, 2001) compared land suitability for present land uses. Using remote-sensing data on Landsat TM, Hossain and Lin (2002) analyzed suitable areas for mangrove afforestation, shrimp farm location, and salt bed development in the Cox's Bazar area. Most suitable areas identified in the study actually coincide with existing land use. In contrast, Quader et al. (2004) found that the Khulna-Satkhira area was more suitable for shrimp farming than Cox's Bazar.

Zoning can also be integrated with time sharing. This will minimize unplanned horizontal expansion of any land use, particularly shrimp farming or salt production, reduce land use conflict, and minimize productivity from smaller areas through vertical integration. Vertical integration was proposed in the Delta Development Project (DDP), which suggested combining rice cultivation and shrimp farming on the same land (in rotation) would have no direct negative effects on rice yield (Nishat, 1988). This integration pattern is emerging in the southwest coastal zone where rice and fish are simultaneously cultivated on the same land. In Cox's Bazar, farms largely alternate between shrimp and salt production. Similarly, trees can be planted on embankments to reduce soil erosion. Both actions could improve the socioeconomic condition of the poor as well as maintain the ecological balance in the coastal region (Islam, 2006).

Using the existing information about the coastal region, it can be divided into eight broader areas: (1) shrimp (brackish-water), (2) shrimp (sweet-water), (3) salt-shrimp, (4) forest, (5) mangrove (including Sundarbans), (6) urban and commercial (industrial, port, export-processing zones, and ship-breaking yards), (7) tourism, and (8) agriculture. Even with many agencies agreeing on appropriate land zoning, only a start has been made. This version of land zoning should be the basis for detailed land zoning in the coastal region. In addition, the government should take the initiative, studying the impact of land use change, reviewing regularly to detect socio-economic and environmental changes, and ensuring a continuous monitoring of land use changes and identifying areas under land use change.

COASTAL MANAGEMENT

Management of the coastal zone of Bangladesh is very important because the zone accounts for about one-fourth of the country's land with one-third of the country's population. Relatively abundant coastal resources, which are vulnerable to both natural and man-made extreme events, are vital to sustainability in economic and social development among its impoverished residents. The coastal zone provides distinctive economic opportunities and development potential, and it contributes 29% to the GDP of the country (Islam, 2008).

However, coastal management in Bangladesh is a challenge because management must address issues related to both natural and socioeconomic systems. Poverty stricken groups account for about three-fourths of the coastal population. Their livelihoods have been adversely affected by a combination of natural and man-made hazards and socioeconomic activities, such as tropical cyclones and associated storm surges, river and beach erosion, lack of adequate domestic water, drainage congestion and water logging, saline water intrusion, increases in soil salinity, conflicts between shrimp and rice farmers, degraded ecosystems, deteriorating water ways, risk of sea level rise, and inadequate infrastructural facilities.

Early Initiatives of Coastal Zone Management

Large initiatives for developing the coastal zone started in the early 1960s when Bangladesh was still part of Pakistan. These initiatives were undertaken to control degradation of coastal resources because of increased salinization, flooding, and erosion, and included such projects as constructing coastal embankments, creating the coastal greenbelt, and reclaiming lands, as well as completing many surveys and studies (Iftekhar, 2006). Between independence in 1971 and before the Bangladesh government formulated the CZPo in 2005, many projects were undertaken for the coastal region. These projects can be divided into two broad categories: single sector/single agency led projects and multiagency participation projects (Huda, 2004).

The single sector/single agency led approach was very common particularly in the early years, and many public departments (e.g., forestry, agriculture, water, fisheries, and environmental) initiated projects focused on their specific sector (Iftekar, 2006). This approach allowed specific sectoral policies for natural resources to be developed, such as agricultural, forestry, and fishery policies. These policies enabled the establishment of different ecosystem protection zones, like reserved forests, protected forests, wildlife sanctuaries, national parks, ecologically protected areas, and safari parks. In many cases, this approach produced desirable results, but some negative effects have also occurred. This approach created conflicts among different public departments and resulted in the mushrooming of divergent plans and regulations. Furthermore, this approach suffered from a narrow, departmentalized administration and unilateral action based on departmental priorities. It failed to resolve complex problems that required the attention of multiple departments (Iftekar, 2006).

As the various shortcomings of single sector/single agency approach became apparent, the Bangladesh government became interested in the multiagency participation approach. This approach was first applied in 1981 with the launching of the DDP, financed by the Netherlands and executed by the Bangladesh Water Development Board. Through this approach each participatory agency processed and maintained separate components of projects reflecting an agency's activities and financial outlays.

Each agency maintains full control over the financing and management of its own components. Because of its flexibility, this approach gained widespread support among many public agencies as the preferred approach (Huda, 2004).

For the early initiatives (1971–2005), many projects were undertaken by a number of government departments. Apart from coastal embankment projects, in 1964, the Forest Department started a massive plantation of mangroves in the coastal areas forming a coastal green belt. Approximately 296,520 acres (120,000 ha) were planted with mangroves from 1965 to 1990 (Saenger and Siddiqi, 1993). Plantations of mangroves in the coastal areas still continue today. The mangrove belt saved many lives and much property from tropical cyclones, tidal surges, and other water actions. The government also began to establish cyclone shelters in remote coastal areas and coastal islands in the early 1970s to protect the lives of coastal residents from natural disasters. The number of cyclone shelters was 449 in 1990, increased to 1921 in 1997, and to nearly 3800 in 2009. Currently, more than 3851 shelters are scattered around the coastal districts; another 2000 shelters are needed (Rami, 2016). Most of these shelters are also vital social infrastructures; schools, clinics, community centers, and storehouses during normal times.[4]

Early cyclone warning systems began to be used in Bangladesh after the devastating cyclone of 1970, and the ability to forecast cyclones and storm surges has improved over the years. However, greater precision in forecasting and better ways of disseminating warnings both need further improvement. These initiatives are part of an integrated and comprehensive coastal management system in Bangladesh. Cooperation and support from development partners, notably the World Bank, the Asian Development Bank, and bilateral donors (e.g., the government of The Netherlands and Denmark) have been crucial in building these systems. The Netherlands shared its experience and skills in coastal management and supported Bangladesh in coastal land reclamation, delta development, polder/embankment management, and estuary studies.

A number of other initiatives have helped develop area-specific management in the coastal zone of Bangladesh (Islam, 2004):

- Creation of the Off-Shore Islands Development Board (1977–82)
- Formation of the Bangladesh National Conservation Strategy (1987)
- The UN/ESCAP-GoB Coastal Environment Management Plan for Bangladesh (1987)
- The Coastal Area Resources Development Plan (1988)

- Formation of the Special Parliamentary Committee on Coastal Area Development (1988–90)
- Assessment of the vulnerability of Bangladesh to sea level rise (1992–94), supported by The Netherlands
- Beginning the Bangladesh Integrated Coastal Zone Management (ICZM) process (1995)

The above initiatives laid the groundwork for the Bangladesh government's initiative in 1999 to embark on a process for managing the coastal zones of ICZM. In 1999, the government published a policy note on ICZM. The note was prepared through a joint mission of the Bangladesh government, The Netherlands, and the World Bank. Eventually, the preparatory phase for ICZM in Bangladesh was carried out between February 2002 and June 2006 with support from the governments of The Netherlands and the UK.

Integrated Coastal Zone Management Plan Project

The lack of clear-cut government policy was a main barrier for initiating an integrated approach for conserving and managing the coastal ecosystem. In response, in 1999, the Bangladesh government adopted a policy note, expressing its support for the CZMP (MoWR, 1999). It is a process for managing the coast using an integrated approach, with all aspects of the coastal region considered in an attempt to achieve sustainability. The Integrated Coastal Zone Management Plan (ICZMP) project was executed by the Water Resources Planning Organization (WARPO) of the Ministry of Water Resources between February 2002 and December 2005. With the goal of developing capabilities and priority actions for coastal development in Bangladesh, the ICZMP process has two main components: the coastal zone policy (CZPo) and the coastal development strategy (CDS).

Coastal Zone Policy

An important component of the ICZMP is the CZPo, approved by the government in January 2005 and given a policy framework for long-term ICZMP. It was created through a process of multilevel consultations over 3 years (2002–04) (Islam, 2005). The CZPo establishes the goal of the ICZMP as "to create conditions, in which the reduction of poverty, development of sustainable livelihoods and the integration of the coastal zone into national processes can take place" (MoWR, 2005, p. 3). One of the objectives of the CZPo was to preserve and enhance critical ecosystems and ecological processes. Other objectives included economic growth, basic needs and opportunities for livelihoods, reduction of vulnerabilities, sustainable management of natural resources, equitable distribution, empowerment of communities, and women's development and gender equity (MoWR, 2005).

The CZPo states that the coastal development process aims to meet the national goal for economic growth, poverty reduction, and social development; the code of conduct for responsible fisheries; and the code of conduct for responsible mangrove management as well as conform to other international conventions and treaties to achieve the targets of the millennium development goals. Strategy and investment programs have been developed to implement policy objectives. To achieve its objectives, the CZPo brought relevant government and nongovernmental institutions and agencies together. Thirty-four different government agencies from 13 different ministries were involved. These agencies were structured into an *Inter-Ministerial Steering Committee*, which provided policy guidelines, and an *Inter-Ministerial Technical Committee*, which contributed technical knowledge and helped resolve intergovernmental conflicts. The technical committee also had representatives from universities, NGOs, the private sectors, and the civil society as a whole. These arrangements and representatives gave a feeling of wider ownership and participation (MoWR, 2005).

Coastal Development Strategy

The CDS is the linchpin of the ICZM process, linking the CZPo to development programs and interventions. It builds on existing national sectoral strategies and plans and on the document on National Strategy for Accelerated Poverty Reduction. The government approved and adopted the CDS in 2006. The CDS adopted nine strategic priorities and three routes of implementation. One of the nine priorities is to use of coastal land optimally, recognizing that land use in the coastal areas is diverse, competitive, and often conflicting. Erosion of land, especially in the Meghna estuary and on off-shore islands, is widespread. Land is also degrading because of increased salinity and water logging. Land in the coastal region has also been accreting and is being gradually developed. This new land should contribute significantly to the regional economy and well-being of coastal residents. The CDS emphasizes land zoning as an effective instrument to optimize the use of coastal lands (MoWR, 2005).

Implementing of both CZPo and CDS was slow because of lack of funds, political instability, and major natural calamities. Neither have been included in key national programs and policies. Only one of the identified ICZM investment programs has entered the Government's Annual Development Programs; the Ministry of Land (MoL) began implementing coastal land zoning, an important element of the program. In the absence of proactive institutional leadership and because of the lack of national focus and funding, development partners have also changed their priorities. However, two cyclones, Sidr in 2007 and Aila in 2009, have again brought the coastal zone to national attention. Global recognition of the impact of climate change, moreover, emphasizes the need for further national and international attention in the coastal zone.

Country Programming Framework

In 2010, the MoA and FAO prepared a Country Programming Framework (CPF) that identified the coastal region of Bangladesh as an ecologically depressed and economically deprived area but also as having tremendous potential for agricultural development, poverty reduction, and food security (MoA, 2012). The decision to revise the CPF was taken jointly by FAO and the government of Bangladesh in late 2012 in response to changes in the planning and strategic framework of both the FAO and the Bangladesh government. This new document replaces the previous one. The initial CPF was revised to realign the FAO program with the most up-to-date national development priorities, fill any gaps that have become apparent, and focus FAO's efforts on a reduced but more concrete set of country priority areas (FAO, 2014); the revised CPF was put into place for the 4 years from 2014 to 2018. Like the initial CPF, the revision recognizes that the coastal region has the least developed agriculture in the country but does have the potential for further integration of crop, fish, and livestock farming.

The objectives and priority areas of the revised CPF are (1) to reduce poverty and enhance food security and nutrition; (2) to enhance agricultural productivity through diversification/intensification, sustainable management of natural resources, use of quality inputs, and mechanization; (3) to improve market linkages, value addition, and quality and safety of the food system; (4) to further encourage generating and adapting technology by better linking producers to research and extension; and (5) to increase resilience of communities so they can withstand shocks like natural disasters, health threats, and other risks to livelihoods (FAO, 2014). All these objectives and priorities are relevant to coastal region of Bangladesh.

Master Plan for Agricultural Development in Southern Region of Bangladesh

Master Plan for Agricultural Development in Southern Region (MASR) was prepared by the FAO for 14 of the 19 coastal districts in Bangladesh and was approved by the Bangladesh government in 2012 (MoA, 2012). The plan has aligned with the Sixth Five Year Plan (2011–15) and other government planning documents. Forests, including all reserve forests, unclassified state forests, coastal plantations, and protected areas, have been excluded from the purview of the Master Plan. This master plan provides a detailed plan, identifying investment needs and priority programs, and is directly linked to the Country Investment Plan (CIP). Both the government and the FAO prepared the CIP in 2011 for developing food security. The MASR further broke down the CIP in eight program areas with a regional focus that would operationalize the CIP (Table 6.2). The CIP has been prepared for a 5-year period (2011–15), while the MASR runs for 10 years (2012–22).

As indicated, the MASR has been developed in accordance with, and as a logical consequence of, several other policies and programs that are already on board. It is both vertically and horizontally linked with a host of public

Table 6.2 Linkages of the MASR With CIP

	Program Areas	
Component	**CIP**	**MASR**
Food availability	Sustainable and diversified agriculture through integrated research and extension	√
	Improved water management and infrastructure for irrigation purposes	√
	Improved quality of input and soil fertility	√
	Fisheries and aquaculture development	√
	Livestock development, with focus on poultry and dairy production	√
Food access	Improved access to markets, value-addition in agriculture, and to nonfarm incomes	√
	Strengthened capacities for implementation and monitoring of national food policy (NFP) and CIP actions	
	Enhanced public food management systems	
	Institutional development and capacity development for more effective safety nets	√
Food utilization	Community-based nutrition programs and services	√
	Orient food and nutrition program through data	
	Food safety and quality improvement	

Source: MoA (Ministry of Agriculture), 2012. Master Plan for Agricultural Development in Southern Region of Bangladesh. Dhaka, p. 23.

plans that promote sustainability, economic growth, and poverty reduction. The plan particularly focuses on (1) increasing agricultural production, (2) improving water management and rejuvenating degraded lands to move them back into production, (3) developing climate resilient infrastructure and improving surface water irrigation system, (4) improving productivity of brackish water shrimp and capture fisheries, and (5) developing smallholder poultry and dairy operations (MoA, 2012).

The MASR is a multilevel and multisectoral phenomenon involving several ministries, departments, and other entities, including local government, and the private sector. Certain programs in the plan are local, while others involve several coastal districts with the recommendation that relevant line agencies will implement their programs following existing administrative norms. However, institutional arrangements must facilitate coordination among stakeholders whose participation is important and relevant to achieve the objectives of the MASR. The total investment need is estimated at US $7225 million (MoA, 2012). With the sincere efforts and appropriate allocations among concerned ministries/divisions, and the generous support of development partners, the goals of the MASR should be realized by 2021.

Bangladesh Delta Plan 2100

With technical and financial support from The Netherlands, the Bangladesh government is also in the process of completing Bangladesh Delta Plan (BDP) 2100. This plan comprises a long-term (50−100 year), integrated, holistic vision and strategy to improve safe living and sound economic development in the Bangladesh delta, while also taking into account climate change. The Bangladesh Delta Plan 2100 (BDP 2100) is an umbrella investment plan interacting with sectoral plans, both absorbing input as well as providing coordination and a long-term plan with the appropriate perspective and strategy. For BDP 2100, the Bangladesh Delta is defined as covering almost the whole (85%) country (Consultant Team BanDuDeltAS, 2014).

The overall objective of BDP 2100 is "to realize sustainable and a commonly agreed upon strategy, with all relevant stakeholders, for an optimum level of water safety and food security as well as sustained economic growth of Bangladesh and a framework for its implementation. As such the BDP 2100 will contribute to the overall development of Bangladesh and needs to be incorporated in the seventh and subsequent Five Year plans and also other related plans" (Consultant Team BanDuDeltAS, 2014, p. 6). BDP 2100 has the following specific objectives:

- Support an enabling sociopolitical climate so BDP 2100 can be formulated and implemented,
- Create a common, inclusive, and documented knowledge base for water, land, and related natural resources, as well as spatial planning in the Bangladesh delta with emphasis on water, land, agriculture, public health, environment, disaster management, food security, economic growth, and spatial and ecological development, and using this knowledge base to undertake integrated analyses and scenario building with the main stakeholders,
- Develop a Delta Framework and prepare a draft Act to establish a Delta Framework encompassing necessary and agreed upon reforms of the current institutional framework as well as the involved governmental organizations so they are prepared to formulate BDP 2100 and implement the plan in an integrated, targeted, inclusive, broadly supported, and transparent manner,
- Create with the main stakeholders a delta vision, goals, and measures that use a policy-making process that is adaptive, inclusive, and transparent, ensure the transparent input of both the public and private sector, and provide the anticipated results in the BDP 2100,
- Facilitate entrepreneurship of the private sector, including multinational companies, small-scale and medium enterprises, and NGOs, in Bangladesh and in The Netherlands for ideas, concepts, and innovations that contribute to the BDP 2100 through links to public institutions and other relevant organizations in Bangladesh and that promote short-term regional and sectoral developments leading to appropriate future governance of water, land, and related resources as well as spatial planning in Bangladesh delta.

Concluding Remarks

In recent years, coastal planning and management have had serious attention from the Bangladesh government. Implementing some of the coastal initiatives, such as the ICZM, was difficult for financial and political reasons, particularly between 2002 and 2006, but once those reasons were resolved, the political climate seemed to favor implementing both policies and programs associated with developing the coastal region. The ICZM and other relevant programs not only help mitigate the effects of natural disasters and other environmental issues, but also provide opportunities to exploit resources beneficially and sustainably. These programs are necessary for the survival of millions of people living in the coastal region at the mercy of nature. Bangladesh has no other alternatives but to develop and implement programs for the region and then to continuously monitor progress toward achieving program goals.

The Bangladesh government seems to be committed to improving the living conditions of coastal residents. The government has also been receiving technical and financial assistance from many external sources: the government of The Netherlands, the World Bank, and the United Nations, as well as other development partners. Additionally, Bangladesh has received funding from the European Union (EU) and other European countries for adapting to climate change and implementing adaptation and mitigation measures. Most of these measures were designed specifically for coastal Bangladesh. The country is well in advance of many other developing countries in mitigation and adaptation as well as in developing the capacity for facing the effects of climate change (Rashid and Paul, 2014).

ENDNOTES

1. Three rice varieties are *aman*, *aus*, and *boro*. There are two types of aman: broadcast and transplanted. Both aus and boro rice are grown in the coastal area as transplanted crops. For more about these rice varieties see the next chapter.
2. The three cropping seasons of Bangladesh and the coastal region are discussed in the next chapter.
3. Cropping intensity is the average number of crops grown on a unit of land over a year.
4. For more details about public cyclone shelters, see Chapter 8.

REFERENCES

Ahmed, A., 2011. Some of the major environmental problems relating to land use changes in the coastal areas of Bangladesh: a review. Journal of Geography and Regional Planning 4 (1), 1–8.

Al Mamun, M.A., Raquib, M., Tania, T.C., Rahman, S.M.K., 2014. Salt industry of Bangladesh: a study in the Cox's Bazar. Banglavision 14 (1), 7–17.

ASB (Asiatic Society of Bangladesh), 2004. Banglapedia – National Encyclopedia of Bangladesh. ASB, Dhaka.

Belton, B., Karim, M., Thilsted, S., Jahan, K.E.M., Collis, W., Phillips, M., 2011. Review of Aquaculture and Fish Consumption in Bangladesh. The World Fish Center, Penang, Malaysia.

Baulies, X., Szejwach, G., 1998. LUCC Data Requirements Workshop Survey of Needs, Gaps and Priorities on Data for Land-Use/Land-Cover Change Research. Organized by IGBP/IHDP-LUCC and IGBP-DIS, Barcelona, Spain, November 11–14, 1997.

Brammer, H., 2002. Land Use and Land Use Planning in Bangladesh. Dhaka University Press, Dhaka.

CPA (Chittagong Port Authority) News, 2016. Historical Background, April 24. www.cpa.gov.bd.

Choudhury, N.Y., Paul, A., Paul, B.K., 2004. Impact of coastal embankment on the flash flood in Bangladesh: a case study. Applied Geography 24, 241–258.

Consultant Team BanDuDeltAS, 2014. Inception Report: Bangladesh Delta Plan 2100, Formulation Project. Dhaka.

Food and Agriculture Organization (FAO), 2014. Bangladesh Country Programming Framework: Towards Sustainable Agriculture and Improved Food Security & Nutrition. Dhaka.

GLP (Global Land Project), 2005. Science Plan and Implementation Strategy. IGBP Report No. 53/HDP Report No. 19. IGBP Secretariat, Stockholm.

Haque, A.K.E., 2004. Sanitary and Phyto-Sanitary Barriers to Trade and its Impact on the Environment: The Case of Shrimp Farming in Bangladesh. IUCN Bangladesh Country Office, Dhaka.

Hossain, M.S., Lin, C.L., 2001. Land Use Zoning for Integrated Coastal Zone Management. ITCZM Monograph No. 3. Aisian Institute of Technology, Bangkok, Thaliand.

Hossain, M.S., Lin, C.K., 2002. Land suitability analysis for integrated coastal zone management in Cox's Bazar, Bangladesh: a remote sensing and GIS approach. In: Proceedings of the Coastal Zone Asia Pacific Conference, Bankok, Thailand.

Huda, A.T.M.S., 2004. Interagency collaboration for integrated coastal zone management: a Bangladesh case study. Coastal Management 32, 89–94.

Iftekhar, M.S., 2006. Conservation and management of the Bangladesh coastal ecosystem: overview of an integrated approach. Natural Resources Forum 30, 230–237.

ICM (Integrated Coastal Management), 2010. Best Practices and Lessons Learned from the South Asian Countries of the Bay of Bengal Large Marine Ecosystem (BOBLME); India, Bangladesh, Maldives and Sri Lanka. Workshop Report, Colombo.

Irfanullah, H.M., Azad, M.A.K., Kamruzzaman, M., Waheed, M.A., 2011. Floating gardening in Bangladesh: a means to rebuild lives after devastating flood. Indian Journal of Traditional Knowledge 10 (1), 31–38.

Islam, M.R., 2008. ICZM initiatives and practices in Bangladesh. In: Krishnamurthy, R., et al. (Eds.), Integrated Coastal Zone Management (ICZM) – the Global Challenge. Research Development Services, New Delhi, pp. 73–91.

Islam, M.R., 2006. Managing diverse land uses in coastal Bangladesh: institutional approaches. In: Hoanh, C.T., Toung, T.P., Gowing, J.W., Hardy, B. (Eds.), Environment and Livelihoods in Tropical Coastal Zones: Agriculture-Fishery-Aquaculture Conflict. IRRI, Manila, pp. 237–240.

Islam, M.R., 2005. ICZM program in Bangladesh: an overview of policy and strategy initiatives. In: AIT (Ed.), Proceedings of the International Symposium on Floods in Coastal Cities under Climate Change Conditions. Asian Institute of Technology, Asia Pacific Network for Global Change Research, Bangkok, pp. 107–113.

Islam, M.R. (Ed.), 2004. Where Land Meets the Sea: A Profile of the Coastal Zone of Bangladesh. The University Press Limited, Dhaka.

Kasperson, J.X., Kasperson, R.E., Turner, B.L. (Eds.), 1995. Regions at Risks: Comparisons of Threatened Environments. UN University Press, Tokyo.

Lambin, E.F., Geist, H.J., Lepers, E., 2003. Dynamics of land-use and land-cover changes in tropical regions. Annual Review of Environmental Resources 28, 205–241.

Messerli, B., Grosjean, M., Hofer, T., Nunez, L., Pfister, C., 2000. From nature-dominated to human-dominated environmental change changes. Quaternary Science Reviews 19, 459–479.

MoA (Ministry of Agriculture), 2012. Master Plan for Agricultural Development in Southern Region of Bangladesh. Dhaka.

MoWR (Ministry of Water Resources), 2005. Coastal Zone Policy. Ministry of Water Resources, Dhaka.
MoWR (Ministry of Water Resources), 1999. Integrated Coastal Zone Management Concepts and Issues: A Government of Bangladesh Policy Note. Ministry of Water Resources, Dhaka.
Mukerjee, R., 1938. The Changing Face of Bengal – a Study in Riverine Economy. University of Calcutta, Calcutta.
Mustard, J.F., Defries, R.R., Fisher, T., Moran, E., 2004. Land use and land cover change pathways and impacts. In: Gutman, G., Janetos, A.C., Justice, C.O., Moran, E.F., Mustard, J.F., Rindfuss, R.R., Skole, D., Turner II, B.L., Cochrane, M.A. (Eds.), Land Change Science: Observing, Monitoring, and Understanding Trajectories of Change on Earth's surface. Kluwer, Boston, pp. 1–20.
Nishat, A., 1988. Review of present activities and state of art of the coastal areas of Bangladesh. In: Coastal Area Resource Development and Management. Part II. Coastal Area Resource Development and Management Association (CARDMA), pp. 23–35. Dhaka.
Quader, O., Islam, Z., Rahman, H., Sarkar, M.H., Khan, A.S., 2004. Suitable site selection of shrimp farming in the coastal areas of Bangladesh using remote sensing techniques (4 S Model). In: Proceedings of the XX ISPRS (International Society for Photogrammetry and Remore Sensing) Congress, July 12–23, 2002, Istanbul, Turkey.
Rahman, M.R., Hossain, M.B., 2015. Changes in land use pattern at Chakaria Sundarbans mangrove forest in Bangladesh. Bangladesh Research Publications Journal 11 (1), 13–20.
Rami, M., 2016. Jiban Bachatey 11 Lakh Manus Cyclone Shelteray (11 Lakh People are in Cyclone Shelter to Save Their Lives). Ittefaq, May 22.
Rashid, H., Paul, B.K., 2014. Climate Change in Bangladesh. Lexington Books, Lanham, Maryland.
Rozario, S., 2002. Gender dimensions of rural change. In: Toufique, K.A., Turton, C. (Eds.), Hands Not Land: How Livelihoods are Changing in Rural Bangladesh. Department for International Development and Bangladesh Institute for Development Studies, Dhaka, pp. 137–149.
Saenger, P., Siddiqi, N.A., 1993. Land from Sea. http://epubs.scu.edu.au/cgi/viewcontent.cgi?article=1646&content=esm_pubs.
Saha, B.K., 2002. Rural development trends: what the statistics say. In: Toufique, K.A., Turton, C. (Eds.), Hands Not Land: How Livelihoods are Changing in Rural Bangladesh. Department for International Development and Bangladesh Institute for Development Studies, Dhaka, pp. 123–136.
Sarker, S., Kuri, K.C., Chowdhury, M.S., Rahman, M.T., 2010. Mangrove: a livelihood option for coastal community of Bangladesh. Bangladesh Research Publications Journal 3 (4), 1187–1192.
Veldkamp, A., Lambin, E., 2001. Predicting land-use change. Agricultural Ecosystem and Environment 85, 1–6.

CHAPTER SEVEN

Nonstructural Adaptation

DEFINITIONS OF RELEVANT TERMS

Besides preparing successive assessment reports on the status of climate change, one of the central goals of the Intergovernment Panel on Climate Change (IPCC) is to promote mitigation of climate change and adaptation to its adverse impacts. "In general, adaptation is aimed at reducing the effects and mitigation is aimed at reducing the causes of climate change, in particular the emissions of the gases that give rise to it" (Houghton, 2006, p. 11). Objectives and methods of mitigation and adaptation have been expanded by IPCC in its Annex I: Glossary (Verbruggen, 2015):

- *Mitigation*: Technological change and substitution that reduce resource inputs and emissions per unit of output. Although several social, economic, and technological policies would produce an emission reduction, with respect to climate change, mitigation means implementing policies to reduce greenhouse gas emissions and enhance sinks.
- *Mitigative capacity*: This is a country's ability to reduce anthropogenic greenhouse gas emissions or to enhance natural sinks, where ability refers to skills, competencies, fitness, and proficiencies that a country has attained and depends on technology, institutions, wealth, equity, infrastructure, and information.
- *Adaptation*: Initiatives and measures to reduce the vulnerability of natural and human systems against actual or expected climate change effects. Various types of adaptation exist, e.g., anticipatory and reactive, private and public, and autonomous and planned. Examples are raising river or coastal dikes, the substitution of more temperature shock resistant plants for sensitive ones.
- *Adaptive capacity*: The whole of capabilities, resources, and institutions of a country or region to implement effective adaptation measures.

The IPCC Working Group II (WGII) assesses the vulnerability of socioeconomic and natural systems to climate change, negative and positive consequences of climate change, and options for adapting to it. The Working Group III (WGIII) assesses options for mitigating climate change through limiting or preventing greenhouse gas emissions and enhancing

Climatic Hazards in Coastal Bangladesh
ISBN 978-0-12-805276-1
http://dx.doi.org/10.1016/B978-0-12-805276-1.00007-7

activities that remove them from the atmosphere (IPCC, 2007a). Although IPCC recognizes that there can be conflicts between adaptation and mitigation in relation to the implementation of specific national policy options, it stresses that both of them must be treated as twin issues (Hamin and Guran, 2009). The IPCC Fourth Assessment Report warns about the peril of not following both of these measures simultaneously (IPCC, 2007b, p. 65):

> *There is a high confidence that neither adaptation nor mitigation alone can avoid all climate change impacts. Adaptation is necessary both in the short term and longer terms to address impacts resulting from the warming that would occur even for the lowest stabilization scenarios assessed...Unmitigated climate change would, in the long term, be likely to exceed the capacity of natural, managed and human systems to adapt. Reliance of adaptation alone could eventually lead to a magnitude of climate change to which effective adaptation is not possible, or will only be available at very high social, environmental and economic costs.*
>
> ***Cited by Hamin and Gurran (2009)***

Despite this warning, our focus in this chapter and in the next chapter (Chapter 8) is exclusively on adaptation, primarily because Bangladesh contributes very little greenhouse gas emissions (less than 0.2% of total global emissions) and, therefore, any mitigation measures by it would not result in a meaningful reduction of global greenhouse gas emissions. Perhaps, more pragmatically, Bangladesh has a long tradition of adaptation to different types of climatic hazards, especially river floods and tropical cyclones (Rasid, 2000; Rasid and Mallik, 1995; Rasid and Paul, 1987; Paul, 1984; Haque and Zaman, 1993; Haque and Blair, 1992; Paul and Routray, 2011; Paul et al., 2010). From planning and epistemological perspectives, such research on human adjustments to natural hazards, which predates the birth of the IPCC by several decades, has direct application to climate change adaptation in coastal Bangladesh. Sociological research on "human adjustment to the risk of environmental extremes" has identified several mechanisms of human adjustment to the natural environment (Mileti, 1980). Among them, the following are relevant to the adaptation process: (1) engineering measures that include technological inventions and their application, (2) regulatory approaches that define public policy and social control, and (3) distributional mechanisms that specify the movement of people, activities, and resources (Mileti, 1980). Gilbert White (1974) pioneered geographical research on adjustment to natural hazards and introduced several terms that are directly relevant to climatic adaptation in coastal Bangladesh:

> Natural hazard: *An interaction of people and nature governed by the coexistent state of adjustment in the human use system and the state of nature in the natural event system.*
>
> ***White (1974), p. 5***

Tropical cyclone is clearly an example of natural hazard. Global warming—induced sea level rise, on the other hand, may not be considered as a natural hazard but its impacts are likely to exacerbate other natural hazards, particularly storm surges associated with tropical cyclones.

> Hazard perception: *Individual organization of stimuli relating to an extreme event. The interest is in finding out how people view the "occurrence or threat of the extreme event and of the opportunities open to them in coping with the event".*
>
> ***White (1974), p. 4***

There are three types of responses to natural hazards (White, 1974):

- *Indigenous adjustments*: "Folk, or preindustrial, adjustments which involve a wide range of adjustments requiring more modifications in behavior in harmony with nature than control of the nature, are flexible and easily abandoned, are low in capital requirements, requiring actions only by individuals or small groups, and can vary drastically over short distances" (White, 1974, p. 5).
- *Modern technological, or industrial adjustments*: These involve "a more limited range of technological actions emphasizing control of nature, are inflexible and difficult to change, are high in capital requirements…" (White, 1974, p. 5).
- *Comprehensive, or postindustrial adjustments*: These combine features of both indigenous and modern technological adjustments allowing "a larger range of adjustments, greater flexibility and variety of capital and organizational requirements" (White, 1974, p. 5).

According to the preceding definitions, the existing and planned climate change adaptation projects in coastal Bangladesh provide an example of comprehensive or postindustrial adjustments. In climate change literature, adaptation has emerged as the preferred term for all of these activities as it is used extensively in both scholarly work and IPCC documents. Therefore, we have incorporated White's (1974) concept of comprehensive, or postindustrial, adjustments and some of its terms like indigenous adjustments under the broader rubric of adaptation. Further, for the sake of clarity and distinction, we have classified all relevant adaptation measures in coastal Bangladesh into two broad categories: (1) nonstructural adaptation (this chapter) and (2) structural adaptation (Chapter 8). Among others, some of the leading nonstructural adaptation projects include the following activities:

- *Indigenous adjustments of housing structures*: These apply mostly to traditional features of rural homes that are designed for living with rising sea levels and for withstanding the effects of cyclones and storm surges.
- *Modern scientific nonstructural measures*, such as agricultural adjustments to salinity through development of salt-resistant crops.
- *Disaster management initiatives* designed to save lives during cyclone disasters.

While we have defined structural adaptation as those projects requiring major engineering constructions, some of the projects are at the border line of indigenous adjustments and engineering interventions. The triple-F platform for forest, fruit/food, and fish is an example of this, as it may or may not require major engineering work (see Chapter 8). Similarly, expansion of forests by planting mangrove and other species is not an engineering project but we have included it under structural measures because large-scale intervention using machinery and manpower may be necessary for its effective implementation.

BROADER CONTEXTS OF ADAPTATION AND ADAPTIVE CAPACITY

The concept of adaptation, deeply rooted in natural hazards and disasters research, encompasses a wide spectrum of activities and practices to minimize vulnerability to hazards. In the context of climate change, adaptation is expected to enhance impacted residents' resilience to climate change, which may be characterized as a form of climate-proofing. Contrary to such a broad approach, the term has been used somewhat narrowly in some of the recent interdisciplinary literature to refer to "structural modification of human practices to accommodate unavoidable climate change" (Page, 2007). Emphasizing its broader application, Smit et al. (2000) suggest that adaptations to climate change are not just discrete technical measures but modification of existing practices with respect to multiple (climatic and nonclimatic) stimuli and conditions. Successful adaptation is expected to reduce a country's vulnerability to climate-related natural disasters and sea level rise and thus enhance its adaptive capacity to combat climate change. Because of the uncertainty of climate change impacts, a more appropriate interpretation of adaptation to climate change entails "adjustment by human systems to the actual or expected physical effects of climate change, variability and extreme conditions" (Smit and Pilifosova, 2003). Using IPCC's broader approach, Adger et al. (2005) have defined adaptation to climate change as an adjustment in ecological, social, or economic systems in response to observed or expected changes in climatic stimuli and their effects and impacts in order to alleviate adverse impacts of change or take advantage of new opportunities. In this context, the adaptive capacity of a society or a system can help in building "the ability of a system to adjust to climate change (including climate variability and extremes) to moderate potential damages, to take advantage of opportunities, or to cope with the consequences" (Wall and Smit, 2005, p. 113). To put it more succinctly, adaptive capacity is the ability of the people or a system to easily and quickly cope with the impact of a disaster (Paul, 2011). To cope effectively with a disaster or to minimize exposure, active strategies are required to manage knowledge and resources in response to a disturbance (Easterling, 2011). Exposure measures the degree to which a system or a unit is vulnerable to a stress or disturbance. In the language of sustainability, the more adaptive capacity a system has, the more resilient it is—the system maintains its integrity despite stresses from internal and external factors and hence is more sustainable (Wall and Smit, 2005). Resilience is the capacity of a system to absorb disturbance and reorganize while undergoing change so as to retain essentially the same predisturbance process, form, identity, and feedbacks (Walker et al., 2004).

Interpreted in such broad contexts, a portfolio of adaptation designed to minimize risks associated with climate change may include a wide range of responses, some of which may be behavioral at individual levels, some may be managerial or technical at community levels, while others may be related to national policies and planning. Strictly from operational perspectives, adaptation to climate change in coastal Bangladesh may be classified broadly into two categories: *autonomous adaptation* requiring very little external input and *planned adaptation* with external intervention (Adger et al., 2003). Adaptation to river flood regimes in major floodplains of Bangladesh provides a classic example of autonomous adaptation. For generations, the floodplain residents of Bangladesh have lived with normal flood regimes by building their homesteads on elevated tracts of lands, such as abandoned natural levees that have been further elevated from time to time by adding excavated earth from surrounding lands (Rasid, 2000). The floodplain residents have taken such initiatives using their own resources without any external help. Similarly, by trial and error perhaps over a period of millennia they have perfected different varieties of rice adjusting them to varied flood depths (Rasid and Mallik, 1995). Most of these experiments with rice crops have been autonomous, i.e., based on farmers' indigenous technical knowledge, while more recently the Bangladesh Rice Research Institute (BRRI) has played a significant role in developing different high-yielding varieties of flood-tolerant rice crops.[1]

One of the constraints of replicating floodplain experience in coastal Bangladesh is that, unlike normal river flood regimes that are periodic in nature and more or less predictable, regimes of future coastal inundation due to sea level rise are largely uncertain. Not only is the sea level rise an incremental process (i.e., a gradual increase in level without foreseeable prospect of recession), the rate of rise is nonlinear. In a nonlinear process of sea level rise if there is a sudden jump in the rate of change in level due to some triggering factors, such as acceleration of glacier melt, this may amount to a significant constraint for autonomous adaptations. Whenever such thresholds are crossed external interventions may be required for designing appropriate planned adaptations. To put it simply, if the sea level continues to rise major engineering interventions may be required to protect coastal residents for a limited time period (maybe up to 2050 at the current rates of rise). With further increases in depth of inundation beyond the capacity of planned structures, coastal residents may not have any choice but to relocate to higher grounds further inland.

CYCLONE ADAPTATION FEATURES IN INDIGENOUS HOUSES

Types of Rural Houses in Coastal Bangladesh

Poor quality of houses is one of the most basic reasons for large numbers of deaths from catastrophic cyclones in coastal Bangladesh, because often these houses are blown away by strong cyclone winds or washed away

[1] Bangladesh Rice Research Institute (BRRI) has played a significant role in rice research and farming innovations in Bangladesh that have boosted rice yield from 1.7 t/ha in 1970 to 4.3 t/ha in 2012, amounting to a "green revolution"! This success is largely the result of genetic improvement in the form of "high-yielding, climate change-ready and short-duration rice varieties". BRRI has also developed flood-tolerant rice varieties. Development of salt-tolerant rice varieties is its latest innovation in adaptation of rice farming to saline soils of coastal Bangladesh (http://irri.org/our-work/locations/bangladesh).

by storm surges that have also been responsible for most of the cyclone deaths under their crushing impacts. Most of the poor-quality houses are called *kutcha* houses (meaning weaker houses), which may vary in quality. Most of them have mud floors and thatched roofs. Their sidings may include various types of wooden and bamboo frames and walls may be made of bamboo mats, often intertwined with thatched materials. Most of the poorer people cannot afford to build stronger cyclone-resistant structures. Better-quality *kutcha* houses may have corrugated iron sheet for both walls and roofs.[2] Only small proportions of coastal residents (about 10% of the total) belong to middle and upper classes. Very few well-to-do rural residents build brick homes, that is, the entire structure is made of brick and cement. These are called *pucca* houses (meaning stronger houses). Most of them can withstand strong cyclone winds, at least up to category 3 storms. During high-magnitude cyclones, often such structures act as informal cyclone shelters for many rural residents, including their original occupants and their relatives, friends, and neighbors. Another category of houses that belong mostly to middle class people are called *semi-pucca*. Normally, these houses have wooden frames and corrugated iron sheet walls and roofs. They may have earthen foundations and floors, like *kutcha* houses, but brick and cement foundations and floors are also common for *semi-pucca* houses. While most of the *kutcha* houses share some of the cyclone-proofing measures with the remaining two categories, greater ranges of traditional cyclone adaptation features are incorporated in *semi-pucca* houses. These features can broadly be classified into two categories based on their intended purposes, i.e., protection from wind damage and from storm surge impacts.

Protection From Wind Damage

Tropical cyclones are not a new problem for coastal Bangladesh. For generations coastal residents of Bangladesh have lived with cyclones and storm surges by incorporating various types of indigenous, most often ineffective, cyclone-resistant measures in their vernacular homes. Some of the common features are described below.

[2] Besides *pucca* houses (brick buildings), *semi-pucca* houses (buildings made of brick walls with corrugated iron sheet roofs), and *kutcha* houses (with mud floor, corrugated iron sheet or bamboo mat walls, and corrugated iron sheet or thatched roofs), the Bangladesh Bureau of Statistics (BBS, 2011) lists a fourth category of housing structure called *jhupri*, a shack made of the poorest quality of materials, such as straw, canvas, polythene, plastic, etc. The *jhupris* are found mainly in urban slums.

Windbreaks and shelter belts: One of the basic cyclone adaptation features in indigenous houses is that many individual houses are surrounded by different types of trees. More commonly, the entire common-property homesteads are surrounded by continuous belts of trees or shelter belts.[3] Among a variety of trees that are used for this purpose, some of the flexible but strong trees include coconut, palm, and betel nut trees. Most of these stronger trees are planted on the south and southwestern sides of the houses since these are the prevailing wind directions during cyclonic storms originating from the Bay of Bengal. In addition, many fruit trees, such as mango, jack fruit, guava, and others, are also planted around individual homes or as a part of the shelterbelts. Most of these trees are planted approximately 25−35 ft (8−11 m) away from home, to avoid damage from uprooted and fallen trees (Islam and Siddiqua, 2014). Most of the respondents of a recent field survey dealing with cyclone-vulnerable houses in coastal Bangladesh indicated that they considered heights of specific trees to determine how far away from home these should be planted (Islam and Siddiqua, 2014). Survey respondents also confirmed that individual trees and shelterbelts provided significant protection to their houses from cyclones, especially from lower-magnitude storms.

Shape and orientation of houses: Two basic architectural characteristics of indigenous houses that are designed to minimize wind damage relate to their shape and orientation. Square form for the house plan is avoided in favor of rectangular designs. All survey respondents indicated that they preferred placing the short-arm side of the rectangular house towards the prevailing wind so that the house could not be turned over easily by strong cyclone winds (Islam and Siddiqua, 2014).

Roof shape: The geometry of roofs of some of the houses reflects an attempt to adjust the roof structures to storm winds. Traditional cyclone-resistant roof structures, the so-called *hip roof* (see Fig. 8.1 for an example), have extra support of the roof ridge, tie between the vertical support and the roof structures, and extra tie for extended roof overhang (Haq, 2007; also see Fig. 9 in Islam and Siddiqua, 2014). Based on their past experience, most of the respondents of the above survey indicated that houses with hip roofs had survived the impacts of most of the cyclones, especially lower-magnitude storms, whereas those with gable roofs had been damaged severely or collapsed entirely.

[3] In rural Bangladesh a homestead is called *bari*, which refers to a cluster of houses, normally built on an elevated tract of land. A village consists of several *baris*. See also endnote 1 in Chapter 8.

Window designs and positions: Both of these features are relevant for minimizing damage from strong winds. Most of the houses have top-hung windows as these are less likely to be blown off during cyclones once they are closed. In contrast, *side-hung* windows are more vulnerable to wind damage during storms (see Fig. 10 in Islam and Siddiqua, 2014). Normally windows are placed symmetrically at the middle positions of the windward facade between the main doors and the corners of the buildings (see Fig. 10 in Islam and Siddiqua, 2014). Some of the survey respondents suggested that windows placed near the corners of facades were more vulnerable to wind damage than symmetrically placed windows.

Anchoring devices: Different types of crude reinforcing devices, such as installing additional pillars to reinforce foundations or tying houses with strong ropes to nearby trees, have been employed to protect homes from the impacts of cyclonic winds (Paul et al., 2010). One of the fundamental problems with such traditional anchoring devices is that the foundation is not firmly anchored to the ground so that the cyclones could lift up or blow away the houses at times of high-magnitude storms (Haq, 2007).

Protection From Storm Surges

Elevated homesteads: Most of the measures to adjust homesteads and individual houses to storm surge impacts have been very similar to those of the river floodplains. As in the case of floodplains, raising the levels of homesteads (yards) by earth-fills borrowed from the surrounding common property lands and elevating the plinth levels of individual houses above yard levels have been one of the most common practices in the coastal region (Rasid, 2000). It may be possible to raise existing homesteads of coastal Bangladesh by an additional 1 m (3 ft) to keep up with rising sea levels, at least in the short and medium terms (i.e., up to 2050). Citing a specific case study (below), the Government of Bangladesh has recognized this potential in its Bangladesh Climate Change Strategies and Action Plan:

> *Sona Mollar Dangi is a small village of 26 households with about 250 people on an island in the middle of the* Padma (Ganges)*. Sona Molla, a respected village elder and his family moved to this newly-formed channel bar (an alluvial island locally called* char*) in search of a new life. Several other families also settled in this island but the entire island was inundated successively by the floods of 1998, 2004 and 2007. As a long-term solution to the problem of living with high flood levels, Sona Molla and other villagers decided to raise the homesteads of all of the 26 households above the 1998 flood level by at least 1.5 ft (0.5 m). The project was supported by the Faridpur Development Agency, a local NGO, with*

external funding from UK's Comprehensive Disaster Management Program. The cost has been nominal at about US $300 per family. Now, all of the infrastructures necessary for sustainable living in a farming community, including barns, poultry sheds, backyard gardens, fruit trees, sanitary latrines, and hand tube-wells are all safely above high flood levels.

GoB (2009), p. 22

Stilt houses: There are only limited examples of indigenous stilt houses in coastal Bangladesh. Some of the stilt houses have been built near the edge of water, i.e., within the limits of daily tide regimes, to avoid inundation by high tide levels. Most of these houses have been built on wooden poles with an elevated wooden platform as the first floor of the house, allowing free movement of water under the house at high tide levels. Perhaps these houses have not necessarily been designed to withstand much higher levels of storm surges. In any future planned adaptation of housing structures to rising sea levels, reinforcing, and redesigning stilt homes seem to be a viable option for coastal Bangladesh (see Chapter 8). However, unlike the Mekong Delta in Cambodia and Vietnam, stilt houses are not popular in coastal Bangladesh among the vast majority of Bengali population. Some of the aboriginal (tribal) homes may be an exception. Hasan (n.d. = not dated) has provided an example of elaborate cyclone-proofing features in indigenous houses constructed by the *Rakhains*, an aboriginal group from Burma/Myanmar who has settled in the coastal zone of Kuakata in Patuakhali district for several centuries. The *Rakhain* architecture includes mostly two-storied stilt houses with provisions for extensive openings throughout the entire structure (such as many windows), to allow free flow of both cyclone winds and storm surges without major obstruction. As a rule, the *Rakhain* houses were built on strong wooden poles and a wooden platform with a minimum opening of 1.6 m (5.5 ft) below the platform. Many residents survived past cyclones by taking shelter in the upper floor allowing storm surges to pass under the house and through the lower floor without major obstruction. Similarly, many of the *Rakhain* houses survived repeated cyclones and storm surges without major damage.

AGRICULTURAL ADAPTATION TO CLIMATE CHANGE

Salinity Challenges for Agricultural Adaptation in Coastal Bangladesh

Throughout the floodplains of Bangladesh farmers have used their indigenous technical knowledge to adjust their cropping practices to different microtopographic levels, varied flood depths, and specific timing of flood regimes (i.e., the beginning and the end of rising and falling flood levels) (Brammer, 2000; Rasid, 2000; Rasid and Mallik, 1995). In more familiar terms, they have adjusted their crop calendars to climate-sensitive flood regimes. One of the challenges of indigenous agricultural adjustments in the floodplains was to maximize crop production in ideal flood depths. In years of abnormal floods crop losses were substantial due to excessive flood depths (Paul and Rasid, 1993), whereas in years of normal floods it is customary for paddy farmers to partition their rice crop fields by building low earth embankments for maintaining desired water levels in their individual plots. Such compartmentalization of crop fields has worked well for the river floodplains but in coastal areas it was not an adequate technique for preventing saltwater intrusion in rice fields. Therefore, extensive coastal embankments and polders were built by the Government of East Pakistan in the 1960s (and subsequently by the Government of Bangladesh), mainly to prevent saltwater intrusion in rice fields and partly to alleviate local (river) flood problems.

Climate change impacts, especially impacts of rising sea levels, have introduced entirely new challenges for indigenous agricultural adaptation in coastal Bangladesh. Overall, climate change impacts on agriculture in Bangladesh are likely to be complex. Theoretically, increasing amounts of atmospheric carbon dioxide concentrations are supposed to increase crop production, but some of the experimental studies (for example, a large-scale field trial of major food grain crops at elevated CO_2 levels using an experiment called Free-Air Concentration Enrichment technology) have shown that elevated CO_2—enhanced crop yields were significantly lower than expected (Long et al., 2006). Besides, any potential gains in crop productivity are likely to be cancelled out by perturbations in temperature and precipitation regimes, which are critical for crop growth. More importantly, in coastal Bangladesh impacts of saltwater intrusion in crop fields are likely to be much more consequential for the agricultural sector than other adverse effects of climate change on agriculture in the rest of the country. Salinity effect is likely to be particularly exacerbated by reduced dry season freshwater flow, which is projected to be one of the outcomes of climate change in Bangladesh. Salinity problem would aggravate during the winter months that constitute a major part of the dry season. In the absence of appreciable rainfall in the winter months soils in the coastal areas would desiccate and salt would be delivered to the surface by the capillary action (CCC, 2009). With rising sea levels, high spring tides and cyclone storm surges have already been flooding low-lying coastal areas increasing soil salinity of these areas. As an evidence of the effect of rising sea levels, saline soils in coastal Bangladesh have recently increased from about 3089 mi^2 (800,000 ha) in 2001 to about 4633 mi^2 (1.2 million ha) in 2011 (Islam et al., 2011). The severity of salinity problem is expected to increase in the future with continued sea level rise. In Khulna region (southwestern Bangladesh), the salinity problem has further been exacerbated by conversion of croplands into shrimp farms (*ghers* or shrimp polders), thus rendering nonsaline croplands to saline soils. Rice crop farmers have very little options for indigenous adjustments of their cropping practices to such saltwater intrusion into their fields. Recent development of several varieties of salt-tolerant rice by BRRI has not only compensated for the lack of scope for indigenous agricultural adjustments in coastal Bangladesh but probably these innovations constitute a major milestone in the continuing "green revolution" in Bangladesh, which has sustained a population of more than 160 million people in a small land area (about the size of Iowa). To understand how these newly developed varieties of salt-tolerant rice would adjust to different types of saline environments of coastal Bangladesh, a description of crop calendars of Bangladesh should be helpful.

Crop Calendars of Bangladesh

Rice is the staple crop of Bangladesh. Dominating its rural landscape, vast expanses of green rice fields provide evidence of this monoculture. Other crops, such as wheat, jute, oilseeds, pulses, tobacco, and large varieties of vegetables are also important in both subsistence and commercial agriculture. However, the dominance of rice is evident from the main crop calendars of Bangladesh. The growing season of the country extends throughout 12 months, which can be further divided into three overlapping cropping seasons (Johnson, 1982):

- Rainy season (*bhadoi* in Bengali): late March to early September
- Late autumn/fall season (*aghani*): August to early December
- Dry season (*rabi*): late November to early April

Although the vernacular names in parentheses are well established in Bengali literature and native traditions, there are some inconsistencies in their definitions. Thus, the *bhadoi* and the *rabi* crops are defined according to their growing seasons (rainy and dry seasons, respectively), whereas the term *haimantic* refers to the harvesting period in autumn (fall), although its growing season is mainly throughout the monsoon (Rasid and Paul, 1987).

Perhaps, it is easier to follow the rice calendar. There are three main varieties of rice in Bangladesh. Their growing seasons are as follows:

- *Aus* may be considered as the premonsoon variety as it is sown throughout March to May (i.e., the premonsoon season), although it is harvested in July and August at the peak of the monsoon.
- *Aman* is the monsoon rice. It has two subtypes, namely transplanted and broadcast *aman*. Although *aman* is the dominant monsoon season rice, it is harvested in the late autumn when it coincides with the dry season (rabi). Several varieties of monsoon vegetables (mostly indigenous varieties that are largely unfamiliar in midlatitude countries) constitute another important group of monsoon crops.
- *Boro* is the dry season (*rabi* season) irrigated rice. Several other *rabi* crops overlap with *boro* cultivation. These include wheat, vegetables, oilseeds, pulses, tobacco, among others. A large variety of vegetables are also grown during the *rabi* season. Most of the *rabi* season vegetables are familiar European and North American varieties, such as cauliflower, cabbage, tomato, beans, potatoes, to name a few.

Coastal Environments for Salt-Tolerant Rice

Successful development of a wide variety of salt-tolerant rice crops by BRRI is a significant success story (indeed a milestone!) since such crops would allow farmers of coastal Bangladesh to continue to live in an environment of shallow seawater flooding at least for next several decades.[4] As we have stressed earlier, coastal Bangladesh is not necessarily a uniformly even low-lying area. Coastal residents have built their homesteads on elevated tracts of lands. Crop fields can be classified into at least three to five or more micro-topographic levels (Brammer, 2000). Some of the rice varieties, especially *aman*, grow well in wide ranges of flood depths. A special variety of broadcast *aman*, called floating rice can grow in deep floodwaters (at least up to a depth of 14 ft or 4 m) (Rasid and Mallik, 1995). Some of the native rice varieties can tolerate certain levels of salinity. Post-*Aila* (2009 cyclone) impacts of storm surge

[4] Besides rice, Bangladesh grows a wide variety of crops, such as jute, tea, sugarcane, cotton, peanut, and others. While BRRI has the mandate to conduct research on rice, Bangladesh Agricultural Research Institute (BARI), the largest public research institute of the country, has been conducting research on the remaining crops, including plant breeding different varieties of these crops. Another state-of-the-art research institute is Bangladesh Institute of Nuclear Agriculture (BINA), which has a plant breeding and biotechnology research program. In addition, there are separate research institutes on jute, sugarcane, tea, cotton, and silk. http://www.fao.org/in-action/plant-breeding/our-partners/asia/bangladesh/en/.

flooding in some of the coastal areas are instructive of future scenarios of early stages of coastal flooding due to sea level rise (Indeed, we are already in the initial stage of sea-water flooding!). Some of the low-lying areas remained flooded for more than one year following *Aila* since surge water could not recede efficiently back to the sea due to higher than normal sea levels (compared to previous years). In such a flooded environment, coastal residents have managed to continue to live in their elevated homes. Although crop production has reportedly suffered from saltwater intrusion, it seems that several varieties have continued to grow. This is the context for the next generation of sea level adaptation in coastal Bangladesh through the development of saltwater-tolerant rice varieties.

Nature of Salinity Measurements

Plant species vary in how well they tolerate salt-affected soils. The relative growth of plants in the presence of salinity is termed their salt tolerance (Agri-Facts, 2001). Salt tolerances are usually measured by electrical conductivity of a saline solution (that is its ability) to transmit an electrical current. The modern metric unit (SI unit) for conductivity is deci-Siemens per meter (dS/m). Electrical conductivity is the reciprocal of resistance (or ohm), which can also be expressed as S = ohm (Ω)/1 (Website: dictionary.reference.com/browse/siemens). Soil salinity levels vary widely from one environment to another. Therefore, their classification may also be site-specific. Based on generic data on electrical conductivity of a salt solution through different soil depths, Alberta (Canada) Agriculture and Forestry classifies saline soils into four or five categories (Agri-Facts, 2001). Compared to nonsaline soils (<2 dS/m), strongly saline soils have conductivity values of 8–16 dS/m. In very strongly saline soils, conductivity values may exceed 16 dS/m (Table 7.1). As electrical resistance increases from the surface to a depth of 2–4 ft (60–120 cm), conductivity values increase correspondingly (mostly double).

Table 7.1 Electrical Conductivity of a Salt Solution Through Different Soil Depths

Soil Depth	Nonsaline Soil: EC: dS/m[a]	Weakly Saline Soil: EC: dS/m	Moderately Saline Soil: EC: dS/m	Strongly to Very Strongly Saline Soil: EC: dS/m
0–60 cm (0–2 ft)	<2	2–4	4–8	8–16 and higher
60–120 cm (2–4 ft)	<4	4–8	8–16	16–24 and higher

[a]EC (electrical conductivity) in *deci-Siemens* per meter (dS/m).
Source: Agri-Facts: Alberta Agriculture and Forestry, www.agric.gov.ab.ca.

Table 7.2 Samples of Soil Salinity Levels in Southwestern Coast of Bangladesh

Sample District: Satkhira Soil Salinity Class	**Electrical Conductivity: dS/m**	**Salinity Level**
S_1	<4	Low
S_2	4–8	Moderate
S_3	8–15	High
S_4	>15	Very high

Source: Prepared by the authors based on data in Islam, M.R., Gregorio, G.B., 2013. Progress of salinity tolerant rice variety development in Bangladesh. SABRAO Journal of Breeding Genetics 45 (1), 21–30.

Soil salinity reduces rice crop productivity significantly. In coastal Bangladesh, several nonaltered (native) varieties of rice have already been growing in weakly saline soils but their productivity is compromised by soil salinity. Many of the native varieties of rice crops are not suitable for growth in strongly saline environments. High salinity levels may result in low yields and sterile grains. Since soil salinity in coastal Bangladesh is likely to increase significantly with saltwater intrusion due to sea level rise, the main challenge for rice breeding scientists in Bangladesh is to develop new varieties that can grow in soils with high salinity levels. Already BRRI has developed several new varieties of salt-tolerant rice. Most of them have been field tested successfully in saline soils in Satkhira, a coastal district in southwestern Bangladesh (CCC, 2009). In general, soil salinity in Satkhira may be classified into four levels, ranging from low (<4 dS/m) to very high (>15 dS/m) (Table 7.2). For field-testing of new varieties of rice crops some of the more specific soil-water parameters were monitored throughout their growing periods. These included salinity levels of:

- Test plot soils
- Pump-water (shallow tube-well water) used for irrigation for *boro* varieties
- Field-water (standing water) throughout the growing period

Salinity levels were measured by an EC Meter (Beckman Coulter Meter/Phi pH/Con/TDS Model 460) (CCC, 2009).

Salt Tolerant Varieties of Boro Rice

Breeding of newer varieties of rice is an ongoing activity of the BRRI. Experimental data on eight varieties of *boro* rice, which were field-tested in 2007 between 22 February and 6 April, are summarized in Table 7.3. All of the test varieties were transplanted as 10-week-old seedlings.

The test plots had medium-high to medium-low elevations, relative to the lowest local levels. No specific data were given on these elevations in the CCC (2009) report, but according to Johnson (1982, Fig. 5.1) medium-low elevations for rice crops in Bangladesh range between 1 and 2 m (3–7 ft) and medium high refers to lands between 2 and 3 m (7–10 ft). Soil salinity ranged from nonsaline (<2 dS/m) to moderately strong saline (up to 6 dS/m). For controlling salinity levels in the test plots, standing water in the field was maintained throughout the growing period by irrigating all plots with shallow tube-well water. The salinity values of pump-water and field-water ranged from very low (0.75 dS/m) to high (up to 8 dS/m). The permissible salinity level for optimum growth is up to 4 dS/m (Islam and Gregorio, 2013).

Among the tested varieties, *BRRIdhan 47* (a dry-season rice) has emerged as the most promising salt-tolerant *boro* variety with a yield of 6.43 tonnes per hectare (t/ha) from a test plot with a salinity tolerance up to 6 dS/m

Table 7.3 Results of Field Tests on Salt Tolerance of *Boro* Rice: 2007 Experiments

Experimental Varieties	**Test Plots With Weak Salt Tolerance Soil Salinity: 0.85–1.6 dS/m Pump-Water Salinity: 0.75–1.5 dS/m Field-Water Salinity: 1–1.9 dS/m**	**Test Plots With Moderately Strong Salt Tolerance Soil Salinity: 3–6 dS/m Pump-Water Salinity: 3.5–8 dS/m Field-Water Salinity: 3.5–8 dS/m**
***Boro* Varieties**	**Crop Yield (t/ha)**	**Crop Yield (t/ha)**
BRRIdhan 28	6–7.4[a]	5.27
China/Shaitta	5.63–6.81[a]	4.36
Kajallata	4.81	
Hira (hybrid)	9.31	
IET		4.59
BRRIdhan 47		6.43
BRRIdhan 45		5.35
BRRIdhan 29		4.67

[a]Range means two different test plots with two different yield values.

Source: Prepared by the authors based on information in text and data in several tables in CCC (Climate Change Cell), 2009. Climate Change Adaptation Research: Adaptive Crop Agriculture Including Innovative Farming Practices in the Coastal Zone of Bangladesh. Government of Bangladesh, Ministry of Environment and Forest, Department of Environment, Climate Change Cell, Component 4b, Comprehensive Disaster Management Project (CDMP), Ministry of Food and Disaster Management, Dhaka. www.bdresearch.org.bd, http://www.climatechangecell.bd.org.

Table 7.4 Results of Field Tests on High Salt Tolerance of Two Varieties of *Boro* Rice: 2010–2011 Experiments

Experimental Varieties	Test Plots With Weak Salt Tolerance: EC: <4 dS/m Crop Yield[a] (t/ha)	Test Plots With Moderate Salt Tolerance: EC: 4–6 dS/m Crop Yield[a] (t/ha)	Test Plots With Strong Salt Tolerance: EC: 7–12 dS/m Crop Yield[a] (t/ha)[a]
BRRIdhan 47	8.3	7.1	5.2 (4.2 under EC: 12–14 dS/m)
BRRIdhan 28	7.2	5.8	2.7

[a]All crop yield data represent the maximum yield out of three test plots in three different villages.
Source: Prepared by the authors based on data in Islam, M.R., Gregorio, G.B., 2013. Progress of salinity tolerant rice variety development in Bangladesh. SABRAO Journal of Breeding Genetics 45 (1), 21–30 (Table 5).

(Table 7.3). Further, in a follow-up experiment in 2010–11 its performance was even more robust in plots with greater ranges of salinity tolerance (Table 7.4). Thus, its maximum yield ranged from 8.3 t/ha in test plots with weak salt tolerance (<4 dS/m), to 7.1 t/ha in plots with moderate salt tolerance (up to 6 dS/m), and to 5.2 t/ha in plots with strong salt tolerance (up to 12 dS/m). Even when the salt tolerance level was further increased to 12–14 dS/m, *BRRIdhan 47* managed to produce 4.2 t/ha (Table 7.4). In contrast, the prospect for adaptation of *BRRIdhan 28* to strongly saline environments seems to be much more limited. Because of its high yield and a short growing period it is one of the most popular *boro* varieties in the test villages. During the 2007 field tests in two separate plots with low salinity (<2 dS/m) its yields ranged between 6 and 7.4 t/ha. Even in a test plot with moderately strong salinity (up to 8 dS/m) it performed fairly well with a yield of 5.2 t/ha (Table 7.3). In the follow-up experiment in 2010–11 its performance was similar up to the salinity level of 6 dS/m, but in strongly saline environments (7–12 dS/m) its yield dropped by about 67% (from the maximum of 7.2 to 2.7 t/ha) (Table 7.4). Another variety, *BRRIdhan 45* also holds some promise for adaptation to highly saline environments (Table 7.3). Although a hybrid high-yielding variety (HYV) named *Hira* produced the highest yield (9.31 t/ha), it is not very popular in test villages. More importantly, it is not tolerant of high salt stress. Development of several other dry-season rice varieties, including *BINAdhan 8, BRRIdhan 29, BRRIdhan 53,* and *BRRIdhan 55,* among several others, has been "a major breakthrough for breeding salt tolerant rice in

Figure 7.1 An experimental BRRI plot displaying *BRRIdhan 29*, a successful salt-tolerant *boro* rice in Bangladesh. *Source: Reproduce from Bangladesh Rice Knowledge Bank. http://www.knowledgebank-brri.org/brri-rice-varieties.php.*

Bangladesh" (Islam and Gregorio, 2013, see Table 6 for a complete list of the new varieties) (Fig. 7.1).

Salt Tolerant Varieties of Transplant Aman Rice

Transplant *aman* is a monsoon-season rice. Due to heavy rainfall during its main growing period (June—September) salinity is not a major issue even in the coastal region. For example, monsoon-season salinity in the test villages (in Satkhira district) ranged between 1.5 and 3 dS/m (i.e., no salinity to low salinity) (CCC, 2009). Salinity becomes a problem only in the late stages of transplant *aman* prior to its harvesting in the late fall (October—November) when rice fields dry out and consequently soil salinity increases significantly. All of the new salt-tolerant varieties of transplant *aman* have been developed to adapt to such seasonal salinity gradients (from very low salinity at early stages to moderately high salinity at the maturity stage).

Since it is a monsoon-season rice, *aman* grows well in naturally flooded rice fields. However, as in the case of the *boro* rice, all test plots were selected at medium-high to medium-low elevations (approximately 1—2 m or 3—7 ft) above the local low levels and they were irrigated with shallow tube-well water to maintain continuous standing water in the field. The following six varieties were field tested in late August, 2007 for their salinity tolerance: *BR23, BRRIdhan 33, BRRIdhan 39, BRRIdhan 40, BRRIdhan 41, and BRRIdhan 44.* About 30-days-old seedlings were transplanted between 22 August and 30 August. The results of the field tests are summarized in Table 7.5.

Table 7.5 Results of Field Tests on Salt Tolerance of Transplant *Aman* Rice: 2007 Experiments

Experimental Varieties	Test Plots With Weak to Moderate Salt Tolerance Soil Salinity: 0.5–4 dS/m Field-Water Salinity: 0.5–0.9 dS/m	Test Plots With Moderately Strong Salt Tolerance Soil Salinity: 1–6 dS/m Field-Water Salinity: 1–4 dS/m
Transplant *Aman* Varieties	**Crop Yield (t/ha)[a]**	**Crop Yield (t/ha)**
BR23	2.80–4.07	3.86
BRRIdhan 33	2.98–3.13	3.49
BRRIdhan 39	3.57–4.15	3.98
BRRIdhan 40	4.37–4.50	4.44
BRRIdhan 41	4.27–4.43	5.51
BRRIdhan 44	4.85–4.93	5.42

[a]Crop yields from two different test plots with two different values for all varieties in low saline environments.
Source: Prepared by the authors based on information in text and data in several tables in CCC (Climate Change Cell), 2009. Climate Change Adaptation Research: Adaptive Crop Agriculture Including Innovative Farming Practices in the Coastal Zone of Bangladesh. Government of Bangladesh, Ministry of Environment and Forest, Department of Environment, Climate Change Cell, Component 4b, Comprehensive Disaster Management Project (CDMP), Ministry of Food and Disaster Management, Dhaka. www.bdresearch.org.bd, http://www.climatechangecell.bd.org.

Perhaps the most convincing evidence of their salt tolerance is that five of the six varieties had higher yields in moderately strong saline environments (field-water and soil salinity up to 4–6 dS/m) than in weakly saline test plots (up to or less than 4 dS/m). Among the tested varieties, *BRRIdhan 41* and *BRRIdhan 44* seem to be the most promising varieties for adaptation to highly saline soils (soil salinity up to 6 dS/m) as their yields were significantly higher (about 25–29% higher) in moderately strong saline soils than in weaker saline soils. Further field tests have shown that they have the ability to tolerate salinity ranges up to 7–8 dS/m during their productive phase (CCC, 2009). The performance of *BRRIdhan 40* is also promising in the sense that its yield remained very similar in both low-saline and moderately strong-saline environments. That means it has also moderately strong salt tolerance. Although the remaining three varieties also performed better in moderately strong-saline environments than in low-saline environments, their yields fell short of 4 t/ha (an arbitrary threshold) making them less preferable for adaptation to saline soils than *BRRI 40, 41,* and *44.*

Salinity Control for Rabi (Dry) Season Crops

The dry season in Bangladesh extends from late October to early April. Crops grown during this period are normally called *rabi*-season crops using vernacular terms from crop calendars. One of the most important *rabi* crops that utilizes extensive irrigation water is *boro* rice, especially high-yielding varieties that produce much higher yields than transplant *aman*. Some of the other *rabi* crops are truly dry-season crops that do not require large-scale irrigation by tube wells. These include sugarcane, tobacco, pulses, mustard, peanut, potato, chilli, and a wide variety of vegetables. Most of the vegetable plants are not salt tolerant. Soil salinity during the dry season tends to be higher than during the monsoon season due to lack of rainfall. Therefore, to enhance yields of some of the popular *rabi* crops, the main focus of agricultural research has been on developing innovative farming practices to minimize or avoid salinity by popular vegetable plants during the dry season. The Bangladesh Agricultural Research Institute (BARI) conducted field tests on tomato, water melon, okra, and aroid (see endnote 4 for BARI's role). Field tests on tomato and watermelon were conducted in a government experimental farm that was located about 2.5 mi (4 km) outside the city of Satkhira. Okra and aroid were tested (monitored) in two designated (experimental) plots that were owned by local farmers. Experiments on watermelons failed due to unexpected rainfalls during the test period, whereas significant amounts of data were obtained from successful experiments on *BARI tomato-3* employing three different farming practices (Table 7.6).

The main farming technique for the experiment on tomato employed three different test beds based on their physical characteristics and irrigation methods (test plot characteristics in Table 7.6). The results were remarkable: the raised bed with mulch and drip irrigation (T3) produced a marketable yield of 42 t/ha, which was more than double the amount produced by a flat bed with can irrigation. Raised bed seems to be a critical determinant of yield as it tends to leach salt downward. Thus, yield from a raised bed without mulch and with water can irrigation (T2) produced 37.86 t/ha, i.e., nearly double the amount of T1. The main success story of this experiment is that both of the tested irrigation methods succeeded in controlling test plot salinity levels below 4 dS/m during their main growing seasons (January and February). It is clear from the salinity data for the month of March, i.e., the maturity stage, that salinity accumulated to excessive levels (above 8 dS/m) in T1 (flat bed with can irrigation). High salinity at the

Table 7.6 Results of Field Tests on *BARI Tomato-3*: Yields Under Controlled Salinity

Test Plot Characteristics	Controlled/Monitored Salinity Ranges (dS/m)			Marketable Yield (t/ha)
Salinity level monitoring dates →	**January 2007**	**February 2007**	**March 2007**	
T1: Water can irrigation on flat bed	3.65–4.70	2.5–4	8–9.65	19.46
T2: Water can irrigation in raised bed without mulch	3.53–4.35	2.22–3.8	5.53–6.42	37.86
T3: Drip irrigation in raised bed with mulch	3.41–3.99	1.95–3.59	4.05–5.38	42.03

Source: Prepared by the authors based on information in text and data in several tables in CCC (Climate Change Cell), 2009. Climate Change Adaptation Research: Adaptive Crop Agriculture Including Innovative Farming Practices in the Coastal Zone of Bangladesh. Government of Bangladesh, Ministry of Environment and Forest, Department of Environment, Climate Change Cell, Component 4b, Comprehensive Disaster Management Project (CDMP), Ministry of Food and Disaster Management, Dhaka. www.bdresearch.org.bd, http://www.climatechangecell.bd.org.

maturity stage (in March) produced the lowest yield. The highest yield from T3 (raised bed with mulch and drip irrigation) was associated with the lowest salinity level in March (4–5.38 dS/m). Addition of mulch helps in downward leaching of salt and also in maintaining moisture in the bed. Drip irrigation, which is a controlled method of low-intensity irrigation, seems to be the ideal method of reducing or avoiding soil salinity, especially during the main growing period of tomato plants. Several other agricultural research organizations in Bangladesh have been conducting similar research for preparing the coastal zone to adapt its *rabi* crops to increasing salt stresses with rising sea levels.

ADAPTATION OF COASTAL FISHERIES TO CLIMATE CHANGE IMPACTS

Bangladesh Fisheries Sectors

Whereas rice is the staple crop of Bangladesh, fish is its main dietary source of animal protein. Coastal Bangladesh is particularly rich in fisheries resources since it has a variety of fishing habitats along its indented coastline, such as the fishing grounds of the Bay of Bengal, several large estuaries (salt-water/freshwater transitional zone), and coastal river channels and freshwater marshes adjacent to them. For adaptation to climate change impacts these habitats can broadly be differentiated as two distinct geographical units: (1) marine fisheries in the Bay of Bengal and (2) inland freshwater fisheries. Out of these two components inland freshwater fisheries constitute by far the most important sector of Bangladesh fisheries. In 2009–10, total fish production in Bangladesh was about 2.89 million metric tonnes (IUCN and GoB, n.d.; cmsdata.iucn.org/downloads/fisheries.pdf). Only about 18% of this amount was produced directly by the sea; the rest came from inland fisheries. The latter, in turn, has two distinct sectors: (1) inland culture fisheries (47%) and (2) inland capture fisheries (35%) (IUCN and GoB, n.d.).

Marine Fisheries

Although marine fisheries account for less than one-fifth of the total fish production of Bangladesh, it is still an important part of the regional coastal fisheries of the Bay of Bengal (Vivekanandan, 2011). The opportunities for adaptation of marine fisheries to climate change impacts in coastal Bangladesh are somewhat limited because of the nature of the problem and are mixed at best because of the diversities of fishing environments. The main problem is that most fish species are highly sensitive to even minor changes in water temperature and pH, which are likely to affect their distribution and life processes. Major shifts in spawning seasons of many species have already been observed in the Indian Ocean with recent increases in water temperature (Vivekanandan, 2011). Overall, productivity in marine capture fisheries is likely to be affected adversely by changes in water temperature and resulting changes in marine ecosystem. The diversities of fishing environments in coastal Bangladesh may provide some opportunities for countering losses in marine capture fisheries by shifting focus on culture fisheries. Adaptation of marine fisheries to climate change impacts is a difficult scientific problem involving manipulation of deepwater marine ecosystems to enhance capture fisheries. While Indian scientists have been active in research on deepwater marine ecosystems, Bangladesh has yet to undertake a major role in such research. Further, any significant outcome of such research is far from certain. The main research topics deal with the effect of iron fertilization (dispersal of microscopic iron particles in ocean water) on acceleration of phytoplankton growth leading to restoration and enhancement of major fishing stocks (because many fish species depend directly on plankton for their food). The downside of this technique is that there are several environmental concerns with such manipulation of the marine ecosystem, including acidification of ocean water, development of "red tide," and unwanted algal bloom. Similar research has also been conducted on planting sea weed and other carbon-fixing plants in ocean water as measures for promoting certain fish species, but with uncertain results.

Inland Fisheries

The estuaries of the Ganges-Brahmaputra-Meghna Rivers and other coastal channels provide a rich fishing ground with diversities of marine and freshwater species. Further inland beyond the reaches of the saltwater ingress, floodplains of coastal rivers, like other floodplains of Bangladesh, provide an important environment for capture fisheries (i.e., wild fish). Among a variety of research programs dealing with climate change impacts on fisheries, the Government of Bangladesh has been studying the effect of saltwater intrusion on aquatic ecosystems and the effect of water temperature increases on inland fisheries and culture fisheries in ponds. Potential impacts of climate change on inland fisheries in Bangladesh include the following among others (IUCN and GoB, n.d.):

Effects on reproduction and growth of fish: Higher water temperature and erratic rainfall may bring changes in physiology and sex ratios of fish species, and major disturbances in timing of spawning and migration of species affecting productivity across marine and freshwater systems.

Effects on species composition, abundance, and distribution: Increased salinity and the resulting changes in water quality are likely to instigate changes in fish species composition and distribution in coastal areas.

Effects on habitat quality: Increased flooding and erosion due to sea level rise are likely to destroy some of the fish habitats and their breeding grounds.

Effects on aquaculture: Changes in water temperature may affect aquaculture adversely by increasing risks of disease outbreaks.

As in the case of marine fisheries, perhaps very little could be done to adapt capture fisheries to climate change impacts. The adaptation research on fisheries in Bangladesh has therefore focused mainly on culture fisheries, which include pond fisheries, shrimp culture (in shrimp polders), river-based and floodplain-based cage aquaculture, spawning of freshwater and saltwater-tolerant species, and development of hatchery and grow-out technologies (GoB, 2009). Culture fisheries are already an important sector of the economy but several production constraints, such as lack of adequate seed and prevention and control of disease, limit future growth.

The Government of Bangladesh has placed a high priority on cultural practices to deal with threats to spawning, growth, and migration of all fish species, particularly of the *Hilsha* (shad), perhaps the most popular fish of Bangladesh (GoB, 2009).

Culture Fisheries

With increasing demand for freshwater fish from an exploding urban population, supplies of capture fisheries (i.e., wild fish) have not been able to keep up with the demand. Consequently, freshwater culture fisheries from ponds now account for most of the fresh fish supplies in large urban markets, especially in Dhaka, the capital of Bangladesh with a population of about 14 million. While culture fisheries seem to have a good potential for growth in Bangladesh because of depleting supplies of wild fish from rivers and swamps, paradoxically saltwater shrimp culture in ponds and enclosed embankments (shrimp polders) have been highly counter-productive because of its adverse environmental impacts. Saltwater shrimp culture expanded very rapidly in the 1980s and 1990s, from about 20,000 ha in 1979 to 130,000–138,000 ha in 1994 (Rahman et al., 2013). The latest available data indicate that in 2005–06 total shrimp culture area in Bangladesh had expanded to 170,000 ha (BSFF: Bangladesh Shrimp and Fish Foundation Website: www.shrimpfoundation.org/statistics.php). The leading shrimp farms are located in southwestern coastal districts, i.e., in Satkhira, Khulna, and Bagerhat. It is also an expanding industry in southeastern coast (Cox's Bazar area). Saltwater shrimp from coastal Bangladesh has a large internal and international market. In particular, a brackish-water large shrimp (*Macrobrachium rosenbergii*), locally known as *Bagda Chingri* (tiger or monster shrimp), attracts a very high price in the international market, such as in Japan, USA, UK, and Canada (Rahman et al., 2013). Because of its high commercial value a profitable industry has emerged with four main subsectors: shrimp farms (*ghers*), shrimp hatcheries, shrimp feed mills, and shrimp processing plants (Rahman et al., 2013). The industry also includes agents of the supply chain, such as shrimp collectors and exporting agents. Proponents of the economically successful shrimp aquaculture have dubbed it as a "Blue Revolution," akin to the Green Revolution of the mid-1960s (Knowler et al., 2009). Rejecting such claims, critics of shrimp aquaculture have criticized it as a "Fake Blue Revolution" because of its destructive environmental impacts (Deb, 1998). Most of these adverse impacts result from the detailed processes of shrimp aquaculture reviewed below.

Conversion of freshwater lands into salt-water shrimp ponds: The main problem with shrimp farms is that they have replaced previous croplands. In coastal areas many of these croplands have already certain amounts of salinity. The shrimp farming processes accelerate salt accumulation in soils and groundwater because shrimp enclosures or *ghers* are inundated with salt water and water-logged for several months prior to the release of shrimp fries. It takes about 3–6 months for the fries to mature for marketing (Rahman et al., 2013). By converting large areas of the coastal belt (formerly rice fields) into salt-water ponds, the process has degraded the overall land quality of the entire area. Shrimps *ghers* have now replaced some of the other environmentally friendly functions, such as agriculture, freshwater ponds for other aquaculture, livestock grazing, and growth of different types of plants.

Deforestation: Constructions of *ghers* often require removal of parts of a forest or individual trees (a form of deforestation), resulting in loss of biodiversity. A major casualty of this process has been that parts of the mangrove forests have been cleared for some of the *ghers*.

Loss of wetlands: Conversion of many freshwater marshes into *ghers* has also destroyed the natural feeding and breeding grounds of many popular freshwater fish, such as different varieties of carps (*ruhi* and *katla*), catfish (*shing* and *tangra*), and perch (*koi*). *Ghers* have also blocked their natural migration routes resulting in a significant decline of these wild fish.

Decline in wild shrimps due to overharvesting shrimp fries: Some of the activities outside the shrimp ponds may also lead to major ecological impacts. Shrimp farmers prefer natural shrimp fries from nearby rivers and other water bodies (in lieu of hatchery fries) because of their superior quality (Rahman et al., 2013). Often they catch fries from rivers and other natural water bodies by using exceedingly fine-meshed nets to maximize their catches. As a result, production of wild shrimp in rivers and marshes has declined.

Adverse effects of chemicals: Shrimp farmers use different types of chemicals for protection against potential disease outbreaks. They also use chemicals for improving water quality. Some of these chemicals might be responsible for killing other living organisms in the pond.

Problems with shrimp feeds: Shrimp feeds are particularly problematic for the environment. Popular shrimp feeds may be made from fishmeal that may use freshwater snail (Rahman et al., 2013). The demand for snails has increased so much that they have disappeared from many local wetlands. This amounts to a major ecological perturbation because snails play a significant role in local food chains. Snail eggs are a potential food for fish, snakes, ducks, rats, and birds (Rahman et al., 2013). Overharvesting snails might thus result in loss of food source for some of these species.

Since shrimp aquaculture has emerged as one of the major sources of foreign exchange earnings for Bangladesh, it is unlikely that the Government of Bangladesh would impose major restrictions on the operations of this industry. However, there is some scope for developing more environment-friendly shrimp ponds that may not contribute to increased salinity. Cultivation of salt-water shrimp in freshwater ponds is a viable option. This has already been tried successfully but we do not have any data on it. Perhaps, more importantly, it makes sense to develop salt-water shrimp ponds in those areas that have already high salinity, since such areas may not be suitable for even some of the new salt-tolerant varieties of rice crops. In southwestern Bangladesh there are many areas (such as in Satkhira) where soil salinity exceeds 8–12 dS/m. Shrimp ponds in such areas seem to be an appropriate adaptation to increasing salt stresses due to sea level rise.

DISASTER MANAGEMENT POLICIES

Third Dimension of Adaptation to Sea Level Rise

Successful adaptation to sea level rise in coastal Bangladesh is a challenging proposition seeking not only sustainable livelihoods in salinity-prone low-lying lands but also measures to save lives during high-magnitude cyclones and storm surges. At the minimum, its success would depend on implementation of at least three complementary initiatives, analogous to the simultaneous support of three legs of a tripod. Flood-proofing rural homesteads and infrastructures above rising sea levels in the short and medium terms is the first leg. Development of salt-tolerant rice varieties is the second supporting leg. Disaster management is the third leg that can save lives during disasters. The main purpose of disaster management is to "plan to reduce losses from hazards, assure prompt assistance to disaster victims, and take steps to recover from the impacts of disasters as quickly as possible" (Paul, 2011, p. 157). The Government of Bangladesh has a good track record of disaster management policies at least during the last three decades following the catastrophic cyclone Gorky (1991). Significant declines in cyclone deaths following *Gorky* provide evidence of this success. Among several interrelated disaster management initiatives, the following four activities played critical roles in saving lives during successive high-magnitude cyclones: (1) cyclone preparedness, (2) cyclone warning, (3) cyclone evacuation, and (4) poststorm relief.

Cyclone Preparedness Program

Preparedness relates to individual and/or community readiness to take precautionary measures in response to threats posed by an impending disaster (Paul, 2011). Preparedness activities may include devising master plans, gathering emergency supplies, training response teams, and educating residents and others about disaster vulnerability and what to do to reduce it (Mileti, 1999). The Government of Bangladesh has established a highly successful Cyclone Preparedness Program (CPP) in collaboration with the Bangladesh Red Crescent Society. First introduced in 1972, the CPP is "a unique institutional arrangement for community preparedness created to mitigate the challenges of catastrophic cyclones that frequently hit Bangladesh coast" (Bangladesh Red Crescent Society: http://www.bdrcs.org/programs-and-projects/cyclone-preparedness-program). The CPP provides a robust early warning system for the coastal population of 13 districts. It also ensures rapid dissemination of the Bangladesh Meteorology Department (BMD) official cyclone warning signals to the coastal communities. Once warned, CPP assists in sheltering, rescuing, and offering immediate medical attention to cyclone victims. The program links national level institutions to at-risk coastal communities through a network of trained and committed volunteers at the local level. As of 2009, the CPP had 28,450 males and 14,225 female trained volunteers in 274 coastal Unions of Bangladesh (Paul, 2009).[5] CPP volunteers are responsible for disseminating cyclone warnings among villagers via megaphone broadcasts and by house-to-house contacts. These volunteers assist people in the evacuation process, execute rescue operations, provide first aid, and assist in distributing relief goods. Volunteers also perform social welfare activities with the help of local administrations, NGOs, *upazila* (subdistrict) disaster management committees, educational and religious institutions, and other entities. Volunteers are equipped with radio receivers, hand sirens, megaphones, microphones, signal lights, raincoats, gumboots, hardhats, life jackets, and signal flags so that they can perform their responsibilities in severe weather. Approximately 5 million

[5] The term Union is the short-form of Union Council, the fourth-tier administrative unit of Bangladesh, following Division (tier 1), District (tier 2), subdistrict or *upazila* (tier 3), and Union Council (tier 4). According to the latest data posted by Wikipedia, Bangladesh has 4553 Union Councils. See also endnote 5 in Chapter 8.

people benefit from their services out of the estimated 15 million coastal area residents who are regularly exposed to cyclonic hazards (Akhand, 2003). The CPP has a network of radio communication facilities along the coast, all linked to the communication headquarters in Dhaka (Paul, 2009). Public awareness is an integral part of cyclone preparedness and is promoted through volunteer activism, cyclone drills and demonstrations, film/video shows, publicity campaigns, radio and television, posters, leaflets and booklets, and staging skits (Paul et al., 2010).

The Government of Bangladesh introduced another preparedness measure by publishing its Standing Orders on Cyclone in November 1985; this was replaced by the Standing Orders on Disasters (SOD) in January 1997 (Paul et al., 2010). This document outlines specific directives and duties and responsibilities regarding disaster management for all concerned entities—for example, relevant ministries, regional and local administrators, and the Bangladesh military (GoB, 1999). These entities must follow the directives listed in SOD during normal times, the precautionary and warning stages, the disaster stage and the postdisaster stage of any given hazard. According to SOD's directives, all special weather bulletins, issued by the Storm Warning Centre (SWC) and broadcast by the radio and television channels, should contain the following information: location of the cyclone, estimated central pressure, direction and speed of movement of the low-pressure system, maximum sustained wind speed, radius of maximum sustained winds, areas likely to be affected, approximate time of commencement of gale force winds (speed more than 30 mi/h or 52 km/h), maximum wind speed, expected storm surge height, and areas most likely to be impacted (GoB, 1999). In reality, the special weather bulletins often fail to provide all of this information (Paul and Dutt, 2010).

Cyclone Warnings

Cyclone warnings are issued to inform people about the imminent cyclone hazard, to advise them to take defensive measures for minimizing potential losses and, sometimes, to mandate evacuation to cyclone shelters or other safer buildings. To fulfill this objective, a warning system encompasses both forecasting the danger and communicating the appropriate warnings. The SWC, a specialized unit of the BMD, prepares all weather forecasts and cyclone warnings for the general public as well as for sea ports, river ports, and all other user agencies, such as health administrators, relief and rehabilitation authorities, NGOs, the media, and local administrative personnel (Paul, 2009; Choudhury, 2001). Forecasting involves continuous observation of various weather parameters over a wide area. Both national and international data are monitored closely for preparing forecasts and warnings. With relatively well-equipped forecasting instruments and an upgraded telecommunication system, the BMD now has a 48-h warning system in place that allows people to evacuate to cyclone shelters and other safer buildings usually hours before a cyclone makes landfall. Cyclone warnings are sent by the SWC directly to the National Coordination Committee, which is chaired by the Prime Minister and includes representatives of both the government of Bangladesh and CPP. The SWC issues special weather bulletins soon after the formation of tropical depressions. Fig. 7.2 illustrates SWC's cyclone warning message dissemination system to the at-risk population.

In addition to informing the electronic media and user agencies, SWC also sends warnings of an approaching cyclone to the CPP headquarters in Dhaka, the capital of Bangladesh. Officials there then transmit the warning

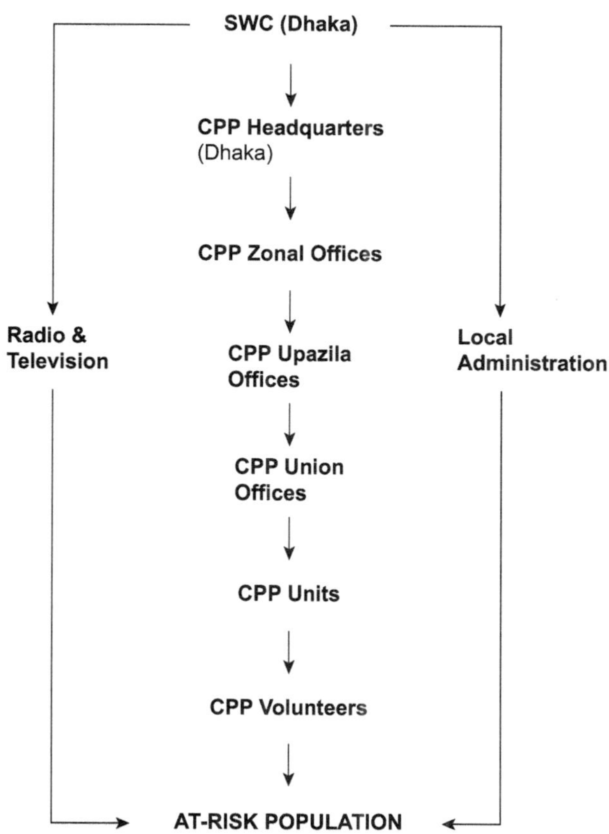

Figure 7.2 SWC's cyclone warning dissemination to at-risk population.

to the six zonal offices. These offices, in turn, relay warnings to 32 *upazilas* and 274 union offices, and from there to 2845 local CPP units, which are distributed throughout coastal villages. Nearly 43,000 trained volunteers in these villages are eventually responsible for directly disseminating cyclone warnings to the at-risk population. Although all radio and television channels continuously broadcast cyclone warnings, such warnings may not reach the targeted population because a majority of coastal residents do not own a radio or a television set (Paul et al., 2010).

Despite such an elaborate scheme, cyclone warning system in Bangladesh has numerous shortcomings. To begin with, forecasting tropical storms is much less accurate than forecasting midlatitude cyclones (storms associated with weather fronts), such as those experienced in higher latitudes in the United States. On top of such forecasting uncertainty, risk communication is another important issue that is not adequately addressed in the forecasting process. Phillips and Morrow (2007) contend that communication of storm risks is a social process that requires receiving, understanding, believing, and personalizing forecasts and warning messages. Akhand (2003) criticizes complications with the SWC storm warning signals as it uses an odd scale of 1–11 for an impending storm without indicating its precise location, intensity, and rates of movement. The system does not specify the potential area likely to be impacted by the storm or the likely depths of storm surge inundation. Despite its repeated use, many coastal residents have a hard time to comprehend the meaning of these signals (Islam et al., 2004). Without a credible and comprehensible warning system during an active storm coastal residents are likely to underestimate the storm magnitude and thus underutilize available mitigation measures, including evacuation to cyclone shelters (Hossain et al., 2008).

Cyclone Evacuation

Real-time responses to cyclone disasters involve a relatively narrow time horizon of an active storm system. These refer to all actions taken immediately prior to, during, and following a storm. To assess how cyclone-vulnerable residents of coastal Bangladesh typically respond to high-magnitude storm events, i.e., cyclone disasters, case studies can provide significant insights into their behavior during this critical period. Based principally on our own surveys (Paul, 2009; Paul et al., 2010), we present two such case studies, comparing coastal residents' responses to cyclones *Gorky* (1991) and *Sidr* (2007), respectively. In our case studies, we have focused on community cyclone preparedness and evacuation initiatives, thus, excluding private, individual indigenous measures. To assess how cyclone-vulnerable residents typically respond to cyclone disasters, the following questions are relevant. How efficient were the cyclone warning systems? How many residents heeded to CPP evacuation initiatives? How efficient were the poststorm relief operations?

In 1991, immediately prior to *Gorky*'s landfall, the Government of Bangladesh weather office had already issued a cyclone danger signal of 10 for the Chittagong and Cox's Bazar areas. By that time, the Bangladesh Red Crescent's team of more than 20,000 cyclone volunteers had been put on full alert through its network of 57 radio contact points. In team of 10, the volunteers set out to warn the teeming coastal population of the impending cyclone, using bullhorns, drums, and radios (*The Washington Post*, May 02, 1991; *The Guardian*, May 06, 1991). Before the storm hit, about 3 million people were evacuated from an area where 7 million lived in 2000 villages (*The New York Times*, May 01, 1991). The number of cyclone volunteers increased to more than double in 2007 prior to *Sidr*'s landfall. Increases in the number of CPP volunteers and public cyclone shelters appear to have had a positive impact on the percentages of coastal residents who received cyclone warning messages and on their evacuation rates. Based on field data collected three months after *Sidr* struck Bangladesh, Paul (2009) reported that about 86% of all households surveyed received cyclone warning and evacuation orders prior to landfall. Chowdhury et al. (1993) reported that in 1991 only about 60% of the respondents surveyed were aware of the warning and evacuation orders. In both instances, there were some problems with the cyclone warning and evacuation procedures. For example, no warnings were issued or announced by responsible authorities in several areas, such as in Mirzaganj (in Patuakhali district) prior to *Sidr* landfall, and in the offshore islands in the case of *Gorky* (Paul, 2009; Haque, 1995). In the former case, in some coastal areas, warnings were disseminated only in the marketplace, which was distant from many rural homes (GoB, 2008a). Consequently, many households did not receive any cyclone warnings or evacuation orders when the household heads were away from home for an extended period of time, for example for fishing. Their female counterparts had no way of receiving warning messages from the distant marketplace.

Prior to *Sidr*'s landfall, many CPP volunteers in the Barguna district were unable to use megaphones and microphones for disseminating warning messages due to loss of power and/or lack of workable batteries. There were no provisions for back-up batteries for a variety of reasons, including inadequate supply. It was not possible to buy replacement batteries within the limited time at hand because they were available only in distant cities. There were some problems also with the nature of warning messages. Most often they were incomplete in the sense that they did not provide some of the crucial pieces of information, such as the probable landfall sites, areas likely to be affected, storm surge height, and magnitude and duration of the cyclone. Several *Sidr* respondents from the Pirojpur district complained that CPP volunteers requested residents via megaphones that they should immediately seek safer shelter, but did not specify either the nature of the shelter or its location. One respondent from the Barguna district complained that he wished he had known how *Sidr* compared with other storms in more comprehensible terms than referring to the signal numbers he and others only vaguely understood (Paul, 2010).

The Government of Bangladesh estimates that the evacuation rates differed significantly between the two events. During *Gorky*, approximately 30% of the cyclone-vulnerable residents heeded to formal evacuation orders/advisories. Out of them, 15–20% took refuge in public cyclone shelters (GoB, 2008a). In contrast, evacuation rates during *Sidr* increased to 35–40% of all coastal residents (GoB, 2008b). Most of the sample surveys indicated similar rates. Thus, according to one study, 15% of the *Gorky* respondents went to public cyclone shelters, whereas another 17% took refuge in neighbors' homes or in public buildings, such as schools, colleges, mosques, and government office buildings (Chowdhury et al., 1993). A similar survey for *Sidr* indicated that approximately 33% evacuated successfully; of them, 20% moved to public shelters and the remaining 13% took refuge in other buildings (Paul, 2009).

Despite significant increases in the number of cyclone shelters in 16 years following *Gorky* and more aggressive CPP evacuation initiatives immediately before *Sidr*'s landfall, nearly two-thirds of the cyclone-vulnerable residents did not heed to evacuation orders/advisories. Probable reasons for why so many coastal residents did not participate in past cyclone evacuation initiatives are presented in Table 7.7. These are based on spontaneous expressions of the respondents, which have been reported in our own studies on *Gorky* and *Sidr* (Paul et al., 2010; Paul, 2010, 2009) and some of the previous studies on *Gorky* (Haque, 1997; Haque and Blair, 1992; Ikeda, 1995). These spontaneous responses provide significant insights into cyclone victims' inability to participate in evacuation initiatives and can be broadly classified into five categories (Table 7.7).

Table 7.7 Potential Reasons for Not Evacuating to Cyclone Shelters

(a) **Cyclone shelter attributes**
- No nearby public shelter
- Public shelters were overcrowded
- Too far to walk to public shelter; no other transportation to shelters
- Unsatisfactory lighting and sanitary conditions
- Fear that cyclone shelter would collapse

(b) **Problems with the warning system**
- Complicated warning signal
- No warning was given
- Disbelief of warning
- Incomplete warning message
- Premature warning
- Warning was too late
- Warning signal changed suddenly
- False warning

(c) **Gender issues**
- Male members were not at home
- Father/husband decided not to leave
- Husband was too old to walk long way
- Could not take along small children
- No separate rooms or toilets for females in cyclone shelters

(d) **Security issues**
- Fear of burglary
- Had to protect shop
- Could not leave behind cattle

(e) **Perception of cyclone hazards**
- No past cyclone experience
- False sense of security
- Felt safe at home
- Thought it would not occur here
- Cyclone in this season would not be severe
- A Northerly wind had never before resulted in a surge
- Cyclone was Allah's will

Sources: Rashid, H., 2011. Interpreting cyclone disasters in Bangladesh and Myanmar from web-based newspaper discourse: media framing of cyclone vulnerability on the Bay of Bengal coast. The Arab World Geographer 14 (1), 1–32; Paul, B.K., Dutt, S., 2010. Hazard warnings and responses to evacuation orders: the case of Bangladesh's Cyclone Sidr. The Geographical Review 100 (3), 336–355; Paul, B.K., Rashid, H., Islam, M.S., Hunt, L.M., 2010. Cyclone evacuation in Bangladesh: tropical cyclones Gorky (1991) vs. Sidr (2007). Environmental Hazards: Human and Policy Dimensions 9, 89–101; Paul, B.K., 2009. Why relatively fewer people died? the case of Bangladesh's Cyclone Sidr. Natural Hazards 50, 289–304; Haque, C.E., 1997. Atmospheric hazards preparedness in Bangladesh: a study of warning, adjustments and recovery from the April 1991 cyclone. Natural Hazards 16 (2), 181–202; Ikeda, K., 1995. Gender differences in human loss and vulnerability in natural disasters: a case study from Bangladesh. Indian Journal of Gender Studies 2 (2), 171–193; Haque, C.E., Blair, D., 1992. Vulnerability to tropical cyclones: evidence from the April 1991 cyclone in coastal Bangladesh. Disasters 16 (3), 217–229.

First, perhaps the most basic factor was the inadequate number of cyclone shelters in the vicinity of the respondents. Thus, many respondents indicated that there were no cyclone shelters around or that the distance to a designated public shelter was too great. One *Sidr* victim from Mathbaria in Barguna district reported that: "There is no cyclone shelter in the village—the nearest one is far away (approximately 2.5 mi or 4 km)... If we could make recommendations to the government, we would tell them to build more cyclone shelters with enough space for both people and livestock" (GoB, 2008b, p. 151). Some of the studies have shown that unless a cyclone shelter is within 1 mi (1.6 km) of a residence, it may be considered too distant (Ikeda, 1995). In the absence of adequate number of shelters, shelter capacity and overcrowding were major constraints of accommodation. Thus, in both 1991 and 2007, many respondents had to return back to their homes when they found out that the shelters were already full or overcrowded. Logistics were a related concern, as some indicated that there were no means of reaching the shelter, especially if they had to cross a river on the way. Some of the coastal residents were also reluctant to use public shelters because of unsatisfactory sanitary conditions, such as inadequate number of toilet facilities and lack of drinking water supply. Most of the shelters were also poorly lighted. Some were concerned that the building might collapse due to overcrowding or due to the impact of the cyclone and storm surges. Often such concerns were justified because many buildings showed signs of fatigue, such as cracked walls.

Second, among factors related to cyclone preparedness, a large number of expressions were related to the credibility of the warning systems. Some felt that the warning systems were overly complicated; others complained that no warning was issued, while many simply did not believe the warning. As an example of the complicated nature of the warning systems, there are two sets of warning signals: one for sea ports and the other for river ports. There are 11 signal numbers for sea ports but only four for river ports. There are no signal systems for inland stations. Another complication is that the signal numbers are constantly updated with changes in the storm magnitude in the Bay of Bengal. It is too much for the coastal residents to keep track of these changes. The decision to or not to evacuate also depends on the timing of warning. Some complained that the warning message was incomplete or that the warning was premature or too late or that it had changed suddenly. As an example of the consequence of premature warning, on November 14, 2007 most of the evacuation to cyclone shelters was completed by midday, particularly in the Bagerhat district, with the anticipation that *Sidr* would make landfall by noon. When the cyclone did not strike the area by 6:00 p.m., many thought that it would not materialize and left the shelter for their homes (Hossain et al., 2008). *Sidr* made landfall around 9:30 p.m. and caught many returnees off-guard, often with tragic loss of lives. Emdad Haque (1997) attributed past false warnings as one of the reasons for noncompliance of evacuation orders in 1991. Similarly, in one of our own studies we found that nearly one-fifth of the respondents did not believe the *Sidr* warnings because of a false tsunami warning issued earlier, just 2 months prior to the cyclone (Paul, 2010). On September 13, 2007, a 7.8 magnitude earthquake off the coast of Sumatra prompted Indonesian authorities to issue a tsunami warning, which it later withdrew. In Bangladesh, however, this warning remained in force, and more than a million coastal residents were evacuated.

Third, among sociodemographic factors, gender was an issue for many female respondents for not evacuating (Ikeda, 1995). All of the five responses listed in Table 7.7(c) are clearly gender related. In a traditional rural society like coastal Bangladesh, females depend heavily on males for all kinds of decisions. Some of the female respondents, thus, indicated that they could not evacuate because male members were not at home at the time of the cyclone. In some cases, male household heads (father or husband) decided not to evacuate for a variety of reasons. Some women could not evacuate because they had difficulties in carrying several small children to the shelter. Also, cultural restrictions for female seclusion, called *purdah* in Bangladesh (in this case, separation of women from men in social gatherings) are an important attribute for conservative Muslim women. Therefore, lack of separate rooms or toilets for females in most of the cyclone shelters was a deterrent for female evacuation in some cases.

Fourth, an important socioeconomic factor that was cited by some of the respondents was that they were worried about the security of things left behind in their homes. The fear of burglary, the need to protect the shop (for a small number of respondents), and to take care of cattle were three such concerns. Nearly 15% of the *Sidr* respondents who did not evacuate stayed home for the fear of burglary (Paul, 2010). Literature on this issue indicates that even in industrial societies disaster evacuees are hesitant to leave behind their belongings (van Duin and Bezuyen, 2000; Alexander, 1993).

Finally, responses related to perceptions, previous experience of the cyclone hazards, and attitudes towards the hazards are consistent with the literature. For example, one of White's (1974) leading hypotheses on natural hazard perception is that variation in hazard perception can be accounted for by the magnitude, frequency, and personal experience of the hazard. Some of the younger respondents who did not experience any catastrophic storm like *Gorky*, consequently, did not evacuate as they had a false sense of security and felt safe at home. Similar responses included statements like: "Thought it would not occur here"; "Cyclone in this season would not be severe"; "A northerly wind had never before resulted in a surge." Another plausible explanation for nonevacuation was some of the respondents' fatalistic attitudes towards the cyclone hazards, typified by such statements that future cyclones were "Allah's will" and that they could not do anything about it. A potential explanation of this pessimistic attitude towards cyclone hazards in coastal Bangladesh has been presented earlier by Islam (1974), based on his post-1970 (i.e., post-*Bhola* cyclone) survey in the southern coast of Bangladesh. According to this study, at least 10% of the respondents who believed that a cyclone would strike the area soon ascribed the probability of the future cyclone to "God's will" (Islam, 1974).

Poststorm Relief

Confronted with death, damage, and destruction of property, poststorm relief and rehabilitation play a significant role in cyclone victims' survival and recovery in the disaster aftermath. Because of a long and repeated experience with catastrophic cyclones, the government and people of Bangladesh have developed a track record of dealing with these disasters in a comprehensive manner. Overall experience with past relief and rehabilitation initiatives has been generally positive, despite myriads of logistical problems with relief operations and alleged inequity in the distribution of relief goods and poststorm reconstruction projects. Thanks to massive international donor assistance and mobilization of substantial internal financial and manpower resources, impoverished coastal residents of Bangladesh have survived repeated cyclone disasters, demonstrating an aura of resiliency in the face of adversities.

As a part of a comprehensive disaster management policy, an extensive relief and rehabilitation program following each major cyclone is one of the highest priorities of the Government of Bangladesh, national and international NGOs, and international donor agencies. The immediate challenges for poststorm relief operations include distribution of food and drinking water and prevention of outbreak of diseases. While loss of field crops and stored food grains is one of the critical effects of storm surge flooding and strong cyclone winds, normally cyclones also cause serious damage to the water supply and sanitation systems. Hand pumped tube wells, which are one of the major sources of drinking water in rural Bangladesh, are often damaged badly and become unusable. Freshwater ponds in surge-affected coastal areas are often contaminated by saline water and sludge. Therefore, after each major cyclone there is usually a serious shortage of drinking water and a high probability of an outbreak of waterborne and other diseases. In the past, major disease outbreaks had been avoided most of the time by proper distribution of relief goods, including food and safe drinking water as well as by timely implementation of health care intervention measures. Most of the relief foodstuffs include dry food, such as rice, pulses, wheat, and canned food. Drinking water is distributed as bottled water or in bulk in different types of containers. Following *Sidr* landfall, medical teams and health workers were sent shortly after the cyclone subsided to provide immediate medical care. The Bangladesh government also opened makeshift hospitals and provided mental health counseling for survivors in need. Initially, medical personnel from public health facilities in neighboring districts were sent. These were followed by nationally formed medical teams. In addition, the Bangladesh military set up its own medical camps to treat the injured and provide health care services (GoB, 2008a).

While efficient relief operations can play a significant role in cyclone victims' survival and recovery in the disaster's aftermath, logistical problems with relief operations can contribute to their misery at a critical time. Relief operations are extremely complex undertakings involving coordination among a variety of government departments, international donor agencies, national and international NGOs and a host of volunteers. The milieu of relief activities is hectic, often chaotic, and uncertain. For cyclone victims who have lost almost everything to the disaster, any uncertainty about the availability of and access to disaster aid in the immediate aftermath of a disaster may operate as a "double whammy," reinforcing their vulnerability to cyclone disasters. Experience from both *Gorky* and *Sidr* has shown that commonly two types of logistical problems may hamper delivery of relief goods. First, some of the logistical problems may be natural, i.e., beyond anyone's control. For example, timely delivery of international aid to *Gorky* survivors was hampered by loss of communication with outlying areas where storm surge flooding had caused road damage and erosion or blockage of navigation channels by sunken boats and ships. Further, relief operations were disrupted significantly by another weaker storm less than a week after *Gorky*.

Second, most of the man-made problems were related to the inadequacy and inefficiency of relief operations, more in 1991 (*Gorky*) than in 2007 (*Sidr*). Prior to *Gorky*, Bangladesh's previous military-backed government under Lt-Gen Ershad (1982–1991) demonstrated efficiency in relief operations during major disaster events, especially during the catastrophic flood of 1988. *Gorky* struck Bangladesh barely six weeks after its democratically elected government took office in early 1991 under the leadership of Prime Minister Khaleda Zia. There were widespread internal and international criticisms of the new government for its apparent inability to cope with the logistics of the disaster. Relief officials and others complained of the "lackadaisical way in which relief distribution was handled" (Johnson, 1991; Rashid, 2011). In contrast, relief operations during *Sidr* (2007) were much more efficient, thanks partly to good coordination between the army and the interim military-backed civilian government (Rashid, 2011). International donors generally play positive roles in relief operations in Bangladesh. The United States, in particular, provided significant and timely logistical support for delivering relief goods in both 1991 and 2007. In 1991, giant US C-130 transport planes (en route from Iraq War) ferried food and other supplies from Dhaka to Chittagong port, headquarters of the task force Operation Sea Angels (*The Independent*, London, May 24, 1991). Similarly, two US C-130 transport aircraft and two naval vessels participated in US–Bangladesh relief operations in 2007, immediately following *Sidr*'s landfall (*The New York Times*, November 20, 2007). Despite concerted efforts by all concerned to distribute aid as fairly as possible, there is a widespread perception among cyclone victims and many others in Bangladesh that often relief is distributed unfavorably among socially marginalized groups and cyclone victims suspected of being supporters of opposition political parties (Paul, 2010).

Media Discourse on Cyclone Disasters in Bangladesh

In addition to the survey data presented above, we have further analyzed information from media discourse to assess how some of the international newspaper reports covered cyclone disasters in Bangladesh, especially in the context of disaster management policies. In particular, we were interested in assessing how the international media interpreted cyclone casualties in Bangladesh as an outcome of inadequate cyclone preparedness. Thus, our analysis was based on an assumption that "Lack of cyclone preparedness or inadequate preparedness may result in significant loss of life, whereas adequate preparedness may save lives" (Rashid, 2011). Discourse analysis of international newspaper reports on *Gorky* and *Sidr* provided an extensive array of information related to subsets of cyclone preparedness in Bangladesh (Tables 7.8–7.10).

Table 7.8 Cyclone Preparedness Program (CPP) Initiatives That Saved Lives During *Gorky* (1991) and *Sidr* (2007)

CPP Initiatives	Number of Reports	
	Gorky	*Sidr*
Early warning systems and evacuation initiatives	8	15
Timely evacuation to cyclone shelters	11	12
New cyclone shelters		5

Source: Rashid, H., 2011. Interpreting cyclone disasters in Bangladesh and Myanmar from web-based newspaper discourse: media framing of cyclone vulnerability on the Bay of Bengal coast. The Arab World Geographer 14 (1), 1–32.

Table 7.9 Causes of Evacuation Failures During *Gorky* (1991) and *Sidr* (2007) That Cost Lives

Gorky[1]:

1. Insufficient numbers of shelters
2. Why some cyclone victims did not evacuate?
    - Defied cyclone warnings
    - Did not believe in warnings because of previous false alarms
    - Afraid of thieves and robbers; reluctant to leave their houses and possessions
    - Did not have a place to go (the shelter was too far)

Sidr[2]:

3. Shelter characteristics
    - Nearest shelter was too far
    - Cyclone shelter swelled out of capacity
4. Why some cyclone victims did not evacuate?
    - Some did not want to move or did not move quickly enough
    - Some ignored cyclone warnings and went fishing into the sea
    - Some were caught at the sea
    - Some could not swim or climb to trees
    - Unaware of the impending storm in remote islands
    - Returned from the shelter assuming that the storm was not coming
    - Some thought that the storm would not hit them
    - Some intended to move later
    - Some were trying to save their belongings

Sources:

1 *The Age* (Australia) (May 7, 1991), *Courier-Mail*, Melbourne (May 7, 1991), *The Guardian* (UK) (May 6, 1991), *The New York Times* (May 11, 1991 and February 16, 1992), *The Sunday Times* (London) (May 5, 1991), *The Toronto Star* (June 5, 1991), *The Washington Post* (May 2, 1991).

2 *The Daily News* (South Africa, November 20, 2007), *London Free Press* (Ontario, November 16, 2007), *The New York Times* (November 20, 2007).

Table 7.10 Problems With the Logistics of Relief Operations: *Gorky and Sidr*

	Number of Reports	
Nature of the Problems	***Gorky*[1]**	***Sidr*[2]**
1. Another storm disrupted relief operations	19	
2. Communication disrupted by damaged and flooded roads	15	12
3. River navigation disrupted by sunken boats and ships; ships struggling in rough seas to reach remote islands	5	3
4. Helicopters, other aircraft, and fast boats were the only means of delivering relief goods but there was an acute shortage of helicopters	24	5
5. US C-130 transport aircraft and naval vessels employed	4	9
6. Slow government response; poor coordination between army and civilian authorities, inefficient relief operations	13	1

Sources:
[1] *The New York Times* (May 2, 3, 5,8, 9, 19, 1991, February 16, 1992); *The Washington Post* (May 2, 3, 6, 7, 11, 1991); *Courier-Mail* (Melbourne, May 3, 7—11, 1991); *St Louis Post-Dispatch* (May 3, 5, 6, 9, 1991); *The Age* (Melbourne, May 3, 1991), *The Toronto Star* (May 4, 6,7,1991), *Globe and Mail* (Canada, May 4, 1991); *The Sunday Times* (London, May 5,12, 1991), *USA Today* (May 6, 16, 1991) and others.
[2] *The Guardian Unlimited* (November 16, 2007), *The Irish Times* (November 21, 2007), *The Pioneer* (India, November 17, 2007), *The Virginia-Pilot* (Norfolk, VA, November 17, 2007), *Edmonton Journal* (November 19, 2007), *The Mirror* (UK, November 19, 2007), *Vancouver Sun* (November 19, 21, 2007); *The Washington Post* (November 17, 2007), *The New York Times* (November 20, 2007); *The Times* (London, November 21, 2007); *Turkish Daily News* (November 21, 2007), *The Star Phoenix* (Saskatoon, SK, November 21, 2007), *The Roanoke Times* (Virginia, November 21, 2007) and others.

Data in Table 7.8 deal with the Government of Bangladesh CPP initiatives that saved significant numbers of lives during *Sidr* and minimized loss of lives during *Gorky*. Since the loss of lives from *Gorky* was catastrophic (amounting to 139,000 or more), the very proposition that the CPP initiatives saved lives seems to be paradoxical without the context of such a claim. The context arises out of the Government of Bangladesh claim that without the CPP initiatives the loss of lives could have been even higher. Most of these claims pointed out that casualties from the 1970 killer storm (*Bhola Cyclone*), with slightly lower wind speeds than that of *Gorky*, were much higher, presumably because of the lack of any cyclone preparedness program by the then Government of East Pakistan. In this context, the Prime Minister of Bangladesh insisted that the death toll would have been far worse than that of the 1970 storm if the government's Cyclone Preparedness Center had not tracked the storm and given 2 days of warning as the winds picked up strength and approached the densely crowded coast (*St. Petersburg Times*, Florida, May 04, 1991). Some of the international newspaper reports agreed with the official assertion that improved cyclone tracking, the construction of shelters and an early warning program persuaded large numbers of people to seek shelter and thus helped to reduce casualties (*The Independent*, London, May 02, 1991, *The Washington Post*, May 02, 1991). Then, an important question is: why so many people still died? Data in Tables 7.9 and 7.10 provide partial explanations. The main problem was the insufficient numbers of cyclone shelters (Table 7.9).

In 1991, Bangladesh had about 300 cyclone shelters (Tathan et al., 2009). A comprehensive cyclone protection plan, conceived 10 years prior to *Gorky*, called for the construction of about 5000 concrete cyclone shelters, to protect the entire coast of Bangladesh. The *Toronto Star* (June 05, 1991) reported that in one such shelter 3000 people found safety from the cyclone, even though the shelters were designed for only 1000 persons. If the cyclone shelters saved so many lives why then the planned structures were not built? The answer is deceptively simple: the cost of construction. Each shelter cost $50,000 to build (in 1991 dollars). The Government of Bangladesh national budgets could not cover the cost of so many structures even during a period of 10 years. In short, the shelter-building program suffered from crippling shortfalls in funds. The program also suffered from the country's priority for flood management programs elsewhere (especially in the Ganges-Brahmaputra floodplains) (*The Sunday Times*, London, May 05, 1991; Brammer, 2000). *The Guardian* (London, May 06, 1991) offered a realistic solution: to give security to the whole coastal population, a vast international effort is needed.

Besides shortage of cyclone shelters and overcrowding in the existing shelters, there were many other reasons for the failure of some of the cyclone victims to seek refuge in nearby shelters (compare these with our survey data presented in Table 7.7). Many reports recounted some of the explanations based on the inputs of the survivors. Distance to the nearest shelter without any means of transportation was a deterrent for evacuation for some of the victims. The security of personal property without protection from potential burglary was a familiar deterrent, as the literature suggests (Alexander, 1993). Other reasons included lack of access to cyclone warnings from remote locations, risk-taking propensity (hence ignoring cyclone warnings), false alarms (many previous warnings did not materialize), and procrastination. In the absence of evacuation to formal shelters, many victims resorted to some of the indigenous adaptation measures to escape from the storm's harm's way (reported in Chapter 2).

While some of the cyclone victims did not or could not evacuate to shelters during *Sidr* (Table 7.7), the casualty figure from this 2007 disaster was remarkably smaller than its 1991 predecessor (3500 vs. 139,000). Considerable amount of scholarly research has been conducted for analyzing the underlying causes of this contrast in death tolls from two similar category 4 storms (e.g., Paul, 2010; Paul, 2009; Tathan et al., 2009). The newspaper reports provided two basic explanations: (1) vastly improved cyclone preparedness, particularly an aggressive drive to evacuate a large number of coastal residents and (2) a significant expansion in the network of cyclone shelters. Based on its previous experience with evacuation initiatives, the Government of Bangladesh, in collaboration with several international NGOs, developed an improved cyclone warning system, organizing simulated cyclone evacuation exercises and educating villagers in the most exposed areas on how to flee. Efficient early warning systems and the widespread use of mobile phones meant that this time even people in remote regions were aware of the impending disaster (*The Observer*, UK, November 18, 2007). A simple early warning system, made up of 40,000 men and women trained in disaster response, was triggered when authorities gave warning that the worst cyclone to hit the country in a decade was about to strike the coast of Bay of Bengal (*The Times*, London, November 24, 2007). An army of these trained volunteers mobilized by the Bangladesh Red Crescent beat drums and used bullhorns or a special flag system to communicate evacuation warnings (*The Irish Times*, November 17, 2007). As a result, an estimated 3.2 million people were evacuated to safety to higher grounds and more than 1 million were moved to cyclone shelters, according to the United Nations and other agencies (*The Times*, London, November 24, 2007). None of the reports provided data on the exact numbers of cyclone shelters in Bangladesh in 2007, but the *Virginia-Pilot* (Norfolk, VA, November 17, 2007) reported that about 2000 new structures have been built since 1991, bringing the total number to about 2400 (Tathan et al., 2009).

National Plan for Disaster Management 2008–2015

One year after *Cyclone Sidr*, the GOB has adopted a comprehensive disaster management plan called *National Plan for Disaster Management (NPDM) 2008–2015* (GoB, 2008c). The NPDM is based on a new paradigm that emphasizes a shift from the conventional reactive approach of relying heavily on postdisaster relief to a greater emphasis on disaster risk reduction (DRR). Three broad goals of the plan are as follows (GoB, 2008c):

First, disaster management would involve the management of both risks and consequences of disasters that would include prevention, emergency response, and postdisaster recovery.

Second, community involvement for preparedness programs for protecting lives and properties would be a major focus. Involvement of local government bodies would be an essential part of the strategy. Self-reliance should be the key for preparedness, response, and recovery.

Third, nonstructural mitigation measures such as community disaster preparedness, training, advocacy, and public awareness must be given a high priority; this would require an integration of structural mitigation with nonstructural measures.

The plan is based on a disaster model to guide disaster risk reduction and emergency response management efforts in Bangladesh. The model has three key elements:

- Defining and redefining the risk environment
- Managing the risk environment
- Responding to the threat environment

The NDMP recognizes that global warming and climate change pose major risks to Bangladesh because of their impending impacts on sea level rise, increased salinity in coastal areas and increasing magnitudes of cyclones and storm surges. To address these challenges, the plan has devoted an entire section on cyclone and flood shelter management plan.

As of 2008, there were 2033 cyclone shelters in 15 districts of coastal Bangladesh (GoB, 2008c). The NDMP calls for construction of additional 2500 shelters to accommodate larger numbers of cyclone victims during future cyclones. The plan recommends construction of multipurpose shelters in primary and secondary schools. Structural aspects of multipurpose cyclone shelters are presented at some length in Chapter 8. Some of the nonstructural measures of the Multipurpose Cyclone Shelter Program (MCSP) include the following (GoB, 2008c):

- Comprehensive risk assessment (Hazard Assessment and Vulnerability Assessment), including tsunami inundation modeling and evacuation mapping;
- Warning guidance, including seismic and sea level monitoring, data evaluation, processing and interpretation, forecasting methods, and warning dissemination.

- Mitigation and preparedness, including education and awareness programs, structural and nonstructural mitigations, and government policy and emergency management procedures.
- Development of rescue, relief, and rehabilitation plan of action based on comprehensive risk assessment.

The Way Ahead: Addressing Hyogo Priority for Disaster Risk Reduction

The DRR elements of the Bangladesh disaster management model are based on the Hyogo Framework for Action (HFA) for building the resilience of nations and communities to disasters. The World Conference on Disaster Reduction held at Kobe, Hyogo, Japan in January 2005 adopted five priorities for action (ISDR, 2005):

1. Ensure that disaster risk reduction is a national and a local priority with a strong institutional basis for implementation.
2. Identify, assess, and monitor disaster risks and enhance early warning.
3. Use knowledge, innovation, and education to build a culture of safety and resilience at all levels.
4. Reduce the underlying risk factors.
5. Strengthen disaster preparedness for effective response at all levels.

The Government of Bangladesh has made a commitment to adopt these priorities in its NPDM 2008–15. Many of these priorities are repeated throughout the NPDM document. The United Nations Office for Project Services (UNOPS), an operational arm of the United Nations, has been assisting the Government of Bangladesh in implementing these priorities (UNOPS, 2016). In a postaudit assessment, UNOPS has elaborated its role in these implementations, which are summarized below.

To address HFA 1 (i.e., DRR is a national and local policy), UNOPS has helped GoB in developing a "project management information system to monitor and report on key program achievements and guided the development or review of key disaster management policies," including the NDMP 2008–15 (UNOPS, 2016). With the help of UNOPS the GOB has created more than 550 union-level community risk reduction action plans, affecting 15–20 million people (consistent with HFA 2 above). "Ten research projects analysing the impact of climate change have been implemented and a climate change library and Website established, as well as an institutional mechanism that predicts climate and flood impact" (UNOPS, 2016). UNOPS has helped in establishing the Disaster Management Information Center and risk information network covering 150 million people in 64 districts and 230 *upazilas* (subdistricts). To build a culture of safety and resilience (HFA 3), UNOPS has helped in disaster management training and professional development programs for government officials and thousands of disaster management committee members. To reduce the underlying risk factors (HFA 4), UNOPS has helped in initiating a local grants program "to provide small-scale risk reductions interventions and livelihood security strategies" (UNOPS, 2016). More specifically, it involved "adaptation research to address agricultural risks." Consistent with HFA 5 (strengthen disaster preparedness), UNOPS provided "early recovery technical support" to GOB in 2007 to address damage caused by two severe floods and *Cyclone Sidr*, which affected nearly 9 million people. This technical support included 200 situation reports on floods and *Sidr*, which were widely disseminated among national and international organizations.

It appears from the preceding UNOPS updates that significant progress has already been made in DRR initiatives in Bangladesh. It remains to be seen if such progress would be reflected in responses to future climatic disasters in Bangladesh, especially in responses to high-magnitude cyclones.

REFERENCES

Adger, W.N., Arnell, N.W., Tompkins, E.L., 2005. Successful adaptation to climate change across scales. Global Environmental Change 15, 77–86.

Adger, W.N., Huq, S., Brown, K., Conway, D., Hulme, M., 2003. Adaptation to climate change in the developing world. Progress in Development Studies 3 (3), 179–195.

Agri-Facts, 2001. Salt tolerance of plants. Agri-Facts: Practical Information for Alberta's Agriculture Industry. Government of Alberta, Canada, Alberta Agriculture and Forestry: www.agric.gov.ab.ca.

Akhand, M.H., 2003. Disaster management and cyclone warning system in Bangladesh. In: Zschau, J., Kuppers, A.N. (Eds.), Early Warning Systems for Natural Disaster Reduction. Springer, New York, pp. 49–64.

Alexander, D., 1993. Natural Disasters. Chapman & Hall, New York.

BBS (Bangladesh Bureau of Statistics), 2011. Population and Housing Census 2011. In: Socio-economic and Demographic Report, National Series, vol. 4. Government of Bangladesh, Ministry of Planning, Bureau of Statistics, Statistics and Informatics Division (SID). http://203.112.218.66/WebTestApplication/userfiles/Image/BBS/Socio_Economic.pdf.

Brammer, H., 2000. Flood hazard vulnerability and flood disasters in Bangladesh. In: Parker, D.J. (Ed.), Floods, vol. I. Routledge, London and New York, pp. 100–115.

CCC (Climate Change Cell), 2009. Climate Change Adaptation Research: Adaptive Crop Agriculture Including Innovative Farming Practices in the Coastal Zone of Bangladesh. Government of Bangladesh, Ministry of Environment and Forest, Department of Environment, Climate Change Cell, Component 4b, Comprehensive Disaster Management Project (CDMP), Ministry of Food and Disaster Management, Dhaka. www.bdresearch.org.bd. http://www.climatechangecell.bd.org.

Choudhury, A.M., 2001. Cyclone in Bangladesh. In: Nizamuddin, K. (Ed.), Disaster in Bangladesh: Selected Readings. University of Dhaka, Disaster Research Training and Management Centre, Dhaka, pp. 61–76.

Chowdhury, A.M.R., Bhuyia, A.U., Choudhury, A.Y., Sen, R., 1993. The Bangladesh cyclone of 1991: why so many people died? Disasters 17 (4), 291–304.

Deb, A.K., 1998. Fake blue revolution: environmental and socioeconomic impacts of shrimp cultures in coastal areas of Bangladesh. Ocean and Coastal Management 41 (1), 63–88.

Easterling, W.E., 2011. Guidelines for adapting agriculture to climate change. In: Hillel, D., Rosenzweig, C. (Eds.), Handbook of Climate Change and Agroecosystems: Impacts, Adaptation and Mitigation (ICP Series on Climate Change Impacts, Adaptation and Mitigation, vol. 1. Imperial College Press, London, pp. 267–286.

GoB (Government of Bangladesh), 2009. Bangladesh Climate Change Strategy and Action Plan 2009. Government of Bangladesh, Ministry of Environment and Forest, Dhaka. http://www.sdnbd.org/moef.pdf.

GoB, 2008a. Cyclone Sidr in Bangladesh: Damage, Loss and Needs Assessment for Disaster Recovery and Reconstruction. Government of Bangladesh, Ministry of Food and Disaster Management, Dhaka. https://www.gfdrr.org/sites/default/files/2275_Cyclone SidrinBangladeshExecutiveSummary.pdf.

GoB, 2008b. Super Cyclone Sidr 2007: Impacts and Strategies for Interventions. Government of Bangladesh, Ministry of Food and Disaster Management, Dhaka. http://www.preventionweb.net/files/9470_cyclonebangladesh.pdf.

GoB, 2008c. National Plan for Disaster Management: 2008–2015. Government of Bangladesh, Ministry of Disaster Management. www.preventionweb.net/files/9472_NationalPlanfor Disastermanagement.pdf.

GoB, 1999. Standing Orders on Disaster. Government of Bangladesh, Dhaka.

Hamin, E.M., Gurran, N., 2009. Urban form and climate change: balancing adaptation and mitigation in the U.S. and Australia. Habitat International 33 (3), 238–245.

Haq, B., 2007. Battling the Storm: Study of Cyclone Resistant Housing. Bangladesh Red Crescent Society/German Red Cross, Dhaka, Bangladesh. http://sheltercentre.org.

Haque, C.E., 1997. Atmospheric hazards preparedness in Bangladesh: a study of warning, adjustments and recovery from the April 1991 cyclone. Natural Hazards 16 (2), 181–202.

Haque, C.E., 1995. Climatic hazards warning process in Bangladesh: experience of, and lessons from, the 1991 April cyclone. Environmental Management 19 (5), 719–734.

Haque, C.E., Blair, D., 1992. Vulnerability to tropical cyclones: evidence from the April 1991 cyclone in coastal Bangladesh. Disasters 16 (3), 217–229.

Haque, C.E., Zaman, M.Q., 1993. Human responses to riverine hazards in Bangladesh: a proposal for sustainable floodplain management. World Development 21 (2), 93–107.

Hasan, M.T., n.d. Surviving Cyclones: The Indigenous Wisdom. Unnayan Parishad, Dhaka. Unpublished manuscript (web document). www.unnayan.org/documents/Climatechange/Surviving_Cyclones_Indigenous_Wisdom.pdf.

Hossain, M.Z., Islam, M.T., Sakai, T., Ishida, M., 2008. Impact of tropical cyclones on rural infrastructures in Bangladesh. Agricultural Engineering International: the CIGR E-journal 10 (no pagination).

Houghton, J., 2006. Global Warming: The Complete Briefing, third ed. Cambridge University Press, Cambridge, UK.

Ikeda, K., 1995. Gender differences in human loss and vulnerability in natural disasters: a case study from Bangladesh. Indian Journal of Gender Studies 2 (2), 171–193.

IPCC, 2007a. Mitigation, adaptation and climate change. In: IPCC Fourth Assessment Report: Climate Change 2007, Section 2.5.2. WMO, UNEP: IPCC. www.ipcc.ch/publications_and_data/ar4/wg3/en/ch2s2-5-2.html.

IPCC, 2007b. Climate Change 2007: Synthesis Report, Fourth Assessment Report. IPCC and Cambridge University Press, Cambridge, UK.

ISDR (International Strategy for Disaster Reduction), 2005. Hyogo Framework for action 2005–2015: building the resilience of nations and communities to disaster. In: World Conference on Disaster Reduction, 18–22 January 2005, Kobe, Hyogo, Japan. http://www.unisdr.org/2005/wcdr/intergover/official-doc/L-docs/Hyogo-framework-for-action-english.pdf.

Islam, M.A., 1974. Tropical cyclones: coastal Bangladesh. In: White, G.F. (Ed.), Natural Hazards: Local, National, Global. Oxford University Press, New York, pp. 19–24.

Islam, M.R., Gregorio, G.B., 2013. Progress of salinity tolerant rice variety development in Bangladesh. SABRAO Journal of Breeding Genetics 45 (1), 21–30.

Islam, S.M., Siddiqua, A., 2014. Using indigenous building patterns to cope with natural disasters: the case of cyclones in Bangladesh. International Journal of Architecture, Engineering and Construction 3 (1), 27–43.

Islam, M.B., Ali, M.Y., Amin, M., Zaman, S.M., 2011. Climate variations: farming systems and livelihoods in the high Barind tract and coastal areas of Bangladesh. In: Lal, R., Sivakumar, M.V.K., Faiz, S.M.A., Rahman, A.H.M.M., Islam, K.R. (Eds.), Climate Change and Food Security in South Asia. Springer Science + Business Media, New York, pp. 479–497.

Islam, M.S., Ullah, M.S., Paul, A., 2004. Community response to broadcast media for cyclone warning and disaster mitigation: a perception study of coastal people with special reference to Meghna estuary in Bangladesh. Asian Journal of Water, Environment and Pollution 1 (1 & 2), 55–64.

IUCN (International Union for Conservation of Nature) and GoB (Government of Bangladesh), n.d. Climate Change and Fisheries & Livestock in Bangladesh: Information Brief. Government of Bangladesh, Ministry of Environment and Forest, IUCN, Embassy of Denmark, UKaid. Web document: cmsdata.iucn.org/downloads/fiheries.pdf.

Johnson, B.L.C., 1982. Bangladesh, second ed. Heinemann Educational Books, London.

Johnson, S., May 16, 1991. God must be very angry; cyclone one more burden for the nation. The USA Today.

Knowler, D., Philcox, N., Nathan, S., Delamare, W., Haider, W., Gupta, K., 2009. Assessing prospects for shrimp culture in the Indian Sundarbans: a combined simulation modelling and choice experiment approach. Marine Policy 33, 613–623.

Long, P., Ainsworth, E.A., Leaky, A.D.B., Nösberger, J., Ort, D.R., 2006. Food for thought: lower-than-expected crop yield simulation with rising CO_2 concentrations. Science 312 (5782), 1918–1921.

Mileti, D.S., 1999. Disaster by Design: A Reassessment of Natural Hazards in the United States. Joseph Henry Press, Washington, DC.

Mileti, D., 1980. Human adjustment to the risk of environmental extremes. Sociology and Social Research 64 (3), 327–347.

Page, E.A., 2007. Fairness on the day after tomorrow: justice, reciprocity and global climate change. Political Studies 55 (1), 225–242.

Paul, B.K., 2011. Environmental Hazards and Disasters: Contexts, Perspectives and Management. Willey-Blackwell, Chichester, UK.

Paul, B.K., 2010. Human injuries caused by Bangladesh's Cyclone Sidr: an empirical study. Natural Hazards 54, 483–495.

Paul, B.K., 2009. Why relatively fewer people died? the case of Bangladesh's Cyclone Sidr. Natural Hazards 50, 289–304.

Paul, B.K., 1984. Perception of and agricultural adjustments to floods in Jamuna floodplain, Bangladesh. Human Ecology 12 (1), 3–19.

Paul, B.K., Dutt, S., 2010. Hazard warnings and responses to evacuation orders: the case of Bangladesh's Cyclone Sidr. The Geographical Review 100 (3), 336–355.

Paul, B.K., Rasid, H., 1993. Flood damage to rice crop in Bangladesh. Geographical Review 83, 150–159.

Paul, B.K., Rashid, H., Islam, M.S., Hunt, L.M., 2010. Cyclone evacuation in Bangladesh: tropical cyclones Gorky (1991) vs. Sidr (2007). Environmental Hazards: Human and Policy Dimensions 9, 89–101.

Paul, S.K., Routray, J.K., 2011. Household response to cyclone and induced surge in coastal Bangladesh: coping strategies and explanatory variables. Natural Hazards 57 (2), 477–499.

Phillips, B.D., Morrow, B.H., 2007. Social science research needs: focus of vulnerable population, forecasting and warning. Natural Hazards Review 8 (3), 61–68.

Rahman, M.M., Giedraitis, V.R., Liberman, L.S., Akhtar, T., Taminiskiene, V., 2013. Shrimp cultivation with water salinity in Bangladesh: the implications of an ecological model. Universal Journal of Public Health 1 (3), 131–142.

Rashid, H., 2011. Interpreting cyclone disasters in Bangladesh and Myanmar from web-based newspaper discourse: media framing of cyclone vulnerability on the Bay of Bengal coast. The Arab World Geographer 14 (1), 1–32.

Rasid, H., 2000. Reducing vulnerability to flood disasters in Bangladesh: compatibility of floodplain residents' preferences for flood alleviation measures with indigenous adjustments to floods. In: Parker, D.J. (Ed.), Floods, vol. II. Routledge, London and New York, pp. 46–65.

Rasid, H., Mallik, A., 1995. Flood adaptations in Bangladesh: is the compartmentalization scheme compatible with indigenous adjustments of rice cropping to flood regimes? Applied Geography 15, 3–17.

Rasid, H., Paul, B.K., 1987. Flood problems in Bangladesh: is there an indigenous solution? Environmental Management 11 (2), 155–173.

Smit, B., Pilifosova, O., 2003. From adaptation to adaptive capacity and vulnerability reduction. In: Smit, B.J., Klein, R.J.T., Huq, S. (Eds.), Climate Change, Adaptive Capacity and Development. Imperial College Press, London, UK, pp. 9–28.

Smit, B., Burton, I., Klein, R.J.T., Wandel, J., 2000. An anatomy of adaptation to climate change and variability. Climatic Change 45, 223–251.

Tathan, P., Spens, K., Oloruntoba, R., 2009. Cyclones in Bangladesh—A case study of a whole country response to rapid onset disasters. In: Proceedings, POMS 20th Annual Conference, Orlando, Florida, USA, May 1–4, 2009 (no pagination).

UNOPS (United Nations Office for Project Services), 2016. Disaster Management in Bangladesh: Comprehensive Disaster Management Programme. UNOPS APO, Bangkok. https://www.unops.org/SiteCollectionDocuments/Factsheets/English.

Van Duin, M.J., Bezuyen, M.J., 2000. Flood evacuation during the floods of 1995 in The Netherlands. In: Parker, D.J. (Ed.), Floods, vol. I. Routledge, London and New York, pp. 350–360.

Verbruggen, A. (Ed.), 2015. Annex I: Glossary. IPCC. https://www.ipcc.ch/pdf/glossary/ar4-wg3.pdf.

Vivekanandan, E., 2011. Options on fisheries and aquaculture for coping with climate change in South Asia. In: Lal, R., Sivakumar, M.V.K., Faiz, S.M.A., Rahman, A.H.M.M., Islam, K.R. (Eds.), Climate Change and Food Security in South Asia. Springer Science + Business Media, New York, pp. 360–371.

Wall, E., Smit, B., 2005. Climate change adaptation in light of sustainable agriculture. Journal of Sustainable Agriculture 27 (1), 113–123.

Walker, B., Holling, C.S., Carpenter, S.R., Kingzing, A., 2004. Resilience, adaptability and transformability in social-ecological systems. Ecology and Society 9 (2), 5.

White, G.F. (Ed.), 1974. Natural Hazards: Local, National, Global. Oxford University Press, New York and Toronto.

CHAPTER EIGHT

Structural Adaptation

LIVING WITH SEA LEVEL RISE THROUGH STRUCTURAL ADAPTATION

Being situated at the edge of a low-lying delta, coastal Bangladesh is a classic example of a local area in a distant third-world country, which is at the forefront of global climate change impacts. The area confronts two climate change—related hazards: (a) gradual sea level rise (SLR) and (b) increasingly more intense tropical cyclones with high-magnitude storm surges. As our estimates have shown (in Chapter 3, Table 3.6), an SLR of 1 m (3 ft) is likely to impact at least 29 out of 64 districts of Bangladesh. Among them 11 districts situated on the Bay of Bengal coastline are directly exposed to the sea (Table 8.1).

In addition to these, several other low-lying inland districts are also highly vulnerable to SLR. However, on a positive note, our review of global SLR during the last 100 years as well as SLR along the Bangladesh

Table 8.1 Districts of Bangladesh Situated on the Bay of Bengal Coastline

Districts (Coastal Zone[a])	Percentage of Area Likely to Be Inundated by a 1 m Rise in Sea Level	Percentage of Population Likely to Be Impacted/Displaced by a 1 m Rise in Sea Level
Satkhira (SW)	100	100
Khulna (SW)	100	100
Bagerhat (SW)	100	100
Barguna (SC)	100	100
Patuakhali (SC)	100	100
Bhola (SC)	100	100
Lakshmipur (SC)	90	90
Noakhali (SC)	90	90
Feni (E)	90	90
Chittagong (E)	10	50
Cox's Bazar (E)	40	50

[a]Coastal zones: *E*, eastern; *SC*, south-central; *SW*, southwest.
Source: Prepared by the authors based on our own estimates (from Table 3.6).

Climatic Hazards in Coastal Bangladesh
ISBN 978-0-12-805276-1
http://dx.doi.org/10.1016/B978-0-12-805276-1.00008-9

coast for the past two to three decades indicate that the sea level has been rising relatively slowly. Further, most of the climate change models project accelerated rates of SLR only after 2050. This implies that the residents of coastal Bangladesh would be able to continue to live in this area with relatively shallow and discontinuous flooding (i.e., flooding in only low-lying areas) for at least another generation, provided a wide range of structural adaptation measures are implemented. It may be stressed here that the entire area of coastal Bangladesh is not necessarily a uniformly low-lying area. There are significant local micro-topographic variations, with higher tracts of land alternating with medium-level and low-level lands (Brammer, 2014). As in the floodplains of Bangladesh, coastal residents have built their homesteads[1] and rural infrastructures on and along relatively flood-free elevated tracts of lands (Rasid, 2000). Given such physical settings and confronted with slowly rising sea levels, coastal residents may still need protection of their homes from impacts of cyclones and storm surges. They may also require additional cyclone shelters during high-magnitude cyclones. With rising sea levels, they also need protection of existing roads and bridges by elevating their levels above projected sea levels. Thus, contrary to popular assumptions, people of coastal Bangladesh are not necessarily in a hopeless situation of outright inundation due to SLR. A number of structural measures may be the short-term and medium-term solutions for confronting SLR in coastal Bangladesh. Such an approach is consistent with the principle of adaptive management, which would allow a wait-and-see approach in an uncertain environment of SLR. The following structural measures, among others, provide opportunities for confronting impending climatic disasters in coastal Bangladesh:

- Improved network of rural roads and bridges
- Cyclone-resistant homes
- Cyclone shelters
- Coastal embankments
- Triple-F Platforms

[1] In Bangladesh, a homestead is called *bari*, consisting of several houses with a common courtyard called *uthan* in the middle of the homestead. Homesteads are built on elevated tracts of land above normal flood levels. In some cases abandoned *natural levees* of migrating meandering rivers provide ideal sites for homesteads. If levees are not available, homesteads are elevated artificially by digging earth from surrounding lands.

COASTAL CLIMATE-RESILIENT INFRASTRUCTURE PROJECT

An ADB-Led Project

The Bangladesh *Coastal Climate-Resilient Infrastructure Project* (CCRIP) is an ambitious infrastructure project in southwestern and south-central Bangladesh that is designed to allow the people of coastal Bangladesh to continue to live with rising sea levels in the short and medium terms, at least up to 2050 (ADB, 2012). This $150 million project is being financed through the Climate Investment Funds, a multilateral financing instrument for "scaled-up financing through ADB, the African Development Bank, European Bank for Reconstruction and Development, Inter-American Development Bank, and World Bank Group" (ADB, 2012).[2] The project area includes 12 districts of southwestern and south-central Bangladesh, which are most vulnerable to an SLR of 1 meter (Table 8.2). Out of these 12, six are coastal districts, that is, they are directly situated on the Bay of Bengal coastline.[3] The remaining six are inland districts but four of them are sufficiently low-lying so that their entire areas are vulnerable to SLR, similar to the coastal districts.

Improvement of Rural Roads

This is the most basic goal of CCRIP. To sustain livelihoods in rural settings of coastal Bangladesh, access to rural market is a basic requirement. While some goods may reach rural markets by indigenous boats bulk of them are transported to the market by roads. It has been estimated that large proportions of rural roads in coastal areas are likely to be inundated by a 3.3 ft (1 meter) rise in sea level. Therefore, upgrading these roads above the projected SLR and, in some cases, rebuilding damaged roads is one of the CCRIP priorities. Two major road-related infrastructure components of this project include improvement of: (a) roads and (b) bridges and culverts.

[2] The Climate Investment Funds (CIFs) have been designed by developed and developing countries to fund different types of low-emissions and climate resilient development projects through the Multilateral Development Banks (MDBs). www-cif.climateinvestmentfunds.org.

[3] Out of the six coastal districts listed in Table 8.2 the following sub-districts (*upazilas*) are situated directly on the coastline of the Bay of Bengal: (Barguna district): sub-districts of Amtali, Barguna, and Patharghata; (Patuakhali district): Kalapara and Rangabali; (Bhola district): Char Fashon and Manpura.

Table 8.2 Bangladesh Coastal Climate-Resilient Infrastructure Project Area

	Percentage of Area Likely to Be Inundated by a 1 m Rise in Sea Level	Percentage of Population Likely to Be Impacted/Displaced by a 1 m Rise in Sea Level
Coastal Districts		
Satkhira	100	100
Khulna	100	100
Bagerhat	100	100
Barguna	100	100
Patuakhali	100	100
Bhola	100	100
Inland Districts		
Pirojpur	100	100
Jhalokathi	100	100
Barisal	100	100
Gopalganj	100	100
Madaripur	80	80
Shariatpur	66	75

Source: Prepared by the authors based on district names taken from the map of Bangladesh in ADB (Asian Development Bank), 2012. Bangladesh: Climate Resilient Infrastructure Project. ADB Technical Assistance Consultant's Report. Final Report, vol. I: Main Report. AECOM Asia Company Limited (in association with several other consultants and the Government of Bangladesh, Local Government and Engineering Department), Hong Kong. Web document available at: www.adb.org/sites/default/files/project-document/73944/4508-001-ban-tacr.pdf. Data on area of inundation and population displacement likely from a 1 meter rise in sea level are based on our own estimates in Table 3.6.

In all, 334 miles (537 km) of roads will be upgraded in three administrative units as follows:

- Upazila (subdistrict): 81 miles (130 km)
- Union: 108 miles (174 km)
- Village: 145 miles (233 km)

One of the basic objectives of road improvement is to increase paved roads or in some cases to strengthen (harden) road surfaces with such construction materials as herring-bone bricks. Other components of road upgrading include:

- Reconstruction or improvement of existing roads
- Widening and raising of embankments
- Slope protection against erosion and wave action

Engineering standards for adaptation to SLR has been characterized in the CCRIP document as climate-resilient standards. To achieve this, the road crest levels will be raised to 2 ft (0.6 m) above the "maximum normal

Table 8.3 Examples of CCRIP Climate-Resilient Engineering Standards

Nature of Engineering Work	New CCRIP Construction Standards (Selected Examples)
Pavements	• All pavement designs to have a full-width drainage layer incorporated into the pavement at either improved subgrade or subbase levels • Special flood roads to have either concrete or mortared brick verges • Subbase widths to be at least 0.25 m (0.82 ft) wider than overlying base • Upazila roads to have a narrow verge (minimum 0.9 m/3.0 ft) to reduce land-take • Village roads to have sand-aggregate subbase rather than a thickened sand cushion
Earthworks	• Minimum of 0.6 m (2.0 ft) above local highest normal flood/tide level for all roads • An additional height for effective sea level rise to be adopted based on local circumstances
Drainage	• Increase of cross-drainage structures to an average of 2 per km and linked into rehabilitated internal polder system

Source: Box 1 (ADB (Asian Development Bank), 2012. Bangladesh: Climate Resilient Infrastructure Project. ADB Technical Assistance Consultant's Report. Final Report, vol. I: Main Report. AECOM Asia Company Limited (in association with several other consultants and the Government of Bangladesh, Local Government and Engineering Department), Hong Kong. Web document available at: www.adb.org/sites/default/files/project-document/73944/4508-001-ban-tacr.pdf, p. 12) (shortened by the authors).

design life flood level" (ADB, 2012). Embankment protection will be achieved through the use of turf and shrubs. In high-risk areas additional protection by geotechnical measures may be necessary. Table 8.3 summarizes some of the CCRIP climate-resilient engineering standards. All of these standards exceed the current Government of Bangladesh (GoB) Local Government and Engineering Department (LGED) standards.

Upgraded Rural Markets

Another major goal of the project is to improve existing rural market centers and promote new growth centers. The road network is a crucial element of this component of the project. Each market will be connected to an existing or to a new paved road. In all, 88 markets will be improved under the project with 15% of space allocated to women with limited income, who will also carry out routine maintenance on rural roads using basic hand tools and materials supplied by LGED (ADB, 2012). One of the new features of market improvement will be to provide paved trading areas. New market sheds will be supported on concrete pillars at a level above the existing maximum normal monsoon level plus the estimated SLR for 2050, with an additional freeboard allowance of 10 inches (250 mm). Other improvements will include provisions of water supply system, drainage facilities, sanitation facilities, market offices, etc (ADB, 2012). Deep hand operated tubewells and sweet water ponds (with pond sand filters) will be installed in selected large markets in southwestern districts of Khulna and Satkhira, where groundwater is polluted with saltwater intrusion (ADB, 2012). Electricity will be provided in some of these facilities. Street lights will be powered by using solar photovoltaic cells (ADB, 2012). If implemented as planned all of these features would constitute a remarkable transformation of rural markets in coastal Bangladesh.

Other Measures and Prospects of CCRIP

The CCRIP has also a component on cyclone shelters. It plans to build 15 new multipurpose shelters, to improve 10 of the existing shelters and to upgrade about 9 miles (15 km) of access roads to shelters. Compared to a large number of cyclone shelters in Bangladesh (nearly 3900), this is a relatively small component of CCRIP. We have devoted an entire section on cyclone shelters, using primary data from the Disaster Management Information Center of the Government of Bangladesh Department of Disaster Management (see below).

While ADB-led $150 million CCRIP is one of the latest show case climate change adaptation projects of the Government of Bangladesh, it has received a mixed review from Ian Burton, one of the world's leading experts in disaster management. In a recent article published in *Dhaka Tribune*, Burton (2015) wrote:

> *If well implemented, the project will bring substantial benefits to the 12 districts in the short to medium term (next 20–30 years). The improved infrastructure will provide greater safety, easier and more reliable access to markets. The regional economy, food security and climate resilience is likely to improve while poverty and disaster risk will be reduced.*

Then, he raises a big question: "Is this a wise investment and a worthwhile project?" Burton's concerns arise from the dominant paradigm in current public discourse on climate change impacts and SLR in Bangladesh. Citing Asian Development Bank's estimates, he has raised three substantive issues:

1. "By 2050, climate change impact could make an additional 14% of the country extremely vulnerable to flooding and dislocate more than

35 million people. Many of these migrants will come from these 12 rural coastal districts."

2. "In my view as a well-disposed external observer, familiar with Bangladesh, this project makes very good sense, but by itself it is not sufficient. The project needs to be thought of in a longer term context incorporating long term measures with a steady process of implementation."
3. "Essentially this means moving the exposed population to safer places."

If we accept Burton's recommendations, it implies that Bangladesh should start preparing for mass relocation of impending climate refugees. Theoretically, such an assumption makes sense, but we have at least three major concerns with the basic assumptions. First, our own research on population displacements from coastal Bangladesh (see Chapter 3) shows that there is no field evidence of significant population movement from coastal Bangladesh (also see Chapter 7 in Rashid and Paul, 2014). Admittedly, Dhaka receives close to 400,000 migrants each year, as Burton (2015) has pointed out, but they are from all parts of Bangladesh and most of them are economic migrants and not necessarily climate refugees. Second, ADB's estimate of displacement of 35 million people from coastal Bangladesh by 2050 is grossly exaggerated. Compare this with World Bank's (2009 and 2000) estimates, which indicate that 3.6 million people are likely to be affected by a 10-inch (25 cm) of SLR along the Bangladesh coast by 2050. This makes sense based on most of the model projections that indicate acceleration in SLR rates after 2050. Third, ADB's projections do not consider the details of the countering effects of river sedimentation on SLR along the GBM delta. One of the latest studies has shown that land level rise in an island at the mouth of one of the large rivers in southwestern Bangladesh has kept up with rising sea levels for the last 50 years (Aurbach et al., 2015). Since there are such uncertainties with the rates of SLR along the Bangladesh coast, an adaptive management approach, i.e., a wait and see approach for both the short term and the medium term, seems to be the most appropriate approach for dealing with SLR in Bangladesh. In this context, CCRIP seems to be a tacit admission of this approach. Having said this, it does not mean that the Government of Bangladesh should be complacent about potential climate refugee issues. Burton (2015) has concluded correctly that "there is a need to prepare places for migrants to move to. This means the creation of new towns, or planned expansion of existing towns, by economic development and job creation." Because of an acute shortage of land in Bangladesh this will be extremely difficult, but there are still some scopes for creating new places for migrants with innovative planning.

CYCLONE-RESISTANT HOMES

Potential Strengthening of Existing Rural Homes

In Chapter 7, we have used an analogy of a tripod representing three dimensions ("legs") of adaptation of coastal Bangladesh to impending SLR due to climate change. Building homes on elevated tracts of lands above the rising sea levels is one of the three supporting legs of the tripod. (Salt-tolerant rice and disaster management are the remaining two measures.). We have characterized traditional homesteads as a form of nonstructural adaptation to "normal flood levels" in coastal floodplains, since their construction does not involve modern engineering work. Floodplain residents throughout Bangladesh have been using indigenous methods of elevating and maintaining their homesteads (a common property for several homes) above normal flood levels by digging earth manually from surrounding lands (Rasid, 2000). Such indigenous adjustments (nonstructural adaptation) are not likely to be adequate any more for confronting rising sea levels and increasing magnitudes of cyclones and storm surges. Modern engineering work would be necessary for flood-proofing and cyclone-proofing existing and new homes in coastal Bangladesh. It may not be feasible to cyclone-proof the entire housing stock of coastal Bangladesh because of the sheer numbers of such houses (several million). However, there are scopes for incorporating some of the cyclone-proofing features in existing structures (during their repair and maintenance), in damaged structures (during their reconstruction), and in entirely new structures.

As we have indicated earlier in Chapters 3 and 7, rural homes in coastal Bangladesh (as in the rest of the country), can be classified into three categories: (a) *kutcha* houses made largely of thatched materials (literally meaning weaker houses, accounting for about 90% of rural homes), (b) *pucca* houses (stronger houses) made largely of brick and cement, and (c) *semi-pucca* houses (moderately stronger houses) combining features of both. Among these, *kutcha* houses need significant strengthening to withstand impacts of cyclones and storm surges. Based on an earlier report on the Multipurpose Cyclone Shelters project and field evidence from coastal Bangladesh, Lewis and Chisolm (1996) have reviewed some of the relatively inexpensive changes in construction methods that have good potentials for strengthening both *kutcha* and *pucca* houses. We have highlighted below some of these changes in construction methods for improving different parts of indigenous homes.

Almost invariably all *kutcha* houses have mud floors, i.e., their plinths are made of earth and floors are polished with mud. Most of the *kutcha* houses are built as a cluster on elevated homesteads with a common courtyard. Plinth levels of individual houses may range from 1.5 to 3 ft (0.5–1 m) in their elevation above the courtyard level. Current homesteads are already situated on elevated tracts of land. Some of them are situated on abandoned natural levees but most of them have been raised artificially to minimum heights of 3–10 ft (1–3 m) above the floodplain level.

> Potential improvements of kutcha houses: (1) Elevate homesteads and plinth levels. *Adaptation to rising sea levels is possible by raising the levels of existing homesteads and of plinth levels of individual homes above the courtyard by digging additional earth (Rashid and Paul, 2014). Scarcity of land for digging additional earth is a major constraint for the suggested measure. For additional digging heavy machinery may be helpful.*

Frames of *kutcha* houses are normally made of bamboo. Wooden frames are rare. Frame elements are lashed together with jute rope (Lewis and Chisholm, 1996).

> Potential improvements of kutcha houses: (2) Improve different features of frames. *These include (a) treatment of bamboo against insect attack, (b) treatment of poles against rot in the ground, (c) better anchoring of poles into the ground, (d) inclusion of cross-bracing, and (e) replacement of jute rope with galvanized wire binding.*
>
> ***Lewis and Chisholm (1996).***

Walls of *kutcha* houses are commonly made of split and woven bamboo panels. In rare cases, *kutcha* houses may have corrugated tin walls and roofs (corrugated galvanized iron sheets).

> Potential improvements of kutcha houses: (3) Standardize door and window locations. *This may be done by: (a) placing the door in the center of the wall, (b) adding a small window in the rear wall, and (c) limiting window openings in relation to the walls.*
>
> ***Lewis and Chisholm (1996).***

Roofs of *kutcha* houses are usually made of thatched materials. Corrugated tin roofs for such houses are rare.

> Potential improvements of kutcha houses: (4) Standardize roof characteristics and anchoring them. *This may be done by: (a) increasing the pitches of roofs to 30–40°, (b) tying down thatch materials, (c) fixing corrugated tin roofs more*

frequently, and (d) tying down of the entire roof, either permanently or on a temporary basis following a cyclone warning.

Lewis and Chisholm (1996).

Although *pucca* and semi-*pucca* structures are stronger than *kutcha* houses they may still require some modifications to make them cyclone resistant. The following improvements have been suggested by Lewis and Chisolm (1996): (a) Use concrete in foundation, (b) employ improved anchoring techniques, (c) use "sawn timber and galvanised steel straps to tie roof structure members to each other and to walls", (d) improve fixings for roof sheets, and (e) provide external timber window shutters.

Engineering Designs of Cyclone-Proof Homes

Some of the information presented in this section overlaps with that of the preceding section but there are significant differences in the details. Whereas Lewis and Chisolm (1996) suggested some of the very basic structural improvements for indigenous houses in cyclone zones of Bangladesh, in this section Agarwal (2007) has outlined the latest advances in architectural designs of cyclone-proof homes that may be applicable in any cyclone zone. The following is a summary of the main engineering design characteristics of cyclone-resistant buildings.

Location: Strong cyclone wind is the main threat to building stability and integrity in cyclone zones. To minimize wind impacts, cyclone-resistant buildings should be sited behind a mound, if available, or behind a shelter belt of trees. The distance of shelter effect of trees is approximately 8–10 times the height of the trees. To avoid damage to the house from broken trees buildings should be sited at a distance of about 1.5 times the tree heights (Agarwal, 2007).

Foundations: Stability of a building is determined significantly by its foundation. The nature of foundation is partly determined by soil characteristics. Normally foundations tend to be shallow in stiff sandy soil, whereas in clay-rich soils deeper foundations are required. Slab or raft foundations are normally used on soft soils as these tend to spread the weight over a wider area (Agarwal, 2007). In areas where soil conditions vary strip foundations are common. In alluvial and clay soils, pile foundations use different types of pile (pole) footings, to strengthen the foundation (Agarwal, 2007).

Building shape: Building shapes may vary widely due to cultural preferences but three shapes are common: square, rectangular, and L-shaped. Among these, square is the preferred design since it allows wind to move around a square building relatively easily. The main problem with the

L-shape, particularly if the L is facing windward, is that it tends to concentrate wind action against the intersection of two wings of the building (L), locally increasing wind pressure against the building. For rectangular buildings the length should not exceed three times the width of the buildings (Agarwal, 2007). If rectangular houses are built as row houses these may generate wind tunnel effect, i.e., winds may accelerate speeds funneling through buildings. To minimize such wind tunnel effect, buildings should be constructed as a cluster with zigzag openings through them.

Types of roofs: Among a variety of roof geometry, three are common: hip roof, gable roof, and flat roof. Hip roofs are more cyclone resistant than gable roofs or flat roofs.[4]

Hip roof consists of four slanting parts of a roof surrounding the house. Two longer rectangular parts cover the length of the house on opposite sides. Two shorter triangular pieces cover the two ends of the houses (along its width). A gable roof has only two rectangular pieces covering the length of the house on opposite sides. In this case, two ends of the house (along its width) are covered by the wall. Flat roofs are not necessarily entirely flat, unless these are brick homes. Often flat roofs are slanted slightly to facilitate drainage of rain water. In all of the cases, high pitched roofs are more cyclone resistant. A minimum pitch of 22° is recommended for flat roofs, whereas higher pitches for both hip roofs and gable roofs provide maximum cyclone resistance (Agarwal, 2007).

Overhangs, patios, and verandas: These are cultural features of many houses. They may be esthetically pleasing but they trap winds relatively easily often causing major damage to homes. Damage is likely to increase if these are constructed as an extension of the roofs of the main house. To avoid damage to the main house, patios and verandas should be constructed with separate roofs slightly below the main roofs. If constructed in this manner, patio roofs may be blown off by winds without affecting the main roofs.

Securing devices for roofs: Engineering specifications for different types of securing devices for roofs are extensive. These are incorporated in building codes of many cyclone zones. Some of these could readily be used for

[4] There are some variations in roof designs of rural homes in Bangladesh, irrespective of thatched roofs or corrugated iron sheet roofs. Two common designs are (a) hip roofs, called *chouchala* in Bengali (literally meaning four sides of roofs) and (b) gable roofs, called *dochala* (two sides of roofs). There are some single slanting roofs of corrugated iron sheets but these are rare. Some of the houses with hip or gable roofs may have separate single slanting roofs for the balcony, immediately below the main roof.

cyclone-proofing roofs of newer structures in coastal Bangladesh. The following are selected examples.

- Roof ridges (upward junction of two sides of a roof) should be secured by *collar ties* (i.e., timbers connecting the rafters) nailing them to the side of the rafters, or by *gussets* (usually made of steel and plywood), or by metal straps over the top of the rafters (Agarwal, 2007).
- If galvanized sheets (corrugated tin roofs) are used, secure the corrugated sheets by using screws instead of nails, which do not hold as well as screws. Use large washers under the screw heads to prevent the roof sheets from tearing when pulled upward by high winds. Secure the corrugated sheets over the ridge, overhang and eaves (underside end of the roof) by screws at every corrugation or at least at two or three corrugations (Agarwal, 2007).
- Corners and edges of the roof are areas of greater wind suction. For cladding, i.e., securing the roof to the frame (especially at roof corners) J bolts may be used but for cyclone-resistant connections U bolts with M.S. washer (square-shaped washers) are recommended (Agarwal, 2007). Alternately a metal strap may be used to fasten the corners.
- Thatched roofs should be properly tied down to wooden framing underneath by using organic or nylon ropes in diagonal pattern (Agarwal, 2007).
- The connection of roof framing to walls or posts should be made by using properly designed anchor bolts and base plates. For masonry walls this should be accomplished by using anchor bolts embedded in concrete cores (Agarwal, 2007).
- To reduce wind-induced vibration (flutter), all members of the truss (roof-supporting structure) and bracing (devices for fastening or clamping) should be connected at the ends with at least two rivets (i.e., bolts) (Agarwal, 2007).

Doors and windows: Openings in the walls (doors and windows) are areas of weakness and wind stress concentration. Therefore, great care should be taken in placing doors and windows symmetrically in relation to the entire wall. Windows should be avoided near the corner within a distance of one-sixth the height of the wall (Agarwal, 2007). Windows and doors should have strong holdfasts as well as closing and locking arrangements.

Wind-Resistant Huts

The Bangladesh Red Crescent Society (BDRCS) is in the forefront of research, innovation, and field-testing of cyclone-resistant homes. In collaboration with the International Federation of the Red Cross Society it has designed and field-tested a prototype of wind-resistant huts (Haq, 2007). Its design is an admirable attempt to blend some of the modern construction features with the traditional housing structures. Thus, it resembles a vernacular home with a high-pitch gable roof, each side slanting at an angle of 27.5°; but unlike the traditional thatched roof it has a corrugated iron roof. The prototype has precast concrete columns for vertical support, steel truss for roof support, and steel rod for bracing between columns (Haq, 2007). Like traditional houses, it uses bamboo mat as its wall. A prototype has been field tested in the east coast (Chittagong-Cox's Bazar area) with mixed results.

First, as a part of its postcyclone relief operations, BDRCS distributed materials for wind-resistant huts to 1000 cyclone victims in the districts of Chittagong and Cox's Bazar who lost their homes to *Cyclone One Bravo* (May 1997). The hut construction was started in late October and concluded in April 1998. The huts, designed to withstand winds up to 156 miles per hour (250 km/h), were built with prefabricated local materials and erected by the beneficiaries themselves, with assistance from community CPP volunteers (BDRCS, 1998). The huts, 10 ft × 12 ft (3 m × 3.6 m) in size, "were roofed with corrugated iron sheets strongly fixed to a braced steel and concrete structure with bamboo fencing anchored in an earthen foundation" (BDRCS, 1998). One of the special features of these huts was that all conceptual structural engineering calculations were verified by computers, ensuring minimum cost and maximum structural strength. This project proved to be a successful model for cyclone-resistant homes in coastal Bangladesh.

Second, despite this apparent success, surprisingly reactions of some of the rural residents have been lukewarm. For example, one of the beneficiaries of the new experiment preferred covering the iron roof with an additional layer of thatched roof, to conform the overall structure to a vernacular look (Haq, 2007). Although the prototype has excellent anchoring to foundation and jointing details of truss to vertical post, the use of steel sections for fabrication of truss posed a problem with many, particularly in a low technology area where people depend on local expertise and locally available materials. Regular maintenance of steel structures with a specialized coating to avoid rust and damage and the additional costs for new materials also posed a challenge for many impoverished residents. Similarly, recommended chemical treatment of timber and wood also increased the construction cost.

Third, although the cost of construction for the prototype wind-resistant hut is higher than that of a typical vernacular *kutcha* house made of bamboo walls and thatched roofs, it is significantly lower than the cost of a *pucca* house (brick and cement building). Based on the lukewarm reception of the wind-resistant hut, as indicated above, cultural acceptance of a hybrid model may be an issue. In contrast, the *pucca* houses, most of which have proven to be cyclone resistant, are preferred by rural residents, but affordability is the main issue as only a limited few wealthy rural residents can afford it. According to a World Bank study, based on an economic model that seems to be overly optimistic, it is possible that the income levels of coastal residents of Bangladesh would reach sufficiently high by 2050 so that 98% of households would be able to afford a *pucca* house (World Bank, 2010). In short, field-testing of the prototype has demonstrated that it is possible to build affordable wind-resistant homes in coastal Bangladesh, but cultural acceptability of a hybrid design may still be an issue.

Stilt Houses

Unlike the Mekong delta in Vietnam and Cambodia where several million houses have been built on stilts (Boateng, 2012; Nuorteva et al., 2010), stilt houses are not very common in Bangladesh, except in the hilly regions (Hafiz, 2000). Even in deep flood zones of major floodplains of Bangladesh very few floodplain residents have adopted this design despite some initiatives by the Habitat for Humanity (HFH, 2007). A bamboo-based design that would raise family homes safely above water levels was a central concept in HFH's proposed response to flood damage to indigenous houses. In the wake of the 2007 flood, which damaged many rural homes in the Brahmaputra (Jamuna)-Meghna floodplains north of Dhaka, the HFH (2007) proposed "building houses on stilts with bamboo columns inserted in a foundation of something like recycled tires and a fill with adobe blocks and sand for stability against running water ... Wall panels would be made of jute attached to bamboo and covered with earth plaster or stabilized mortar mix." A prototype was built for field testing. The idea was particularly welcomed by many residents of riverbank and *char* (river islands) communities that were liable to bank erosion. One of the advantages of the stilt design is that such structures could be dismantled and relocated relatively easily following flood or erosion damage.

One of the major barriers to large-scale adoption of stilt houses in Bangladesh is a cultural perception that these are low-end structures that are not parts of the mainstream indigenous houses of rural Bangladesh.

This perception is reinforced by recent emergence of rows of bamboo stilt houses in many squatter settlements (slums) of Dhaka. As early as the mid-1980s, unauthorized settlements of bamboo stilt houses appeared in water-logged areas along the Dhaka City flood control embankment (Rasid and Mallik, 1996). Despite many environmental and logistic challenges (such as sanitation and water supply problems and problems of access to houses), these stilt houses provide evidence of their adaptation to high-magnitude floods.

With rising sea levels and increasing impacts of cyclone storm surges stilt houses have been gaining popularity throughout the coastal belts in many parts of the world, including the eastern seaboard of the United States. Many single- and multistoried buildings have been built on concrete stilts (normally called piers or pilings in USA) along the coasts of Florida, the Carolinas, New York, New Jersey, California, and Hawaii. In some areas building codes would specify the height and type of piling or stilt foundations. Typically, stilt home heights in hurricane zones along the US coasts range from a few feet above the ground to as much as 10–20 ft (3–6 m) (http://www.topsiderhomes.com/piling-pier-stilt-house-hurricane-home-plans.php).

Most of the stilt houses along the US coast are luxury homes. Obviously, impoverished residents of coastal Bangladesh cannot afford to build such homes. Yet, such structures evoke images of potential adaptation of rural homes in coastal Bangladesh to rising sea levels. A Google search using the term "stilt houses in coastal Bangladesh" provided pictures of several stilt houses (https://www.google.ca/search?q=stilt+houses+in+coastal+Bangladesh). Their locations were not specified but they seemed to be site specific. So far, stilt houses are not very common in coastal Bangladesh with the exception of indigenous Rakhain homes (see Chapter 7). Some of the examples of experiments with wind-resistant and flood-resistant houses in Bangladesh, reviewed earlier, provide valuable lessons (both positive and negative) for potential adaptation of stilt houses in coastal Bangladesh. First, one of the weaknesses of HFH's bamboo stilt houses for floodplains is that they are not suitable for the coastal belt because these are not designed to be cyclone resistant or storm surge resistant. Put it simply, bamboo stilts are too weak to withstand impacts of cyclone winds. Even sturdy wooden posts may not have adequate strength to resist wind and surge impacts. Besides, bamboo and wooden stilts are likely to rot rapidly in a saltwater environment even if the poles are not placed directly in saltwater. During cyclones they are likely to come in contact with saline surge water. Second, one of the attractive characteristics of wind-resistant huts reviewed earlier is that many of their cyclone-proofing architectural features could readily be adopted for stilt houses in coastal areas. In other words, the wind-resistant hut may be adapted in coastal areas by placing it on a wooden or other flood-resistant platform as a base over a strong stilt structure, instantly transforming a vernacular house or a modern wind-resistant hut into a stilt house (Hafiz, 2000). Third, stilt house technology is focused on the details of the stilt structures. The use of reinforced concrete posts as stilts, substituting typical timber and bamboo stilts, has the advantage of being water resistant and hence more durable (Biswas et al., 2015). Some of the most successful cyclone shelter designs in Bangladesh are based on this principle of building these multistoried shelters on reinforced concrete posts. Free flow of storm surge water beneath the first floor of a shelter through the open structures of the stilts/posts facilitates in dissipating surge impacts. Further, there is some scope for designing wind-resistant huts by emulating aerodynamically compatible shapes and orientation of some of the cyclone shelters. Finally, even the indigenous practice of building floodplain homes on elevated tracts of lands may play an important role in such adaptation. Since indigenous homes are already built on elevated lands there is no reason for placing stilt houses directly on shallow floodwater. Instead, these could be placed on elevated homesteads. This would minimize their direct contact with saline seawater and would help in elevating many stilt houses above the cyclone storm surge levels. Assuming that the majority of the homesteads are already about 10 ft (3 m) above the coastal floodplain, if stilt houses could be raised an additional 10 ft (3 m) then they could avoid the bulk of the surge impacts. The storm surge depths in coastal Bangladesh have remained up to or below 20 ft (6 m) above sea level during most of the high-magnitude cyclones (with the exception of the 1970 Bhola Cyclone when surge water reached up to 34 ft or 10 m). Thus, stilt houses could be built in coastal areas by combining architectural features of wind-resistant huts, conventional stilt houses, cyclone shelters, and floodplain homes. Lack of affordability is likely to be the main downside of wind-resistant stilt houses in coastal Bangladesh since such structures are likely to be significantly more expensive than vernacular homes.

"Build Back Better": A Case Study on Cyclone-Resistant Homes

Some of the design characteristics of wind-resistant huts described above have now gained widespread acceptance in coastal Bangladesh. The Core

Shelter Program (CSP) of the GoB and a UNDP model home called Disaster Resilient Home were based on a similar model of a small but strong cyclone-resistant house, which was built on concrete pillars anchored deeply into the ground. Most of the buildings had earth/mud floors. All of them had corrugated iron sheet roofs. While some of the buildings used corrugated iron sheets for their walls, other construction materials including wood and bamboo introduced variations in wall materials. As a part of its post-*Sidr* shelter strategy, the GoB adopted a "Build Back Better" (BBB) campaign to promote construction of cyclone-resistant homes using its CSP model. Besides GoB and UNDP, the Shelter Working Group (which included several NGOs and governments of Saudi Arabia, India, and Kuwait) was a major partner of the project. The leading NGOs who participated in different components of the program included (in alphabetic order): Bangladesh Rural Advancement Committee (BRAC), CARE, Dustha Shasthya Kendra (DSK) (meaning distressed health center), Concern Worldwide, Habitat for Humanity, Muslim Aid, Oxfam International, and World Vision. Nearly 1.5 million homes were partially or totally destroyed by *Sidr* (Nadiruzzaman and Paul, 2013). Therefore, the demand for reconstruction was very high. The GoB promised to build nearly 79,000 new homes. This was followed by a commitment to build another 78,000 homes by the Shelter Working Group (data on Saudi Arabia in Table 8.4 may overlap with this number). Up-to-date data on commitment to build and on completed projects are not readily available, but limited data in Table 8.4 indicate that at least 115,000 homes were committed for construction. As of 2010, only 10,400 houses were completed and many of the committed projects were ongoing. The total numbers of committed and completed

Table 8.4 Proposed and Completed Cyclone-Resistant Houses in Coastal Bangladesh

Proposed or Completed by	Proposed After *Sidr* (2007)	Completed by 2010
Government of Bangladesh	78,519	No data
UNDP	15,600	9000
Saudi Arabia	21,000	No data
Muslim Aid	No data	1000
BRAC + Oxfam	No data	400
Habitat for Humanity Bangladesh	280	No data
TOTAL	115,399	10,400

Source: Prepared by the authors based on data in Nadiruzzaman, M., Paul, B.K., 2013. Post-sidr public housing assistance in Bangladesh: a case study. Environ. Hazards 12 (2), 166–179.

houses constituted less than 10% of the total numbers of damaged and destroyed homes.

To evaluate performance of the BBB program approximately two years following *Sidr*, we conducted a case study in Gabtola, a sample village in the district of Bagerhat in southwestern Bangladesh, using nine months of field studies (2009–2010) that included formal and informal discussions and interviews with residents, local leaders, and government officials (Nadiruzzaman and Paul, 2013). The case study represents a village that experienced perhaps some of the heaviest damage and destruction of homes by *Sidr*: reportedly only four out of 900 homes were not damaged (Nadiruzzaman and Paul, 2013). It also represents a village where large proportions of its post-*Sidr* houses were built under the CSP construction projects. For the same reason Gabtoli also experienced many of the problems of the CSP projects. These may be classified roughly into four categories of problems, which are outlined below briefly.

First, one of our findings was that there were substantial reductions in disbursement of allocated funds for individual recipients due to administrative procedures and some corruptions at the grassroots levels. Eligible residents of Gabtoli received 41 tin sheets (corrugated iron sheets) and eight concrete pillars from GoB to build a 9-feet (2.74 m) high 150 square-feet (14 sq. m) house (Nadiruzzaman and Paul, 2013). The government appointed external vendors to buy tin sheets and make concrete pillars. A total of taka 62,875 (approximately US$900) was allocated to build a house, which included a budget of taka 10,000 (US$150) for developing an earthen plinth and installing a house on the plinth. From this construction cost, taka 6500 (approximately US$100) was given to the recipient of the house and the remaining taka 3500 (US$50) went to hire outside laborers, ignoring local available labor force. Moreover, a substantial number of recipients, particularly the poorest ones, had to bribe a minimum of taka 1000 (US$14) just to keep their names in the public house recipient list (Nadiruzzaman and Paul, 2013).

Second, we found that the problems with procurement and delivery of construction materials delayed the construction projects. A single core shelter required 4500 bricks, up to 20 bags of cement, and a roof truss that weighed almost 1600 kg (3520 lbs) (Nadiruzzaman and Paul, 2013). Compiling these materials and bringing them to construction sites in remote and isolated locations proved to be challenging. Transporting these materials was not only time consuming but also expensive. In addition, in some areas there were severe shortages of masons, carpenters, and welders. All of these factors were responsible for delaying the construction of many of these new, cyclone-resistant core homes. Third, another startling finding was that some of the houses were abandoned either at the construction stage (i.e., as an incomplete structure) or following the completion of the project.

Some of the respondents complained that the buildings were too small with only one room, two windows, and a front door. Some contended that the lack of an additional room for privacy of women and the absence of a back door for their exit constituted a culturally insensitive design (Nadiruzzaman and Paul, 2013). Finally, another discouraging finding was that some of the houses sustained substantial damage from *Cyclone Aila*, a 2009 category 1 storm. Thus, it appeared that most of the post-*Sidr* homes were not adequately cyclone resistant. To make them resistant to a super cyclone like *Sidr* (a category 4 storm), attention should be given to improving the basic design, especially to anchoring devices, so that such an improved model could withstand a future high-magnitude cyclone. Despite these limitations, undoubtedly the BBB initiative was a promising start for building cyclone-resistant homes in coastal Bangladesh as an appropriate strategy for adaptation to climate change and SLR.

CYCLONE SHELTERS

Symbol of Saving Lives

Cyclone shelters in coastal Bangladesh have emerged as a symbol of saving lives during catastrophic cyclones. In a report on *Cyclone Gorky*, Allen-Mills (1991) of *Sunday Times* (London) described a cyclone shelter in coastal Bangladesh as: "ugly but effective anti-cyclone shelters—a squat two-storey structure built on thick concrete stilts and shaped like an arrow to deflect the coming storm." Data on cyclone deaths provide evidence of effectiveness of cyclone shelters in reducing human casualties. Whereas *Gorky*, a category 4 storm in 1991, was responsible for nearly 139,000 deaths, less than 4000 people lost their lives due to *Cyclone Sidr*, another category 4 storm in 2007. Most of the post-*Sidr* analyses attribute this drastic reduction in cyclone-related deaths to improved cyclone preparedness, especially to initiatives for evacuating people from their homes and moving them to cyclone shelters (Paul, 2009). Recent introduction of a cell phone SMS warning system has been helpful in relaying information about an impending cyclone (Rahman and Islam, 2011). In 1992, there were only 507 cyclone shelters distributed along the entire coast of Bangladesh

(Ikeda, 1995). As of 2009, this number increased to nearly 3800 (see section below on numbers of shelters).

Shelter Designs

A cyclone shelter is basically a solid building (mostly made of reinforced concrete), an elevated construction that provides security from cyclones and associated storm surges (Rahman and Islam, 2011). The foundations, comprising of concrete posts (stilts) and open areas in the ground floor through these posts, are such that they dissipate energy from upcoming cyclonic storm surges from the sea. Above it, at some feet/meters from the ground, one or more floors are constructed to provide shelter (Rahman and Islam, 2011). The description of a cyclone shelter by Allen-Mills (1991) as a "squat two-storey structure built on thick concrete stilts and shaped like an arrow to deflect the coming storm" refers to some of the shelters in Cox's Bazar area (during *Cyclone Gorky*) that were designed aerodynamically in the shape of a ship to be more wind resistant. Sometimes shelters may be built in a "semi-pyramidal shape to withstand the storm surges better" (Rahman and Islam, 2011). It should be stressed here that such designs are not necessarily common along the Bangladesh coast. The vast majority of them have rectangular shape and some have L-shape. A recent Google search of "pictures of cyclone shelters in Bangladesh" showed many variations in cyclone shelter designs (https://www.google.ca/search?q=cyclone+shelter+pictures+in+bangladesh/). Most of the shelters are two-storey buildings on posts/stilts, but some are single-storey structures (Figs. 8.1 and 8.2). In areas above the reach of storm surges some of the shelters have been built on elevated concrete plinths, i.e., without stilts, but these are very rare. In most cases, corners of the buildings (analogous to the points of an arrow) point to the windward side of the prevailing winds. Most of the shelters have access to the buildings by gently sloping long external stairways. Since rapid access to the buildings during an impending storm is a major concern increasing attention is being given to providing access roads.

Numbers of Cyclone Shelters

In 1970 when the "killer storm of the century" struck Bhola Island (hence it is called Bhola Cyclone) there was no cyclone shelter in this island. Immediately following the independence of Bangladesh, the GoB initiated a shelter construction program in 1972 but the pace of the program was extremely slow largely because of acute shortage of funds. The project gained momentum only when many international donor agencies participated in the

Figure 8.1 A two-storied cyclone shelter in Bagerhat district, southwestern coast of Bangladesh. *Photograph by Md Nawfel Huda and Khairul Matin, reproduced with permission.*

Figure 8.2 A single-storied cyclone shelter in Bagerhat district, southwestern coast of Bangladesh. *Photograph by Md Nawfel Huda and Khairul Matin, reproduced with permission.*

program following Cyclone Gorky (1991). By that time, there were only 300 cyclone shelters in Bangladesh (Shamsuddoha and Chowdhury, 2007). Data on total numbers of cyclone shelters in Bangladesh vary widely from one source to another, partly because most of the numbers cited in the literature are based on secondary sources and partly because these numbers refer to different years of completion of the projects. For example, Ikeda (1995) cited a total number of 537 cyclone shelters along the entire coast of Bangladesh, but the dateline for this number was not clear. There are similar uncertainties in the literature on the latest numbers of cyclone shelters in Bangladesh.

Currently, the most reliable primary source of information on cyclone shelters in Bangladesh is a GoB digital platform called Cyclone Shelter Information Database (GoB, 2016). This site is maintained by the Comprehensive Disaster Management Program of the GoB Ministry of Food and Disaster Management (in collaboration with UNDP, Sida, UKaid) (http: www.kmp.dmic.org.bd/). This database maintains hundreds of thousands of entries dealing with different types of information on all of the cyclone shelters in Bangladesh. Among others, entries include location of each shelter in a village (GPS coordinates of each shelter) and many of its attributes such as its building design, capacities, multiuses, and different types of facilities (see Table 8.7 for examples). Some of these data have been analyzed and summarized in a comprehensive report by CEGIS (2009). This report differs significantly from other sources as it has drawn its data from the primary source of the GoB Cyclone Shelter Information Database. In addition to drawing data from CEGIS (2009) in Tables 8.5 and 8.6, we have provided latest data (as of 2016) in Table 8.7 dealing with some of the available facilities in selected samples of cyclone shelters in three coastal zones of Bangladesh.

As of 2009, there were 3753 cyclone shelters in Bangladesh (columns 2 + 3 + 4 in Table 8.5). These included three categories of buildings:

1. Cyclone shelters, constructed for shelters during cyclone emergencies, are also used for other purposes, especially as schools, during normal (noncyclone) time.
2. Primary Education Development Program buildings are cyclone-resistant school buildings that are also used as cyclone shelters during cyclone emergencies.
3. Unusable shelters have been badly or partly damaged so that it is not safe to use them as cyclone shelters any more. Most of the damaged structures had major fractures or failures of parts of the posts (stilts). There were also signs of weaknesses in the walls (cracks) and stairways.

The total numbers of shelters increased to 3841 (i.e., the final tally) when totally destroyed and abandoned structures (88 buildings) were added to the preceding three categories. These buildings were completely or partly

Table 8.5 Distribution of Cyclone Shelters in Bangladesh, 2009

District (# of Sub-Districts) Population: Million	Cyclone Shelters (+PEDP)[a]	Capacity (Number of Persons)	Unusable Shelters	Destroyed Shelters
Bagerhat (8): 1.79	98 (54)	86,159	11	0
Barguna (5): 1.01	147 (59)	147,590	10	2
Barisal (9): 2.65	37 (15)	41,050	0	0
Bhola (7): 2.05	429 (195)	390,050	53	49
Chandpur (6): 2.71	21 (58)	26,350	1	0
Chittagong (22): 8.65	573 (71)	683,010	29	11
Cox's Bazar (7): 2.31	504 (104)	607,310	10	6
Feni (4): 1.45	57 (5)	61,275	12	3
Jhalokati (4): 0.89	12 (3)	7650	2	0
Khulna (6): 3.07	77 (46)	76,541	2	0
Lakshmipur (4): 1.76	106 (119)	118,000	10	5
Noakhali (3): 3.03	245 (16)	266,112	33	5
Patuakhali (7): 1.79	165 (103)	157,675	72	7
Pirojpur (6): 1.25	36 (33)	32,300	1	0
Satkhira (6): 2.43	65 (16)	55,071	0	0
Shariatpur (4): 1.28	11 (27)	14,375	0	0
Number of districts (sub-districts): 16 (108) Population (million): 38.19	**2583 (924)**	**2,770,518**	**246**	**88**

[a]Cyclone shelters refer to buildings constructed for shelters during cyclone emergencies. During noncyclone time most of them are used for other purposes, especially as schools. In contrast, PEDP refers to buildings constructed under the Government of Bangladesh Primary Education Development Program phase II (PEDP-II). These serve as schools + cyclone shelters.
Source: Prepared by the authors based on data in Table 3.1 in CEGIS (Center for Environmental and Geographic Information Services), 2009. Report on Cyclone Shelter Information for Management of Tsunami and Cyclone Preparedness. Ministry of Food and Disaster Management, Dhaka.

destroyed by erosion and some of them have been dismantled following major damage.

Geographical Distribution and Capacities of Cyclone Shelters

In general, geographical distribution of cyclone shelters reflects areas that are most vulnerable to cyclones. East coast is most directly exposed to cyclones. Historically, it has been the site of several high-magnitude cyclones, such as Cyclone Gorky (1991). Therefore, largest numbers of shelters have been constructed in the districts of Chittagong and Cox's Bazar (Table 8.5). If the total population of a district was divided by the numbers of shelters in that district, we might have some ideas about the density of shelters

Table 8.6 Distribution of Cyclone Shelter Facilities in Bangladesh, 2009

District	Water Supply: % of Total Shelters	Toilet (Women's) % of Total Shelters	Women's Separate Space: % of Total Shelters	Storage Facilities: % of Total Shelters
Bagerhat	36	99 (19)	1	3
Barguna	39	96 (41)	9	8
Barisal	86	97 (70)	19	19
Bhola	58	94 (24)	3	14
Chandpur	71	95 (57)	10	5
Chittagong	57	97 (30)	5	9
Cox's Bazar	42	97 (37)	9	15
Feni	64	98 (52)	21	22
Jhalokati	83	100 (42)	0	0
Khulna	62	96 (62)	12	13
Lakshmipur	71	95 (42)	32	33
Noakhali	50	98 (52)	30	29
Patuakhali	72	96 (25)	4	10
Pirojpur	44	81 (36)	6	17
Satkhira	58	98 (52)	5	8
Shariatpur	91	100 (100)	27	0

Source: Prepared by the authors based on data in Table 3.12 in CEGIS (Center for Environmental and Geographic Information Services), 2009. Report on Cyclone Shelter Information for Management of Tsunami and Cyclone Preparedness. Ministry of Food and Disaster Management, Dhaka.

(i.e., per capita distribution). However, such data might not be representative of all districts because most of the shelters were concentrated along the coastal subdistricts. Therefore, data by subdistricts (or better by Unions) would be more representative.[5] Data by districts seemed to be representative of only two districts. Because of its elongated shape all of the seven subdistricts of Cox's Bazar are situated in a relatively narrow coastal zone. Thus, the density of shelter for Cox's Bazar, which seemed to be truly representative of the entire district, was 1 shelter per 4600 people, accounting for the highest density of shelters. This was followed by Bhola, an island district with a relatively small area and the site of the notorious Bhola Cyclone (1970), which killed at least 300,000 people. All of its seven subdistricts are directly exposed to cyclones, some directly from the Bay of Bengal in the south and some through large estuaries on both west and east sides of

[5] According to the latest update posted on Wikipedia (as of February 7, 2016), Bangladesh is divided into eight divisions, 64 districts, and 488 subdistricts (called upazila in Bengali). Upazilas are further divided into 4553 Union Councils (often short-formed as Unions).

Table 8.7 Samples of Cyclone Shelter Facilities

Sub-District (Union): Village	Year Built (Funded by)	Floors (Capacity)	Main Use	Water Supply	Toilets	Separate Women Space
(a) Sample District: Cox's Bazar (Southeast Coast)						
Teknaf (Teknaf): Moheshkhalipara	1996 (Saudi)	3 (1500)	School	No	4	No
Teknaf (Teknaf): Moheshkhalipara	2006–2007 (GOB)	2 (825)	School	Yes	0	No
Teknaf (Teknaf): Khonkar Para	1999 (Swiss Red cross)	2 (1750)	Comm. Center	Yes	4	No
Teknaf (Teknaf): Lengurbil	1996 (Saudi)	3 (1500)	School	No	4	No
Teknaf (Teknaf): Mithapanirchara	1994 (Caritas)	2 (850)	School	No	2	No
Teknaf (Teknaf): Palianpara	1997 (Saudi)	3 (1500)	School	Yes	4	No
Teknaf (Teknaf): Rajarchara	2001 (Saudi)	2 (700)	School	No	2	No
Teknaf (Teknaf): Tulatoli	1991 (Swiss Red cross)	2 (1750)	Cyclone shelter	Yes	2	No
(b) Sample District: Bhola (East-Central Coast)						
Bhola Sadar (char Shibpur): Shibpur (1/2)	2007–2008 (GOB)	2 (825)	School	No	2	No
Burhanuddin (Deula): Gangapur (1/9)	2005–2006 (GOB)	3 (2025)	Office	No	5	No
Char Fasson (Aminabad): Uttar Shiba (1/12)	2000–2001 (OPEC)	2 (450)	School	No	2	No
Daulatkhan (char Khalifa): Kalakopa (1/12)	1986 (ADB)	2 (600)	School	Yes	2	No
Lalmohan (Badarpur): Char Kachapia	1994 (ADB)	2 (700)	School	No	2	No
Manpura (Sakuchia): Rahmanpur (1/25)	1994 (ADB)	2 (725)	School	Yes	4	Yes
Tajumuddin (Sonapur): Indra Narayanpur	2002 (German KFW)	2 (1800)	School	Yes	4	Yes

Source: Prepared by the authors based on primary data in GoB (Government of Bangladesh), 2016. Cyclone Shelter Information Database-DMIC. Government of Bangladesh, Ministry of Food and Disaster Management, Comprehensive Disaster Management Program (CDMP) (in collaboration with UNDP, Sida, UKaid). http: www.kmp.dmic.org.bd/.

the island. The density of shelters in Bhola was second only to Cox's Bazar with 1 shelter per 4780 people.

Data on cyclone shelter capacity (column 3 in Table 8.5) refer to the official estimate of space that is capable of accommodating the maximum number of people in a shelter during a cyclone emergency. In reality, most often cyclone shelters are overcrowded far exceeding the official capacity, implying the need for expanding the network of cyclone shelters in coastal Bangladesh. As of 2009, all of the cyclone shelters in Bangladesh had the total capacity to accommodate 2.77 million people, representing only about 7% of 16 cyclone-vulnerable districts (Table 8.5). According to an estimate by a World Bank group, at least 7000 cyclone shelters (i.e., an addition to 3000 shelters) would be required by 2050 to provide nearly full protection to storm surge—vulnerable population of coastal Bangladesh (World Bank, 2010). Besides an initial investment of $1.8 billion for their construction, their annual maintenance cost has been estimated at $39 million.

Official capacities of individual cyclone shelters vary widely from one shelter to another (see Table 8.7 for examples). Therefore, the ratio between total capacities of all cyclone shelters in a given district and the district population (i.e., shelter capacity in a district/district population x 100) provides a better measure of cyclone shelter protection in that district, compared to the data on shelter density. Calculated in this manner, the following three districts had significantly higher capacity to provide cyclone shelters for the district population:

- Cox's Bazar: 26% (i.e., capacity to accommodate 26% of district population)
- Bhola: 19%
- Barguna: 15%

Shelter capacities were at the minimum for the following three districts, each providing shelter for only 1% of its district population: Chandpur, Shariatpur, and Jhalokathi. For the remaining 10 districts shelter capacities ranged between 2% and 9% of the district population. It may be stressed here that, as in the case of shelter density (indicated above) determining shelter capacity by district population may not be similarly representative of actual cyclone shelter protection for cyclone-vulnerable population of the district. Based on estimates of storm surge—vulnerable population of several districts, Mahmood et al. (2014) have provided an improved method of determining shelter capacity. For example, shelter capacity for Noakhali amounted to only 9% based on its district population (Table 8.5), whereas Mahmood et al.'s estimates increased this proportion to 15% of the surge-vulnerable population of the district. The main limitation of this method is that such estimates would require further research and interpretations of surge-vulnerable population of a given district. In contrast, more objective and reliable data may be obtained by analyzing the distribution of cyclone shelters by subdistricts or preferably by Unions. Such data are now readily available from the GoB Cyclone Shelter Information Database (which includes information on the names of the village, Union, subdistrict, and district of each cyclone shelter).

Whereas the GoB cyclone shelter database provides a valuable resource for analyzing the mesoscale distribution of cyclone shelters in coastal Bangladesh (i.e., by subdistricts, Unions, and villages), microscale distribution is a much more challenging problem. One such microscale case study was conducted by Mallick (2014) in 12 villages in two Unions in Satkhira district (southwestern coast), by employing a household survey a geographic information system (ArcGIS 9.3) software and one of the latest versions of the SPSS statistical packages (SPSS 16). The study found that the location of cyclone shelter(s) in a specific village was determined by "neither a right-based nor a demand-based planning approach." Instead, their location was influenced more by "socio-political affluence of local-level decision-makers" (Mallick, 2014, p. 654). The study further found that the distance from home to the nearest shelter was a critical determinant/deterrent to take refuge in a shelter. The average distance for the total sample was reported as 1.53 km (about 1 mile). Data showed that increasing percentages of people who were closer to a cyclone shelter took refuge in the shelter (Mallick, 2014).

Functions and Facilities of Cyclone Shelters

We have analyzed two sources of data on cyclone shelter facilities: (a) Table 8.6: a summary table based on the CEGIS (2009) report and (b) Table 8.7: samples of primary data on cyclone shelter facilities in two high-risk cyclone districts (Cox's Bazar and Bhola), based on the GoB cyclone shelter database. All cyclone shelters have been constructed for multipurpose use. Among 2583 shelters 82% was used as education centers, 8% as offices, 1% as health centers, and 1% as community centers (CEGIS, 2009). Some of the shelters included animal/cow sheds, called *killa*, which were elevated grounds to provide shelters to the cattle.

Water supply and sanitation facilities are two of the most basic requirements for human beings even during periods of disasters. Therefore, lack of these features in anyone of the cyclone shelters is difficult to comprehend.

Contrary to our expectation, data in Table 8.6 indicate that water supply was available in only one-third to one-half of the shelters. Only in five districts water supply was available in 71%–86% of the shelters. Lack of toilet facilities in some of the shelters was more troubling. Approximately 1%–5% of the shelters in 13 out of 16 districts did not have any toilet facilities, implying an additional risk for the cyclone victims, in case they had to venture out for the call of nature (Table 8.6). In Pirojpur, one out of every five shelters lacked this facility; only in the districts of Jhalokati and Shariatpur all of the shelters had toilet facilities. Among the sample villages, all of the shelters, with the exception of only one, had toilet facilities but their numbers seemed to be too inadequate (Table 8.7). For example, in Tulatoli village in the subdistrict of Teknaf (Cox's Bazar district) there were only two toilets in a shelter for 1750 people (Table 8.7). The total lack of any toilet facility in one of the cyclone shelters in the village of Maheshkhalipara (Teknaf subdistrict, Cox's Bazar district) is difficult to comprehend especially because it was one of the relatively new shelters constructed by the GoB in 2006–2007 (Table 8.7). Cyclone shelters in sample villages of Bhola district had larger numbers of toilets per person than in Teknaf (Cox's Bazar), but surprisingly in this culturally conservative district only one in seven shelters had separate space for women (Table 8.7). Lack of separate toilet facilities for women and the absence of separate space for women were two most common reasons for women's reluctance to move to cyclone shelters. Consequently, women and children who had normally stayed with them accounted for larger proportions of cyclone deaths. Increasing and improving both of these facilities would go a long way for increasing women's evacuation to cyclone shelters. Provision of additional storage facilities for cyclone victims would also assist them in saving their valuable documents and critical grain seeds for postcyclone agricultural recovery. As data in Table 8.6 indicate, storage facilities were available in only small percentages of shelters and some of the shelters did not have any storage facilities.

Despite shortcomings, existing cyclone shelters have saved lives. Based on this experience, an expanded network of shelters seems to have enormous potentials for saving lives in the future with rising sea levels at least in two ways. First, increasing shelter capacity by building additional cyclone shelters seems to be an interim solution for saving lives during recurrent storm surges until other adaptation measures are in place or in progress. As indicated above, in this regard, the progress so far has been impressive. Second, as we have already suggested earlier, some of the architectural designs of multipurpose cyclone shelters, such as concrete posts (stilts) to allow free flow of storm surges, may be experimented for residential housing structures in high-risk coastal zones as an adaptation measure to live with the rising sea levels.

COASTAL EMBANKMENTS AND POLDERS

The Dutch Model

"God created the earth but the Dutch made Holland", says a popular proverb in a blog (cited below). The Dutch, living on a seacoast, have saved "disappearing land by building protective dikes, and reclaiming marshlands for agriculture, and have been doing so for a very long time" (UMB blog): http://blogs.umb.edu/buildingtheworld/waterworks/protective-dikes-and-land-reclamation-the-netherlands/.

For some optimists the Netherlands provides a model for living with rising sea levels by building high embankments along the Bangladesh coast. There are several problems in realizing such a lofty goal. First, Bangladesh is too poor to build such high embankments. Second, the Dutch are renowned for their excellence in hydraulic engineering. Although Bangladesh has also developed significant expertise in water resources engineering, following Dutch models and technical training, it has a long way to go before matching the Dutch success stories in confronting the sea. Third, coastal morphology of Bangladesh is much more complex than that of the Netherlands. Unlike the relatively stable flows of the Rhine and its distributaries, the flows of the GBM rivers and morphology of their estuarine channels are much more unstable and extreme (Burton, 2015). In addition, Bangladesh coast is too indented for building a continuous line of coastal defense system.

Embankment Traditions in Bangladesh

Despite the preceding challenges, Bangladesh has a long tradition of experimenting with embankments and polders for dealing with both river floods and coastal inundation. Rulers and large landowners (*zamindars*) have built embankments for flood protection in Bangladesh for a long time (Brammer, 2000). Historical records indicate that coastal embankments were constructed as early as the 17th century (Islam and Bari, 2006). More recently, a large number of coastal polders were constructed in the 1960s by the East Pakistan Water and Power Development Authority, the predecessor of the current Bangladesh Water Development Board, to prevent intrusion of saltwater from the sea and tidal channels into enclosed interstream areas

(Rasid and Mallik, 1993). The estimates on the total length of coastal embankments in Bangladesh vary widely, depending on the categories of embankments in each estimate, including those of roads and highways that often serve the secondary purpose of flood protection. The latest official estimate indicates that for the last 25 years much of the coastline of Bangladesh has been protected by at least 4350 miles (7000 km) of earthen embankments, mostly in the form of polders (GoB, 2009). The polders have been constructed by using the Dutch engineering technology in which smaller intertidal channels are closed off by embankments but larger streams are kept open. Sluice gates, especially those with automatic flap gates, have been used to prevent intrusion of saltwater during high tides but to allow drainage of excess monsoon rainwater from the polders during low tide (Rasid and Paul, 1987).

Environmental Impacts of Polders and Embankments

Postconstruction environmental impacts of polders and embankments include (a) reductions in tidal water discharges and velocities due to blockage of flows by these structures, (b) siltation in tidal channels resulting in raising of the bed levels of these channels, (c) drainage congestion and water logging problems affecting performance of the polders (Choudhury et al., 2004; SWMC and BWDB, 2000), and (d) major breaches in embankments. The latter in particular has been a problem with rising sea levels as waves at higher levels have been pounding these structures leading to frequent embankment failures and breaches. We expand here on four events responsible for damage to embankments. First, in late September 2000, after an interval of 62 years, the western part of the western coastal zone experienced an unexpected and devastating flood that caused significant damage to crops, dwellings, roads, and other properties. One of the main reasons for this flood was breaching of embankments as well as the shallowness of local rivers caused by rapid sedimentation of riverbeds. The breaching was the result of poor maintenance of the embankments. Residents of flood-affected areas also deliberately breached dams and embankments to release floodwater from their crop fields and shrimp farms (SWMC and BWDB, 2000). Second, in September 2, 2006 high tides eroded parts of embankments in Patuakhali, Barguna, Bagerhat, and Satkhira districts and a total of 168 villages had been under water for weeks. The following year embankment failure occurred in 12 coastal districts, which brought immense miseries to affected population (Ahmed, 2008). Third, the problem of embankment breach has reached its peak following *Cyclone Sidr*, which struck the western coastal zone on November 15, 2007. This category 4 storm was accompanied by tidal waves up to 5 m high and surges up to 6 m in some areas, breaching over 1250 miles (1875 km) of coastal and river embankments. The event also caused damage to 1687 bridges and culverts in affected districts (Ahmed, 2008). Fourth, Cyclone Aila, which also hit the same area on May 21, 2009, damaged 172 miles (258.5 km) of roads in the 14 districts and destroyed and damaged 339 miles (509 km) and 1437 miles (2155 km) of embankments, respectively (Mallick and Vogt, 2012).

Reconstruction Projects

The polders have so far been used principally for rice cultivation and commercial shrimp culture. Protection from low storm surges of tropical cyclones has been one of the secondary benefits of these projects. Coastal polders have been receiving greater attention in the context of SLR, which would undermine their intended purpose unless the elevations of the embankments were raised adequately. Recently, the Climate Change Cell of the Government of Bangladesh Ministry of Environment and Forest has conducted a benefit-cost analysis for engineering works necessary for raising the embankment elevations of existing 138 polders. Savings from expected damage of agricultural crops due to overtopping embankments under the IPCC Third Assessment Report (TAR) scenario of 2 ft (62 cm) of SLR by 2080 indicated a benefit-coast ratio of 2.17 at an investment of about US $500 million (Climate Change Cell, 2009). In the absence of such engineering reconstruction, according to this study, a net SLR of 2 ft would result in inundation of 741 square miles (192,000 ha), damaging at least 541 square miles (140,000 ha) of agricultural land and about 10.4 square miles (2700 ha) of shrimp ponds. Most of the original embankments in high-risk areas—with crest widths ranging from 8 to 15 ft (2.5–4.5 m), seaward slopes ranging from 3:1 to 10:1, and maximum freeboard allowance of 5 ft (1.5 m) above the high tide level were designed with a 1% risk (probability) of over-dike flooding (Rasid and Paul, 1987). An SLR of 2 ft would increase this risk up to 3.33%. To restore the original probability at 1%, according to one option of this study, reconstruction needs would include raising the crest height of about 435 miles (700 km) of highways and feeder roads in high-risk areas by about 3 ft (1 m) and another 77 miles (124 km) of secondary roads by about 1.6 ft (0.5 m) above the original level (Climate Change Cell, 2009).

Compared to this internal assessment by the Government of Bangladesh, a recent study by a World Bank Group has estimated much higher reconstruction costs for these polders—at a total investment of US $5.5 billion by 2050 and $112 million recurrent annual maintenance costs—to provide improved protection from increased storm waves associated with climate change–induced SLR. Of this, much needed strengthening of 43 polders alone would require an investment of $2.46 billion and recurrent annual maintenance costs of $49 million (World Bank, 2010). The World Bank is in the forefront of embankment construction in Bangladesh. It has already launched a Coastal Embankment Improvement Project (Phase I) in 2015. By 2020, the project is expected to reconstruct and/or improve 623 km of embankments and replace or upgrade 129 drainage structures and 134 drainage regulators (Forni, 2015). Another study by two independent water resources engineers, using a hydrodynamic model to simulate floods in rivers, floodplains, and coastal plains under a range of climate change scenarios, indicates that most of the areas vulnerable to SLR are concentrated in deep flood zones in southwestern districts of Bangladesh (Bhuiyan and Dutta, 2012). In particular, this study has identified the deep flood zones downstream of the Passur River in Bagerhat district as areas of concern for future SLR.

Secondary Role of Polders

Another secondary role of polders that has received very little attention is its potential for sediment accretion, leading to induced or planned build-up of floodplain levels. A *New York Times* report (March 20, 2009) by Somini Sengupta provides evidence of this potential. This report talks about an "accidental discovery" of an indigenous method of raising land levels in Beel Bhaina, a low-lying 1-square mile (250 ha) bowl-shaped depression (called *beel* in Bengali), situated on the banks of the Hari River, a tributary of the Ganges, about 55 miles (90 km) upstream from the Bay of Bengal. The main findings of this report are summarized (paraphrased) below:

> *A devastating flood 10 years ago left this* beel *inundated with deep water so that no shallow-water rice paddy could grow here. One night, desperate to drain the water, the nearby floodplain residents punched a hole through the mud embankment that encircled the depression. They watched as the water rushed out. Then the high tide began to haul in sediment, and the depression swiftly filled with silt. When the chief engineer of the local water board came to measure it, he saw that in four years, Beel Bhaina had risen by as much as three feet or more near the river*

bank, and almost as much farther inland. Today, it is a quilt of green and gray square patches of paddy, cut by square ponds to cultivate fish and shrimp

Sengupta (2009).

The major rivers of Bangladesh, particularly the Ganges, the Brahmaputra, and the Meghna carry huge amounts of sediments into the Bay of Bengal. With the rise in sea level, the lower reaches of these rivers are expected to experience decreased velocities due to progressive backwater effect, resulting in accelerated sedimentation rates. Consistent with the findings of the preceding report, Khalequzzaman (2012) has stressed the need for further research on major engineering interventions, including river engineering, for diverting river sediments from lower reaches of the rivers of Bangladesh into floodplains and coastal tidal plains to keep up with the rising sea level. Similarly, emphasizing on the nature of microtopographic characteristics of the floodplains of Bangladesh and the hydrodynamic processes shaping them, Brammer (2000) has suggested that adverse consequences of SLR in deeper floodplain basins of coastal Bangladesh could be alleviated by arrangements (i.e., engineering devices) "to let silty river water into polders periodically to deposit sediments (a process known as 'warping'), which could raise land levels pari passu with the rise in external levels" (Brammer, 2000, p. 114).

THE TRIPLE-F MODEL

What Is the Triple-F Model?

The short-form FFF stands for a UNDP model for growing "forest, fruit and fish" on an artificially constructed earth platform in low-lying coastal areas of Bangladesh (UNDP Bangladesh, 2011). The model has the following physical components:

- An earth platform by excavating earth from surrounding lands
- Several excavated ditches and ponds within the platform for fish and duck
- Several mounds (embankments) along excavated ditches for fruit trees and vegetable crops
- Mangrove trees planted along the perimeter of the platform to provide protection from cyclone winds and storm surges
- Other trees for timber among or adjacent to mangrove forest

By raising the platform significantly above the surrounding low lands to avoid shallow seawater flooding, the project is intended to be sustainable through rainwater irrigation for growing fruit trees and vegetable crops on earth mounds and fish and duck in rainwater ponds and ditches. According to one version of this model, a 1-ha (2.5 acre) triple-F platform would accommodate 15 families (with houses and backyards) and would provide an opportunity to grow different types of vegetables in backyard gardens, such as beans, cucumbers, gourds, to name a few, and different varieties of fish in ditches and ponds. Similarly, quick-growing and early yielding fruit trees (e.g., papaya, banana, guava, and many other indigenous varieties of fruits) would be able to produce 10–20 kg (22–44 pounds) of fruit per tree within two years of planting, providing a sustained source of income (about $700 per mound per year), besides satisfying the family needs (Rawlani and Sovacool, 2011). Fish production, amounting to approximately 150 kg (330 pounds) of fish per ditch, is also likely to exceed family needs leaving behind left-over amounts for sale with an annual income of $300 per ditch (Rawlani and Sovacool, 2011; UNDP Bangladesh, 2011). The protective mangrove forests, other trees for timber and fruit trees could be pruned regularly for fuel. Once matured, some of the trees would have good commercial (timber) values.

Community-Based Adaptation Project

The FFF model is one of the components of a five-year $5 million climate change adaptation project called "Community Based Adaptation to Climate Change through Afforestation Program." The project is being implemented jointly by the UNDP, Global Environmental Facility, and the GoB. The project aims to reduce vulnerability of coastal communities to the impacts of climate change through afforestation in the districts of Barguna and Patuakhali (southwestern coast), Chittagong (east coast), Bhola (south-central coast), and Noakhali (central coast) (Rawlani and Sovacool, 2011). Since the emphasis of the project is on coastal forestry and community involvement the triple-F model is largely a nonstructural adaptation. The main reason for including it here under structural adaptation is that there is some potential for involving heavy machinery for the construction of the earth platforms and for digging the necessary ditches and ponds. The emphasis of the original project, however, is on manual earthwork by a group of household members (UNDP Bangladesh, 2011). Plant saplings, vegetable seeds, and fish fingerlings would be supplied by the GoB. Thus, this project lies at the borderline between indigenous and modern as well as between structural and nonstructural adaptation to SLR.

Limitations of the FFF Model

As Brammer (2000, 1990) has emphasized repeatedly, the floodplains and coastal plains of Bangladesh are not necessarily uniformly level. There are many elevated tracts of lands that could be further raised artificially for preparing the so-called FFF platforms. Many such platforms may be constructed throughout the coastal plains but the amount of land resources required for excavation is limited and the total capacity of all platforms may fall far short of the need for accommodating millions of coastal people. Besides, potential environmental impacts and challenges include saltwater intrusion through leaching contaminated groundwater and/or through contamination of freshwater ponds and ditches by storm surge impacts (as it may be impossible to design all platforms high enough to avoid storm surge impacts of all magnitudes). This $5 million experimental project was initiated in 2013 and was supposed to be completed by 2015. No post-hoc assessment of the project is available as yet.

COASTAL FORESTRY

The Sundarbans

The southwestern coast of Bangladesh is the site of the world's largest mangrove forest in the world. Popularly known as the *Sundarbans* in Bangladesh and West Bengal (Indian state), covering an area of about 3861 square miles or one million ha, it is the largest such coastal wetland ecosystem in the world (Gopal and Chauhan, 2006).

The Sundarbans derives its name from its single most important tree species called *Sundari* in Bengali (*Heritiera fomes*). Among at least 22 families of trees, some of the other popular species include (vernacular names followed by Latin names) *Gewa* (*Excoecaria agallocha*), *Passur* (*Xylocarpus mekongensis*), *Keora* (*Sonneratia apetala*), *Goran* (*Ceriops decandra*), *baen* (*Avicennia officinalis*), and *Golpata* (*Nypa fruticans*) (Iftekhar and Islam, 2004). The latter has a palm-like large fond, which is used in *kutcha* houses for thatched roofs. The Sundarbans is also the home of 453 animal species, 120 species of fish, 290 species of birds, 42 species of mammals, 35 reptiles, and 8 amphibian species (Iftekhar and Islam, 2004). The Royal Bengal tiger (*Panthera tigris tigris*) is of course the most famous animal species of this ecologically fragile mangrove habitat. Its ecological vulnerabilities are partly natural and partly human induced. Different types of diseases have been depleting many species, including the *Sundaris*. Perhaps, its greatest threat is related to illegal poaching of many animal species, especially tiger and deer, and clearing of forests for other land uses, especially shrimp culture.

Role of Forestry in Confronting Climate Change

Protection of the *Sundarbans* and its further expansion by a deliberate plan of plantation can play two complementary roles in confronting climate change in Bangladesh; first, by sequestering atmospheric carbon dioxide and thus mitigating greenhouse gas emissions and, second, by offering a variety of climate change adaptation features. Currently only about 7% of the land area of Bangladesh is under forest cover. It has the lowest density of forests in South Asia—0.023 square mile (6 ha) per 1000 people, compared to 0.227 square mile (59 ha)/1000 for India and 0.39 square mile (101 ha)/1000 for Sri Lanka (Ravindranath et al., 2011). Most of the natural forests of Bangladesh have been cleared for agricultural land use due to high density of population. Yet, there are significant opportunities for reforestation because many government-owned forest reserves have very little forest cover and are thus ideal sites for large-scale reforestation. Even at individual levels, homestead and social forestry programs hold great promise for the entire country. Most of the rural homesteads in Bangladesh have certain types of fruit trees and other trees for providing protection against storms and winds. Reinforcing this traditional practice, the GoB (2009) has identified homestead and social forestry as one of its programs for providing protection against storms and waves in coastal areas.

Perhaps the greatest potential for adaptation to climate change–induced SLR in coastal Bangladesh is provided by mangrove forests, which are very important tropical coastal tidal ecosystems that grow on nutrient-rich muddy substrates that are low in oxygen and that undergo variations in salinity (Ferreira et al., 2007). Mangroves have the appearance of thick bushes, consisting of a large variety of sturdy trees of medium height (mostly 30–50 ft or 9–15 m) and an abundance of shrubs. Mangroves have a proven record of retarding the impacts of storm surges along the coast of Bangladesh. Several studies from southeastern India, Sri Lanka, and the Andaman Islands, following the catastrophic Indian Ocean tsunami in 2004, indicated that mangroves offered a significant defense against the full impact of the tsunami, protecting some of the man-made structures directly behind the most extensive mangroves (Alongi, 2008). Mangrove stands provide protection from wave action by creating a physical barrier, stabilizing the sea floor and affecting the angle of slope of the sea bottom (Kelly and Adger, 2000). However, these "protector" forests require protection for themselves from overexploitation and mismanagement (Khan, 2009). Current deforestation rates for global mangrove forests, at about 1%–2% per year, are a greater threat to their sustainability than the maximum global loss of 10%–15% due to projected climate change and sea level rise by 2100 (Alongi, 2008). Utilization of potential benefits of mangrove forests would therefore hinge upon their protection from overexploitation and appropriate management practices for promoting their sustainability, including their adaptation to rising sea levels.

Substantial areas of the Sundarbans along the coast of Bay of Bengal are expected to be inundated by rising seawater, resulting in intrusion of salinity landward. The annual maximum tidal range along the southern coast of Bangladesh has already increased by about 30 inches (75 cm) during the last two decades (Wahid et al., 2007). According to the same source, about 60% of the mangrove area remains in high-salinity conditions for at least 45 days per year. Yet, these coastal forests are highly resilient to SLR because of their special biotic and abiotic characteristics. Their biotic resilience is related to their biodiversity, characterized by at least 350 species of vascular plants (Gopal and Chauhan, 2006). Many of these species are salt resistant and those less tolerant of salinity can grow further inland in areas with lower salinity under the influence of freshwater river flow. Using this basic ecological principle, the BCCSAP has already set up research and extension programs for planting salt-tolerant species along the coast and for developing/selecting other appropriate species for varied mangrove environments (GoB, 2009).

The abiotic role of mangrove forests is even more promising for their adaptation to SLR and perhaps for countering the effect of SLR because of their ability to build lands upward (Ali, 1999, 1996; Thorne, 1997; Richards, 1990). The mangroves with their thick structure (i.e., high tree densities) act as a geomorphic agent of accretion (i.e., accumulation or deposition of sediment by a barrier) leading to build-up of the forest floor. Thus, mangroves act as the dominant geomorphic agent in the evolution of tidal shoals and their accretion to the main landmasses along the Bay of Bengal coast (Chakrabarti, 1975). The role of mangroves in building coastal landmasses has been probed in different parts of the world. Based on an analysis of stratigraphic sequence in mangrove peat deposits in northeastern Australia, Alongi (2008), an Australian ecologist with expertise in mangrove forests, has postulated six possible patterns of mangrove development in relation to SLR. In almost all of the scenarios, mangroves play a constructive role in building lands by accreting sediments on the forest floor, irrespective of the nature of SLR. Whether it is a constant rise in sea level or an irregular rise, or a rise in levels concurrent with episodic land subsidence, the overall pattern is one of sediment accretion on the forest floor leading to rise in land level above the tidal range (Alongi, 2008). All of these scenarios assume a relatively slow rise in sea level because the stratigraphic sequence does not include any period of accelerated rise in sea level. Based on this finding, it is reasonable to assume that at the IPCC-projected rates of relatively gradual rise in sea level by the end of this century, mangroves are likely to survive and shift their positions with rising sea levels. It is not clear whether this shift would be landward or seaward because of the contradictory evidence from different mangrove sites. For example, the mangroves in Key West, Florida, have shifted inland by about 0.9 mile (1.5 km) since the mid-1940s with SLR of about 0.09–0.1 inch/year (2.3–2.7 mm/year) (Alongi, 2008). On the other hand, recent historical records from Bangladesh indicate that its coastline has migrated 50 miles (80 km) seaward (southward) during a period of 200 years (1780-1980) (Rob, 2008; cited in Khalequzzaman, 2012). Another study supports a similar trend indicating that high rates of sediment deposition in the Sundarban area have resulted in an increase in land area of Bangladesh by more than 300 square miles (800 square kilometers) between 1793 and 1870 (Richards, 1990). At least two factors have been working in favor of seaward shift of the mangroves in tandem with coastal migration. First, a complex network of river channels throughout the mangrove forests—mostly distributary channels of the Ganges—contribute large quantities of sediments, which are deposited at their mouths as deltaic islands (locally called *chars*), gradually building land seaward. The main mangrove landscape, confined largely in the districts of Khulna and Bagerhat, consists of a large number of islands separated by rivers and creeks. At times of high water levels some of the river sediments overflow the forest floor, leading to sedimentation by accretion and build-up of land. This source of river sedimentation is now under threat from upstream engineering interventions in the Ganges basin in India. Several studies have shown that construction of the Farakka Barrage on the Ganges and other dams and barrages on the Ganges and the Damodar Rivers have reduced sediment load and increased salinity in most of the distributary channels draining through the Sundarbans (Gopal and Chauhan, 2006; Mirza, 1998). Second, based on Alongi's (2008) analysis of sea-level impacts on global mangrove forests, the Sundarbans of Bangladesh coast, situated within a "macro-tidal range" or at least a moderate tidal range, are less vulnerable to disturbance due to sSLR compared to many other mangrove sites throughout the world.

Coastal Plantation Projects

The role of the Sundarbans in protecting coastal Bangladesh from cyclones and storm surges and its importance as an ecological and economic resource had been recognized well before it was postulated that climate change posed a major challenge for coastal Bangladesh. To compensate for the continuing loss of mangrove forests, a comprehensive multiyear coastal plantation program (commonly called afforestation projects) started in the mid-1960s (i.e., prior to the independence of Bangladesh). The World Bank (1980–1986) financed this program in several coastal districts, including Patuakhali, Barisal, Noakhai, and Chittagong (Mangrove Afforestation Project: 1980–1986: http://www.worldbank.org/projects/P009404/mangrove-afforestation-project?). The leading planted species included, among others, *Gewa, Keora, Baen*, and *Golpata*. The initial objective of the project was to protect coastal areas from impacts of cyclones and storm surges. Subsequently, its goal was expanded to include, among others: (a) conservation and conversion of newly accreted land for agriculture, (b) production of timber for fuel wood and industrial uses, and (c) creation of local employment opportunities (Iftekhar and Islam, 2004). By 1996, the completed projects covered an area of approximately 765 km^2 (295 square miles) (Iftekhar and Islam, 2004). Additional amounts of plantation in the coastal area have been implemented under the following afforestation projects (Iftekhar and Islam, 2004):

- Coastal Greenbelt Project (CGP)
- Forest Resources Management Project (FRMP)
- Sundarban Biodiversity Conservation Project (SBCP)
- *Jhau* (*Tamarix gallica*) Plantation in the Cox's Bazar Sea Beach Area
- Coastal Embankment Rehabilitation Project (CERP)
- *Char* (islands) Development and Settlement Project (CDSP)

By 2001, the total plantation area increased to 1485 km^2 (573 square miles), but more than one-third (39%) of the projects failed due to disease, insect infestation, erosion, wind damage, and other disturbances (Iftekhar and Islam, 2004). *Keora* (*Sonneratia apetala*), in particular, was vulnerable to disease infestation, wind damage, and storm shocks (Nandy et al., 2013).

Community-Based Coastal Forestry

Failure of a substantial portion of coastal plantation provides an example of a major shortcoming of a narrow technical approach for addressing a much broader issue of sustainability of mangrove ecosystem in coastal Bangladesh. Biodiversity is a critical element of ecosystem health because it provides protection from shock in case certain species collapse due to disease, infestation, and other disturbances. More recently, the GoB has adopted an ecosystem approach for managing coastal mangrove forests in collaboration with NGOs and local people. By involving local people in participatory forestry the objectives are to protect and improve local tree cover and to generate supplemental income for the local people from managed forest resources (see endnote 6). As an example of participatory forestry under the Coastal Embankment Rehabilitation Project (CERP), *khash* lands (public lands) along embankments and roads have been allocated to "embankment settler groups" (poor landless people) in exchange of routine embankment maintenance and planting trees along the embankments and in "strip or roadside plantations" (Iftekhar and Islam, 2004). Various NGOs have played active roles in mobilizing and motivating the settlers and in training them in plantation techniques. Similarly, NGOs have been involved in participatory forestry under several other projects, notably the Coastal Greenbelt Project (CGP) and the Sundarban Biodiversity Conservation Project (SBCP).

One of the most recent GoB/UNDP coastal forestry programs that is based on a comprehensive ecosystem approach for protecting and expanding coastal forests through participatory forestry ("social forestry") is the "Community Based Adaptation to Climate Change through Coastal Afforestation Program" (2010–2015) (introduced in the previous section under FFF model). As already indicated, this project was funded by the UNDP. The anticipated and proven benefits of the project include "improving physical responsiveness by deploying soft and indigenous infrastructure through the creation of green coastal belts, the enhancement of institutional effectiveness by training policymakers, and the strengthening of community knowledge by providing alternative livelihood options and establishing early warning systems" (Rawlani and Sovacool, 2011, p. 854). Challenges include (1) ensuring that adaptation efforts are enough to adequately respond to climate vulnerability, (2) lack of coordination, and (3) constrained institutional capacity to ensure successful implementation (Rawlani and Sovacool, 2011). A survey among the coastal residents—farmers, fishers, and others—found that past projects did not adequately develop a sense of ownership of coastal mangroves among the local population (Rawlani and Sovacool, 2011).[6] The average per

[6] Besides farming, fishing, and aquaculture (especially shrimp/prawn farming in polders), the Sundarbans also supports several specialized livelihoods, such as the *bawalis* (wood coolectors), *mawalis* (honey collectors), and crab collectors (Iftekhar and Islam, 2004).

capita income in many parts of the coastal forests is far below the national average, making people completely dependent on the resources of the mangrove forests for making a living. Consequently, overharvesting has continued contributing to significant reductions in mangrove forests, making coastal Bangladesh highly vulnerable to climate change impacts.

CONCLUSION

In public discourse on climate change Bangladesh is frequently cited as the most vulnerable country for climate change impacts. Besides exacerbating other climatic hazards, such as river floods, tropical cyclones, and storm surges, global warming–induced SLR poses the greatest climatic hazard for Bangladesh since it is likely to displace millions of people from its low-lying coastal areas. However, contrary to such a gloomy prospect for coastal Bangladesh, one of the encouraging components of the IPCC projections is that the rates of SLR are expected to be relatively slow up to 2050. We have framed this chapter based on this assumption, demonstrating that several structural and semistructural measures provide a basis for adaptation to rising sea levels in the short and medium terms.

Our interpretations were based partly on our previous experience with flood adaptation in Bangladesh. For generations floodplain residents of Bangladesh have lived with floods by building their homesteads on elevated tracts of lands. Although SLR is an incremental process, unlike cyclic flood regimes, our interpretations were based on the IPCC projections that assume an SLR of up to 1 m (3 ft) by the end of this century. This approach is consistent with most of the ADB's infrastructure projects in coastal Bangladesh (CCRIP, for example), which are designed to upgrade roads, embankments, market places, and other infrastructures by at least 1 m (3 ft) above the projected SLR. Upgrading coastal embankments (making them stronger and higher) is a part of the GOB and World Bank infrastructure projects.

Besides raising homesteads and roads above shallow seawater flooding, another major challenge for continued human settlements in coastal Bangladesh is to design cyclone-resistant homes. Significant progress has been made in this project and several model homes have been tested in the field with some success. Some of the model homes have incorporated concrete posts (stilts) emulating successful experience with cyclone shelters. The network of life-saving cyclone shelters has already expanded significantly and plans are under way for building many more shelters.

Protection of the coastal belt by improving and increasing mangrove forests is a high priority of the GoB, several NGOs, and many donor agencies. As Rawlani and Sovacool (2011) have put it, community-based adaptation through coastal forestry is a "soft indigenous" approach. We have included coastal forestry in this chapter because there is some scope for large-scale intervention (involving heavy machinery) for tree plantation. Besides, the distinction between structural and nonstructural adaptation to climate change and SLR in Bangladesh is largely pedagogical. The main implications of both Chapters 7 and 8 (this chapter) is that, contrary to popular assumptions, Bangladesh is not necessarily in a desperate situation to relocate millions of people from the coastal area. By reviewing several nonstructural and structural adaptation measures, we have demonstrated that it might be possible to continue to live in coastal Bangladesh at least up to 2050 by adopting some of these measures.

REFERENCES

ADB (Asian Development Bank), 2012. Bangladesh: Climate Resilient Infrastructure Project. ADB Technical Assistance Consultant's Report. Final Report, vol. I: Main Report. AECOM Asia Company Limited (in association with several other consultants and the Government of Bangladesh, Local Government and Engineering Department), Hong Kong. Web document available at: www.adb.org/sites/default/files/project-document/73944/4508-001-ban-tacr.pdf.

Agarwal, A., 2007. Cyclone Resistant Building Architecture. Government of India. UNDP Disaster Risk Management Program. Web document: www.unisdr.org/files/11711_CycloneArchitecture1.pdf.

Ahmed, A.U., 2008. Assessment of Vulnerability to Climate Change and Adaptation Options for the Coastal People of Bangladesh. Practical Action, Dhaka, Bangladesh.

Ali, A., 1996. Vulnerability of Bangladesh to climate change and sea level rise through tropical cyclones and storm surges. Water Air Soil Pollut. 92, 171–179.

Ali, A., 1999. Climate change impacts and adaptation assessment in Bangladesh. Clim. Res. 12, 109–116.

Allen-Mills, T., May 12, 1991. Deluge Worsens Misery in Ravaged Bangladesh. The Sunday Times, London.

Alongi, D.M., 2008. Mangrove forests: resilience, protection from tsunamis, and responses to global climate change. Estuar. Coast. Shelf Sci. 76, 1–13.

Auerbach, L.W., Goodbred Jr., S.L., Mondal, D.R., Wilson, C.A., Ahmed, K.R., Roy, K., Steckler, M.S., Small, C., Gilligan, J.M., Ackerly, B.A., January 5 , 2015. Flood risk of natural and embanked landscapes on the Ganges-Brahmaputra tidal plain. Nat. Clim. Change 2472. http://dx.doi.org/10.1038/nclimate2472.

BDRCS (Bangladesh Red Crescent Society), 1998. Bangladesh Cyclone One Bravo. Situation Report No. 7 (Final Report). http://www.ifrc.org/docs/appeals/97/119707f.pdf.

Bhuiyan, M.J.A.N., Dutta, D., 2012. Analysis of flood vulnerability and assessment of the impacts in coastal zones of Bangladesh due to potential sea level rise. Nat. Hazards 61, 729–743.

Biswas, S., Hasan, M., Islam, M.S., 2015. Stilt housing technology for flood disaster reduction in the rural areas of Bangladesh. Int. J. Res. Civ. Eng. Archit. Des. 3 (1), 1–6.

Boateng, I., 2012. Gis assessment of coastal vulnerability to climate change and coastal adaptation planning in Vietnam. J. Coast. Conser. 16 (1), 25–36.
Brammer, H., 1990. Floods in Bangladesh: I. Geographical background to the 1987 and 1988 floods. Geogr. J. 156 (1), 12–33.
Brammer, H., 2000. Flood hazard vulnerability and flood disasters in Bangladesh. In: Parker, D.J. (Ed.), Floods, vol. I. Routledge, London and New York, pp. 100–115.
Brammer, H., 2014. Bangladesh's dynamic coastal regions and sea-level rise. Clim. Risk Manag. 1, 51–62.
Burton, I., October 31, 2015. Adaptation to Climate Change—moving from Local to National, and Planning for the Long Term. Dhaka Tribune. www.dhakatribune.com/feature/2015/oct/31/adaptation-climate-change-moving-local-national-and-planning-long-term.
CEGIS (Center for Environmental and Geographic Information Services), 2009. Report on Cyclone Shelter Information for Management of Tsunami and Cyclone Preparedness. Ministry of Food and Disaster Management, Dhaka.
Chakrabarti, P., 1975. Evolutionary history of the coastal quaterneries of the Bengal Plain, India. Proc. Indian Natl. Sci. Acad. 61A, 343–354.
Choudhury, N.Y., Paul, A., Paul, B.K., 2004. Impact of coastal embankment on the flash flood in Bangladesh: a case study. Appl. Geogr. 24, 241–258.
Climate Change Cell, 2009. Economic Modeling of Climate Change Adaptation Needs for Physical Infrastructure in Bangladesh. Ministry of Environment and Forest; Ministry of Food and Disaster Management, Dhaka, Bangladesh. http://www.bdresearch.org.bd/home/climate_knowledge.
Ferreira, T.O., Vidal-Torrado, P., Otero, X.L., Macias, F., 2007. Are mangrove forest substrate sediments or Soils?: Study in Southeastern Brazil. Catena 70 (1), 79–91.
Forni, M.S., 2015. Bangladesh – Coastal Embankment Improvement Project – Phase I (CEIP-I): P128276 – Implementation Status Results Report: Sequence 05 (English). World Bank Group, Washington, DC. http://www-wds.worldbank.org/external/default/.
GoB (Government of Bangladesh), 2009. Bangladesh Climate Change Strategy and Action Plan 2009. Government of the Peoples Republic of Bangladesh, Ministry of Environment and Forest, Dhaka, Bangladesh. http://www.sdnbd.org/moef.pdf.
GoB (Government of Bangladesh), 2016. Cyclone Shelter Information Database-dmic. Ministry of Food and Disaster Management. Comprehensive Disaster Management Program (CDMP) (website managed in collaboration with UNDP, Sida, UKaid). http:www.kmp.dmic.org.bd/.
Gopal, B., Chauhan, M., 2006. Biodiversity and its conservation in the Sundarban mangrove ecosystem. Aquat. Sci. 68 (3), 338–354.
Hafiz, R., 2000. Adapting traditional housing design to cope with natural hazards: stilt houses for flood-prone areas of Bangladesh. In: Seraj, S.M., Hodgson, R.L.P., Ahmed, K.I. (Eds.), Village Infrastructure to Cope with the Environment, pp. 151–164. Dhaka, Bangladesh and Exeter, UK: The Proceedings of H & H 2000 Conference. http://salekseraj.com/TP16.pdf.
Haq, B., 2007. Battling the Storm: Study of Cyclone Resistant Housing. Bangladesh Red Crescent Society/German Red Cross, Dhaka, Bangladesh. http://sheltercentre.org.
HFH (Habitat for Humanity), 2007. Habitat Responds with 'Houses-on-stilts' Concept to Bangladesh Floods as Team Visits Island that Is Eroding Away. Web-Based News Report. http://www.habitat.org/asiapacific/news/2007/08_23_2007_Habitat_Responds_Bangladesh_Floods_as_Team_Visits_Island.aspx.
Iftekhar, M.S., Islam, M.R., 2004. Managing mangrove in Bangladesh: a strategy analysis. J. Coast. Conserv. 10, 139–146.

Ikeda, K., 1995. Gender differences in human loss and vulnerability in natural disasters: a case study from Bangladesh. Indian J. Gend. Stud. 2 (2), 171–193.

Islam, M.A., Bari, M.F., 2006. Embankment. In: Islam, S. (Ed.), Banglapedia: National Encyclopedia of Bangladesh. Asiatic Society of Bangladesh, Dhaka, Bangladesh. http://www.banglapedia.org.

Kelly, P.M., Adger, W.N., 2000. Theory and practice in assessing vulnerability to climate change and facilitating adaptation. Clim. Change 47, 325–352.

Khalequzzaman, M., 2012. Geological aspects of climate change: an alternative policy option for the Bengal delta. In: Islam, M.M., Zingel, W. (Eds.), Climate Change Effects and Energy Development. Bangladesh Environmental Network (BEN), Germany, pp. 82–88.

Khan, S.I., 2009. Protecting the protectors: lessons for adapting strategies of mangrove forests from Bangladesh. In: IOP Conf. Ser. Earth Environ. Sci., vol. 6, p. 382025. http://iopscience.iop.org.

Lewis, J., Chisholm, M.P., 1996. Cyclone resistant domestic construction in Bangladesh. In: Hodgson, R.L.P., Seraj, S.M., Choudhury, J.R. (Eds.), Implementing Hazard-Resistant Housing (Proceedings of H&H Dhaka 1996), pp. 29–38. Dhaka: Proceedings of H&H Dhaka 96 workshop held at BUET. Web document: salekseraj.com/pf4.pdf.

Mahmood, M.N., Dhakal, S.P., Keast, R., 2014. The state of multi-purpose cyclone shelters in Bangladesh. Facilities 32 (9/10), 522–532.

Mallick, B., Vogt, J., 2012. Cyclone, coastal society and migration: empirical evidence from Bangladesh. Int. Dev. Plan. Rev. 34 (3), 217–240.

Mallick, B., 2014. Cyclone shelters and their locational suitability: an empirical analysis from coastal Bangladesh. Disasters 38 (3), 654–671.

Mirza, M.M.Q., 1998. Diversion of the Ganges water at Farakka and its effects on salinity in Bangladesh. Environ. Manag. 22 (5), 711–722.

Nadiruzzaman, M., Paul, B.K., 2013. Post-sidr public housing assistance in Bangladesh: a case study. Environ. Hazards 12 (2), 166–179.

Nandy, P., Ahammad, R., Alam, M., Islam, A., 2013. Coastal ecosystem based adaptation: Bangladesh experience. In: Shaw, R., Mallick, F., Islam, A. (Eds.), Climate Change Adaptation Actions in Bangladesh. Springer, Tokyo, Japan, pp. 277–304.

Nuorteva, P., Keskinen, M., Varis, O., 2010. Water, livelihoods and climate change adaptation in the Tonle Sap Lake area, Cambodia: learning from the past to understand the future. J. Water Clim. Change 1 (1), 87–101.

Paul, B.K., 2009. Why relatively fewer people Died?: The case of Bangladesh's cyclone Sidr. Nat. Hazards 50, 289–304.

Rahman, A., Islam, R., 2011. Shelters and schools: adapting to cyclonic storm surges: Bangladesh. In: Misdrop, R. (Ed.), Climate of Coastal Cooperation. Coastal & Marine Union-EUCC, Leiden, the Netherlands, pp. 169–172. Internet version (pdf of the complete book). www.eucc.net. www.coastalcooperation.net/part-0/ccc.pdf.

Rashid, H., Paul, B., 2014. Climate Change in Bangladesh: Confronting Impending Disasters. Lexington Books, Lanham, MD, Boulder, CO, New York, Toronto, Plymouth, UK.

Rasid, H., Mallik, A., 1993. Poldering vs. compartmentalization: the choice of flood control techniques in Bangladesh. Environ. Manag. 17, 59–71.

Rasid, H., Mallik, A.U., 1996. Living on the edge of stagnant water: an assessment of environmental impacts of the construction–phase drainage congestion along the Dhaka city flood control embankment, Bangladesh. Environ. Manag. 20 (1), 89–98.

Rasid, H., Paul, B.K., 1987. Flood problems in Bangladesh: is there an indigenous solution? Environ. Manag. 11 (2), 155–173.

Rasid, H., 2000. Reducing vulnerability to flood disasters in Bangladesh: Compatibility of floodplain residents' preferences for flood alleviation measures with indigenous adjustments to floods. In: Parker, D.J. (Ed.), Floods, vol. II. Routledge, London and New York, pp. 46–65.

Ravindranath, N.H., Murthy, I.K., Swarnim, S., 2011. Options for forest management for coping with climate change in south Asia. In: Lal, R., Sivakumar, M.V.K., Faiz, S.M.A., Rahman, A.H.M.M., Islam, K.R. (Eds.), Climate Change and Food Security in South Asia. Springer Science + Business Media, New York, pp. 344–358.

Rawlani, A.K., Sovacool, B., 2011. Building responsiveness to climate change through community based adaptation in Bangladesh. Mitig. Adapt. Strat. Glob. Change 16 (8), 845–863.

Richards, J.F., 1990. Agricultural impacts in tropical wetlands: rice Paddies for mangroves in south and Southeast Asia. In: Williams, M. (Ed.), Wetlands: A Threatened Landscape. Basil Blackwell, Oxford, UK, pp. 217–233.

Rob, M.A., 2008. Changing morphology of the coastal region of Ganges delta. In: Amin, S.M.N. (Ed.), Studies on Coastal Environments in Bangladesh. A.H. Development Publishing House, Dhaka, pp. 54–69.

Sengupta, S., March 20, 2009. Silt, Bangladesh Sees Potential Shield against Climate Shift. The New York Times.

Shamsuddoha, M., Chowdhury, R.K., 2007. Climate Change Impact and Disaster Vulnerabilities in the Coastal Areas of Bangladesh. COAST Trust, Dhaka.

SWMC (Surface Water Modeling Center) and BWDB (Bangladesh Water Development Board), 2000. Late Monsoon Flood in the Southwest Region of Bangladesh 2000. Ministry of Water Resources, Dhaka.

Thorne, C., 1997. Book review. In: Warrick, R.A., Ahmad, Q.K. (Eds.), 1996. Implications of Climate and Sea Level Change for Bangladesh. Kluwer Academic Publishers, Dordrecht, Netherlands. The Holocene 7, 251.

UNDP Bangladesh, 2011. A New Land Use Model: Forest Fruit Fish. UNDP Bangladesh, Dhaka. Community-Based Adaptation to Climate Change through Coastal Afforestation. (Public education website pamphlet). http://www.bd.undp.org/content/bangladesh/en/home/library/environment_energy/a-new-land-use-model–forest-fruit-fish.html.

Wahid, S.M., Babel, M.S., Bhuyian, A.R., 2007. Hydrological monitoring and analysis of the Sundarban mangrove ecosystem, Bangladesh. J. Hydrol. 332 (3–4), 381–395.

World Bank, 1980–1986. Mangrove Afforestation Project (US $11 million). http://www.worldbank.org/projects/P009404/mangrove-afforestation-project?lang=en.

World Bank, 2000. Bangladesh: Climate Change and Sustainable Development. World Bank, Washington, DC.

World Bank, 2009. World Development Report 2010: Development and Climate Change. World Bank, Washington, DC.

World Bank, 2010. Economics of Adaptation to Climate Change: Bangladesh. The World Bank Group, Washington, DC.

CHAPTER NINE

Conclusion

VULNERABILITIES OF THE COAST

Coastal zones are the interface between the land and water. These zones are important because much of the world's population inhabit such zones (Naylor and Coombes, 2015). Additionally, coastal regions are a rich ecosystem with estuaries, mangroves, beaches, tidal flats, and offshore islands providing great biodiversity. Coastal zones change continually because of the dynamic interaction between the oceans and the land. These zones are highly vulnerable to both natural and anthropogenic perturbations. Over exploitation of the abundant natural resources as well as effects of natural disasters like tropical cyclones and associated storm surges have degraded coastal environments almost everywhere. Constant stress and strain from both natural processes and human interference affect the living conditions of the coastal residents, creating physical, socioeconomic, and ecological vulnerabilities and challenging their ability to cope, an ability that has evolved over the centuries. Rampant pollution, coastal erosion, changing patterns of land use, and salinization also threaten the resilience of coastal environments, ecology, and the livelihoods of the residents.

With climate change emerging as the greatest concern for the future in many countries, coastal regions across the globe have received increasing attention from researchers, policy makers, general public, and many others because coastal residents are more vulnerable than anyone in the world to climate change. These residents are on the front lines; because they will be the first victims of climate change—induced sea level rise, which will invariably displace many residents from their ancestral homes. Rising sea levels are likely to inundate large areas of coastal regions particularly in countries located in the tropics. Such inundation, among other things, will have severe consequences on coastal water resources; fresh water, for instance, will be less available because of increasing salinity in soil and water. Both will destroy coastal resources, ecology, and the livelihoods of its impoverished residents.

In most coastal areas of the world, sea levels are rising, storminess is increasing, the ocean is warming and acidifying, and coastal habitats are

Climatic Hazards in Coastal Bangladesh
ISBN 978-0-12-805276-1
http://dx.doi.org/10.1016/B978-0-12-805276-1.00009-0

shrinking (Naylor and Coombes, 2015; Cubasch et al., 2013). These combined pressures mean that coastal areas are particularly vulnerable to natural hazards and climate change risks. Excepting small island countries like Tuvalu and Maldives, no other country in the world is as vulnerable to climate change and subsequent sea level rise as Bangladesh. The level of vulnerability differs greatly within Bangladesh from one region to another, but among the most vulnerable in the country is the coastal region. While exposure to rising sea levels and other natural hazards has serious implications for both human and natural systems, other socioeconomic factors, including widespread poverty, inequality, poor infrastructure, and inadequate government capacity, surely exacerbate this risk situation. Realizing this, Bangladesh government has recently turned its attention to reducing coastal vulnerability to climate change and its frequent exposure to climatic and other natural hazards.

Since the early 2000s, the Bangladesh government has adopted several plans and policies with the overall goal of creating conditions in which the reduction of poverty, development of sustainable livelihoods, and the integration of the coastal zone into national processes can take place. The integrated coastal zone management policies also emphasize the preservation of critical ecosystems and ecological processes, as well as enhanced productivity, balanced growth, increased employment, and increased access to food and coastal resources (MoA, 2012). In 2012, the government prepared a master plan for agricultural development in the coastal region (MoA, 2012). Climate change is likely to severely affect agriculture over the entire country, but saltwater intrusion into the groundwater and topsoil has already reduced crop productivity in coastal areas of Bangladesh. This poses a serious threat to food security in the country as a whole.

Many coastal channels suffer from excessive sedimentation and drainage congestion because of ongoing rise in sea level. Despite limited human, institutional, and financial capacity, the government of Bangladesh has identified the direct and indirect effects of events associated with climate change, as well as other physical and anthropogenic stresses. Accordingly, the government outlined management strategies and plans to mitigate these effects and stresses. With financial assistance from the EU and other European countries, the government has already developed effective policies and implemented, or is in the process of implementing, several proactive climate change adaptation and mitigation measures for its coastal region. In fact, the country is well ahead of many other developing countries in implementing such measures as well as in developing the capacity to deal with the impacts of climate change (Rashid and Paul, 2014).

To sustain coastal resources, confront the impacts of climate change–induced sea level rise, and develop appropriate strategies for the survival of residents under constant threats and stress, an intensive study of both physical and socioeconomic activities, including climatic variability and irregularities, is essential for optimal use of resources in the coastal region in Bangladesh. Based on available empirical data, first-hand field experiences of the authors, and appropriate literature, this book has attempted to focus exclusively on the Bangladesh coast and examine the varied environmental and economic challenges and problems of the coastal region, particularly those associated with climatic hazards, including anticipated sea level rise. As noted, public and private efforts have already undertaken or to be undertaken to overcome these problems.

In addition to tropical cyclones and storm surges, coastal Bangladesh is also prone to other hazards and disasters like floods, earthquakes, tsunamis, water logging, soil salinity, saline water intrusion, coastal erosion, and above all, climate change (MoWR, 2005). Many of these hazards will intensify as sea levels rise due to climate change caused by human-induced global warming. Both the livelihoods of coastal residents and the natural environment of the coastal zone will be severely affected by the forecast increase in sea level. All or part of the coastal zone will be permanently inundated as early as middle of this century, and according to the Intergovernmental Panel on Climate Change (IPCC) prediction, sea level rise in Bangladesh will displace an estimated 20 million coastal residents from their ancestral homes (Rashid and Paul, 2014). All these extreme events have direct and indirect effects on coastal resources. Most vulnerable are coastal agriculture, fisheries (including shrimp cultivation), forestry, salt industries, tourism, settlement, urban development, and infrastructures. Degradation of coastal land has already been reported from the effects of increased salinization, flooding, and erosion. This has adversely affected lives and livelihoods of coastal residents as well as slowed the pace of social and economic development in the region (MoWR, 2005).

In addition to the coastal zone, the national economy and food security, biodiversity, ecosystem, and the ecosystem will also be adversely affected by these extreme events. Availability of land has consistently decreased because of increasing population pressure, and different uses of coastal land have created land use conflict. The coastal region contains Export Processing Zones, urban centers, harbors, airports, land ports and tourism complexes, opportunity for other industries, as well as diverse natural resources. This region has high potential for both onshore and offshore natural gas, and exploration for exploiting that resource continues. More, several ecosystems in the region are important for conservation. The Sundarbans is known for its wide range of flora, fauna, and aquatic life. Coastal Bangladesh also provides an important common property resource for many of its poor residents. Any persons can exploit this resource without owning it or paying anything. Coastal fisheries have provided food, particularly much needed protein, employment, and income to coastal residents for centuries. Another common property includes the abundant forest resources of the Sundarbans. This mangrove forest is a major source of subsistence for many poor people (Sarwar, 2005).

CHALLENGES AND RECOMMENDATIONS

Agriculture

Given the multiplicity of challenges that coastal agriculture faces, the region must increase crop production. Although this seems to be difficult, it may be possible using suitable high yielding variety cultivars and newly developed salt-resistant crop varieties, better agronomic management practices, and on-farm water control. The Master Plan for Agricultural Development claims that rice productivity in all three crop seasons can be increased more than 2 tons/ha (MoA, 2012). The Bangladesh Rice Research Institute (BRRI) and Bangladesh Institute of Nuclear Agriculture (BINA) have developed new varieties of salt-tolerant rice, fruits, and vegetables. Most of these crops have been successfully field-tested in saline soils in coastal districts and subsequently released for cultivation at farm level. Because salinity is increasing in the coastal region, these institutions have been developing crop varieties that can tolerate increased salinity.

Water resources in the coastal region are abundant, the salinity level in many areas is less than the rice threshold limit, indicating that nearly 2.5 million acres (1 million ha) of land can be brought under surface water irrigation for rice cultivation. The districts where surface water irrigation can be expanded include Barisal, Jhalokati, Pirojpur, Barguna, Patuakhali, Bhola, Bagerhat, and Lakhshmipur. Groundwater can be used to cultivate different types of pulses, oilseeds, maize, chili, sugar beets, soybean, jute, and sugarcane. As with rice varieties, both BRRI and BINA have already developed salt-tolerant pulses and vegetables. A good number of fruits (e.g., guava and velvet apple) and vegetables grow well in the coastal region. The authorities should emphasize increasing the yield of these fruits and vegetables.

Farmers of coastal region practice some innovative production techniques for vegetables and spices. As indicated, National Agricultural Research System institutes, like BRRI and BINA, also developed and modified an appreciable number of growing methods suitable for vegetables and spices in the region. Some of these innovative techniques include cultivating vegetables in mounds or raised beds in waterlogged areas, floating gardens in permanently submerged areas, growing vegetables in hanging baskets, and year round vegetables production on *gher* boundary. The nontidal and nonsaline areas of the coastal region have huge potentials for producing summer vegetable (Indian spinach, string bean, brinjal, and okra). By removing existing constraints on surface water irrigation, expanding minor irrigated areas and through extension services, summer vegetable production can be expanded to ridges and adjacent higher elevations (MoA, 2012). Because vegetables are a high-value crop, expanding their cultivation can boost income and enhance agricultural biodiversity.

Although water resource management is a challenge for increasing agricultural production in coastal Bangladesh, improving the surface water supply for irrigation is possible in many parts of the region. One main reason the surface water supply is less than adequate is that internal canals (*khals*) and rivers are silted up and running dry. Therefore, reexcavation to reactivate the canals and rivers would increase the surface water supply for irrigation. In the greater Barisal area, dry season production can be greatly facilitated through irrigation using surface water, which remains fresh almost year round. In other areas, particularly in the northern part of southwestern coastal zone and in the eastern coastal zone, groundwater irrigation can be selectively exploited from the shallow aquifer. Note that groundwater is not everywhere available or suitable for crop production. In the most exposed districts, groundwater irrigation is nearly impossible because the water is saline. Potential areas for intervention are listed below:

- Surface water conservation through construction of rubber dam/cross dam regulators for developing small irrigation projects.
- Excavation of reservoirs/ponds for water conservation to develop surface water irrigation.
- Restoration of severely waterlogged areas.
- Dredging silted up and dry river beds to improve drainage and flood management and to prevent water logging.
- Integrated farm water management.
- Repair and rehabilitation of damaged polders.

Apart from river dredging, fresh water from other areas of the country could possibly be diverted to the coastal region. Preserving and using rain water must be encouraged and rewarded.

Although land is generally scarce in Bangladesh, it is even more scarce in the coastal region. Land is also highly concentrated in few hands. Data show that the top 7% households own 36% land and the 1% own 10% of the land, while the bottom 46% households together own only 13% of land (MoA, 2012). Moreover, the proportion of landless farmers is increasing. Given this situation, introducing coastal land tenure policies will not only reduce land use conflict and the incidence of widespread poverty, but also increase food production and cropping intensity. The same is true for shrimp cultivation. Unfortunately, no public document seriously advocates any change in the existing land tenure system in the coastal region. In this context, the importance of common property regime (CPR) in making natural resource management policy and creating sustainable development can hardly be overemphasized.

As things stand now, ownership, access to management of natural resources in coastal region, especially land, water, and forestry, lie at the heart of at least three fundamental problems in the region: (1) poverty, (2) resource degradation, and (3) resource use conflict. Although it will be difficult to enhance the concept of CPR and make it widely acceptable to policy makers and other social forces, CPR has high potential for alleviating poverty in the coastal region as well as makes development more sustainable.

In essence, cropping intensity in the coastal region must be increased, and coastal agriculture must produce staple foods. Intercropping should be encouraged to intensify help diversify crops and produce more nonstaple food crops. Agricultural production can also be increased by reducing postharvest loss and encouraging postharvest processing, preservation, and storage. The neglected subsector of agriculture is livestock, which suffers from a shortage of fodder. Grazing land is either not available or available only in limited areas. Therefore, establishing and expanding feed mills should be a priority, along with encouraging private veterinary practices by training of professionals and paraprofessionals (MoA, 2012). The coastal region does not have a single meat-processing plant. As urbanization increases, demand for milk, meat, and egg will also increase, as will the potential of processing plants and marketing systems. Unfortunately, rural producers have no organized marketing linkage. Normally farmers sell their products on the local market. Recently, in some areas of the region, organizations have established milk collection centers and chilling points.

Fisheries

After agriculture, the second most popular occupation for coastal residents in Bangladesh is fishing. It is the primary source of income for 14% of all households in the coastal region. Additionally, this is the most common source of secondary income for more than a quarter of the rural households of the region (Rashid and Paul, 2014). Fisheries resources in the coastal region are of three types: fresh water, brackish water in the estuary and sometimes in the deep sea, and saline water of the sea. The brackish waters of the coast are famous for diversified fisheries and aquaculture production. Although shrimp farming expanded rapidly over the past two decades, productivity is still very low and needs effective encouragement for improvement through better technology, better scientific management, and more investment.

Because of its rapid expansion, and social and environmental consequences, shrimp farming has garnered more attention from policy makers and researchers alike. Shrimp farming has not only created huge employment opportunities for coastal residents, it has also accelerated trade and commerce in the region. While such employment and export gains seem impressive, they have been achieved at a considerable cost, beginning with inequitable shares between *gher* owners and land owners, adverse environmental spillovers like loss of vegetation and biodiversity, and decreases in rice yield, all of which increased social chaos and complexities.

The losers in ecological backlash from shrimp farming are not the shrimp farm owners but the owners of the land under shrimp culture. The shrimp *gher* owners represent the elite and privileged class with easy access to urban economic resources, while the land owners represent the powerless rural people who suffer the social and environmental consequences. Therefore, if forced to abandon shrimp farming because of some ecological disasters, the shrimp *gher* owners can recover easily but not the land owners and the landless people who depend on the farms for employment. Thus, the benefit of shrimp farming is concentrated in the hands of a few specific groups of people, while the present development process poses a challenge to coastal sustainability; shrimp farming has a great impact on coastal environment as it actively acts on different coastal processes (e.g., salinity, water logging, and sedimentation).

Because the shrimp sector is export-oriented, shrimp production fails to guarantee food security for many poor coastal residents. The future of shrimp farming depends on whether the coastal environment will continue to support shrimp culture and how long the international market demand will continue. If the international market falls, a backup plan will be needed for farmers to survive. Additionally, the integrated shrimp-rice farming has proved more sustainable than just shrimp farming. Thus, the interculture of shrimp and rice would not only be more environmental friendly than shrimp farming alone, such a practice will increase food security, contribute to eradicating coastal poverty, enhance social inclusion, improve human welfare, and create employment opportunities for the poor. As noted, shrimp yield is relatively low in Bangladesh. Yield can be improved by enforcing land zoning and by changing current management practices. Restricting unplanned expansion of shrimp culture may be necessary in areas where the potential is low, but it can be encouraged in areas where suitability for shrimp farming is high.

Coastal fishing resources face serious threats from overfishing, lack of enforcement of conservation regulations, morphological changes in the estuary, and increasing pollution in coastal waters. Despite these challenges, fish production could be increased by enhancing productivity of pond culture and expanding cage/pen culture in the central coastal zone. Currently, these practices are negligible. Expansion will also be a viable source of employment for coastal residents. In particular, seasonally water-logged areas of Noakhali and Lakhshmipur districts have good potential for cage/pen culture. Coastal fisheries will also benefit from establishing shrimp and fish processing and preservation plants, and by supporting environmentally friendly, high-quality fish drying plants as well as modern fish hatcheries. All-out effort is necessary to encourage production and consumption of fish through developing capture fisheries and fish feed multiplication farms to supply more fish fry as well as to increase production of fish feed for fish fry and pisciculture. Maintaining marine and coastal ecosystems and keeping the sea area free of pollution and marine debris are also needed.

Forestry and Other Resources

The coastal region is highly suitable for forestry. Agro-forestry, homestead plantations, strip plantations, and mangrove forestry have progressed significantly. These forests help preserve coastal habitats and provide employment. Mangrove forests in particular are an important bioprotective system in the tropics and subtropics. They reduce the impact of wind and waves and offer protection to coastal residents against cyclones and storm surges. Therefore, maintaining ecological balance and sustainable development, conservation, and preservation of coastal forests are essential. The Forest Department has been entrusted with the responsibility of managing this precious resource for more than a century. So far it has seen considerable success. However, efficiency and mangrove forestry practices should be standardized. This necessitates knowledge of ecological process, cooperation among related national and international organizations, and adequate financial support. Also, the question of ecosystem sustainability must be pursued more rigorously.

Climate change, however, will change the rainfall pattern, which may intensify as predicted in the fourth IPCC report. Climate change will lead to increasing frequency and intensity of cyclones and tidal surges. These changes are likely to affect the phenology of the forests. Flowering and fruiting seasons may change. Thus timing seed collection may require change, and nursery and afforestation schedules will consequently require reformulation. The honey collecting seasons may have to be rescheduled. These factors need to be borne in mind, and research into ways of integrating these changing parameters will help in developing to assist in planning for the forestry sector in coastal Bangladesh.

Climate change and other coastal hazards like saline intrusion will change coastal landforms and livelihoods and intensify coastal erosion and land use competition. To reduce the impacts of such changes requires many adaptations and mitigation measures at both public and private levels. Measures already initiated by the Bangladesh government consist of a mix of both nonstructural and structural adaptation measures. Although nonstructural measures are generally preferred over structural measures, given the nature of the threats coastal region of Bangladesh faces, a combination of adaptation and mitigation measures is more appropriate to confront the effects of rising sea levels and other coastal hazards. Nonstructural measures are generally much less expensive than structural measures, typically support more biodiversity, and so provide more ecosystem services. However, the government should closely monitor the effectiveness of both nonstructural and structural adaptations and change its adaptation and mitigation policies as needed. Unfortunately, many of the structural mitigation measures (e.g., sea walls, embankments, and polders) do not work properly without maintenance, strong institutional arrangements, and proper construction. The government should oversee structural mitigation to overcome these problems.

Bangladesh has received funds and expects to receive more from western countries to combat climate change. A significant part of these funds were spent implementing mitigation measures in the coastal region. The funds are not, however, sufficient to adequately protect the region from the effects of climate change. However, the use of these funds by the government was not without controversy (Rashid and Paul, 2014). The government must be more transparent and accountable in spending these funds. Another critical problem is the lack of proper and effective governance, which is required to receive, properly manage, and effectively use climate funds from donor countries. The government also must integrate adaptation and mitigation responses to climate change with overall planning and strategies for development. To insure transparency and accountability for all ongoing and future projects, monitoring and evaluation should be strengthened at all levels.

Many estimates have been published on the number of people who will be displaced from coastal Bangladesh due to rising sea levels. Although these estimates vary widely, no one denies such displacement will occur. Where will displaced people resettle? The Bangladesh government has no comprehensive plan for their resettlement. The bulk of these people will resettle within the borders of the country. The government consistently demands that western countries responsible for emission of greenhouse gases should agree to take a significant portion of these displaced people. However, the government has yet to develop criteria for choosing who will emigrate to foreign countries.

Apart from climate change, coastal land use competition has intensified with rapid population growth; urbanization and industrialization; and expansion of recreational, aquaculture, and agricultural activities. This competition often creates conflicts and social unrest, which must be carefully resolved to avoid severely disrupting coastal ecosystems. Bangladesh does have coastal land use zoning regulations, but unfortunately, these regulations are not strictly enforced. Problems are also associated with ownership of newly created lands, with powerful elite groups taking them. Once again, poor governance and widespread public corruption block allocation of these lands to the landless farmers and agricultural laborers of the coastal region.

The role of marine resources in alleviating poverty, acquiring self-sufficiency in food production, protecting environmental balance, facing adverse effects of climate change, and other economic possibilities are virtually unlimited. But with the potential and possibilities come the challenges. Although the government has outlined public plans and policies to manage the coastal ecosystem, implementing these policies depends on political stability, government's sincerity, and the availability of adequate domestic and external funding, among other things. In addition to poor governance, lack of data and research also create problems in enforcing existing plans and projects for comprehensive and sustainable development of the coastal region. Extensive empirical research must continue to facilitate monitoring the progress of all existing projects in coastal Bangladesh.

REFERENCES

Cubasch, U., Wuebbles, D., Chen, D., Facchini, M.C., Frame, D., Mahowald, N., Winther, J.-G., 2013. Introduction. In: Stocker, T.F., Qin, D., Plattner, G.-K., Tignor, M., Allen, S.K., Boschung, J., Nauels, A., Xia, Y., Bex, V., Midgley, P.M. (Eds.), Climate Change 2013: The Physical Science Basis. Contribution of Working Group I to the Fifth Assessment Report of the Intergovernmental Panel on Climate Change. Cambridge University Press, Cambridge, UK.

MoA (Mnistry of Agricultre), 2012. Master Plan for Agricultural Development in Southern Region of Bangladesh. Government of Bangladesh, Dhaka.

MoWR (Ministry of Water Resources), 2005. The National Coastal Zone Policy. Government of Bangladesh, Dhaka.

Naylor, L., Coombes, M., 2015. Bioprotection: working with nature tc manage coastal hazards. Geography Review 28 (4), 37–41.

Rashid, H., Paul, B., 2014. Climate Change in Bangladesh: Confronting Impending Disasters. Lexington Books, Lanham.

Sarwar, M.G.M., 2005. Impacts of Sea Level Rise on the Coastal Zone of Bangladesh (Master's thesis). Lund University, Sweden.

INDEX

'*Note:* Page numbers followed by "f" indicate figures, "t" indicate tables.'

C

T

FSC
www.fsc.org
MIX
Papier | Fördert gute Waldnutzung
FSC® C083411

Kolibri 360 GmbH
Ferdinand-Jühlke-Straße 7
99095 Erfurt, Deutschland
produktsicherheit@kolibri360.de